EVOLUTIONARY PATTERNS AND PROCESSES

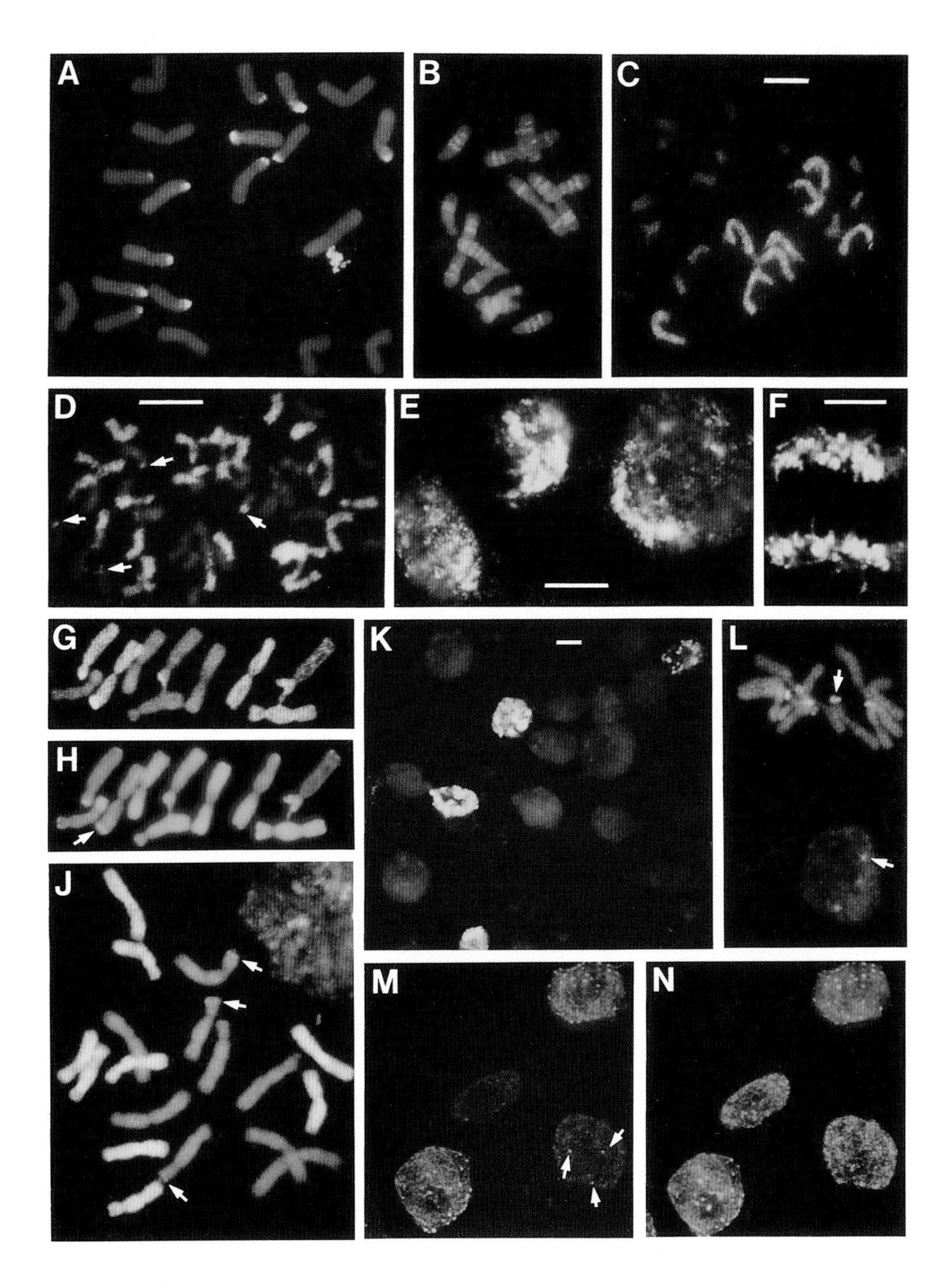

Plate 1 *In situ* hybridization. A: *Gibasis karwinskyana* probed with pTa71, showing 12 sites for NORs (yellow signal); B: *Anemone blanda* probed with the cloned repeated sequence H1; C: *Milium montianum* probed with biotinylated DNA from *M. vernale*. Only the 8 L chromosomes (yellow) are hybridized; D: *Nicotiana tabacum* probed with biotinylated *N. sylvestris* DNA, arrows indicate cross-hybridized areas; E: as D, interphases, showing distinct domains for *N. sylvestris* chromatin (yellow); F: as E, anaphase; G, *G. karwinskyana* × *G. consobrina* ($2n = 10$) probed with biotinylated DNA from *G. consobrina*. Five chromosomes (yellow) are labelled; H: same cell as G, uniform staining with DAPI. The arrowed chromosome is a marker for *G. consobrina*. J: *G. karwinskyana* × *G. consobrina* ($2n = 20$) probed with DNA from *G. karwinskyana* (two chromosomes are missing), cross-hybridized areas are arrowed; K: nuclei of *G. karwinskyana* and *G. pulchella* probed with *G. karwinskyana* DNA (yellow signal); L: *G. schiedeana* probed with DNA from *G. consobrina*, cross-hybridized areas are arrowed; M: as L, interphases without red counterstain, *G. consobrina* nuclei are labelled green, *G. schiedeana* nuclei have cross-hybridizing regions (arrowed); N: same cell as M, to show uniform DNA staining with DAPI. Bar = 10 μm. Unless otherwise indicated, scale bar is the same as for C.

LINNEAN SOCIETY SYMPOSIUM SERIES | NUMBER 14

EVOLUTIONARY PATTERNS AND PROCESSES

edited by
D.R. Lees and D. Edwards

Papers presented at an International
Symposium held at the University of Wales College of Cardiff, Cardiff,
24–26 September, 1991

**Published for the Linnean Society of London
by Academic Press**

ACADEMIC PRESS
Harcourt Brace & Company, Publishers
London San Diego New York Boston
Sydney Tokyo Toronto

ACADEMIC PRESS LIMITED
24/28 Oval Road
London NW1 7DX

United States edition published by
ACADEMIC PRESS INC.
San Diego, CA 92101

A catalogue record is available from the
British Library

ISBN 0-12-440895-8

Typeset by Photo·graphics, Honiton, Devon
Printed in Great Britain by T. J. Press (Padstow) Ltd, Padstow, Cornwall

Contents

Contributors vii
Preface ix

Part I Patterns in the Fossil Record **1**

1 The inexorable logic of the punctuational paradigm: Hugo de Vries on species selection 3
S.J. Gould

2 Making sense of microevolutionary patterns 19
P.R. Sheldon

3 How big was the Cambrian evolutionary explosion? A taxonomic and morphological comparison of Cambrian and Recent arthropods 33
D.E.G. Briggs, R.A. Fortey & M.A. Wills

4 Adaptive radiation: definition and diagnostic tests 45
P.W. Skelton

5 Depth habitats and the Palaeocene radiation of the planktonic foraminifera monitored using oxygen and carbon isotopes 59
R.M. Corfield

6 Patterns of evolution in Quaternary mammal lineages 71
A.M. Lister

Part II From Pattern to Process **95**

7 Postglacial distribution and species substructure: lessons from pollen, insects and hybrid zones 97
G.M. Hewitt

8 Holocene mollusc and ostracod sequences: their potential for examining short-timescale evolution 125
J.G. Evans & H.I. Griffiths

9 Stress, metabolic cost and evolutionary change: from living organisms to fossils 139
P.A. Parsons

10 The importance of rare events in polyploid evolution 157
C.A. Stace

11 Rapid evolution in plant populations 171
M.S. Davies

12 Does genome organization influence speciation? A reappraisal of karyotype studies in evolutionary biology* 189
A. Kenton, A.S. Parakonny, S.T. Bennett & M.D. Bennett

Part III Species and Speciation **207**

13 Animal species and sexual selection 209
H.E.H. Paterson

14 A reaffirmation of Santa Rosalia, or why are there so many kinds of *small* animals? 229
G.L. Bush

15 Phylogenetic patterns of behavioural mate recognition systems in the *Physalaemus pustulosus* species group (Anura: Leptodactylidae): the role of ancestral and derived characters and sensory exploitation 251
M.J. Ryan & A.S. Rand

16 Acoustic signals and speciation in cicadas (Insecta: Homoptera: Cicadidae) 269
T.E. Moore

17 Speciation in insect herbivores – the role of acoustic signals in leafhoppers and planthoppers 285
M.F. Claridge

18 Speciation of the *Enchenopa binotata* complex (Insecta: Homoptera: Membracidae) 299
T.K. Wood

* Plate 1 appears as a frontispiece to this volume.

Contributors

M.D. Bennett, Jodrell Laboratory, Royal Botanic Gardens, Kew, Richmond, Surrey TW9 3DS, U.K.

S.T. Bennett, Jodrell Laboratory, Royal Botanic Gardens, Kew, Richmond, Surrey TW9 3DS, U.K.

D.E.G. Briggs, Department of Geology, University of Bristol, Wills Memorial Building, Queen's Road, Bristol BS8 1TR, U.K.

G.L. Bush, Department of Zoology, Michigan State University, East Lansing, MI 48824-1115, U.S.A.

M.F. Claridge, School of Pure and Applied Biology, University of Wales College of Cardiff, P.O. Box 915, Cardiff CF1 3TL, U.K.

R.M. Corfield, Department of Earth Sciences, University of Oxford, Park's Road, Oxford OX1 3PR, U.K.

M.S. Davies, School of Pure and Applied Biology, University of Wales College of Cardiff, P.O. Box 915, Cardiff CF1 3TL, U.K.

J.G. Evans, School of History and Archaeology, University of Wales College of Cardiff, P.O. Box 909, Cardiff CF1 3XU, U.K.

R.A. Fortey, Department of Palaeontology, Natural History Museum, Cromwell Road, London SW7 5BD, U.K.

S.J. Gould, Museum of Comparative Zoology, Harvard University, Cambridge, MA 02138, U.S.A.

H.I. Griffiths, Department of Genetics, School of Biological Sciences, University of Leeds, Woodhouse Road, Leeds LS2 9TJ, U.K.

G.M. Hewitt, School of Biological Sciences, University of East Anglia, Norwich NR4 7TJ, U.K.

A. Kenton, Jodrell Laboratory, Royal Botanic Gardens, Kew, Richmond Surrey TW9 3DS, U.K.

A.M. Lister, Department of Biology, University College London, London WC1E 6BT, U.K.

T.E. Moore, Exhibit Museum, Museum of Zoology and Department of Biology, The University of Michigan, Ann Arbor, MI 48109-1079, U.S.A.

A.S. Parokonny, Jodrell Laboratory, Royal Botanic Gardens, Kew, Richmond, Surrey TW9 3DS, U.K.

P.A. Parsons, Waite Institute, University of Adelaide, Glen Osmond, S.A. 5064, Australia.

H.E.H. Paterson, Department of Entomology, University of Queensland, Brisbane, Old. 4072, Australia.

A.S. Rand, Smithsonian Tropical Research Institute, Apto. 2072, Balboa, Panama.

M.J. Ryan, Department of Zoology, University of Texas, Austin, TX 78712, U.S.A.

P.R. Sheldon, Department of Earth Sciences, The Open University, Walton Hall, Milton Keynes MK7 6AA, U.K.

P.W. Skelton, Department of Earth Sciences, The Open University, Walton Hall, Milton Keynes MK7 6AA, U.K.

C.A. Stace, Department of Botany, University of Leicester, Leicester LE1 7RH, U.K.

M.A. Wills, Department of Geology, University of Bristol, Wills Memorial Building, Queen's Road, Bristol BS8 1TR, U.K.

T.K. Wood, Department of Entomology and Applied Ecology, University of Delaware, Newark, DE 19717, U.S.A.

Preface

This volume is based on the papers presented at the second regional meeting of the Linnean Society held in Cardiff on 24–26 September 1991. The theme of the meeting, and the title of this book, was intentionally a broad one, bringing together botanists, zoologists and palaeontologists whose wide spectrum of interests splendidly reflect that of the Linnean Society itself. However, in considering the crux of evolution – the processes of speciation – it was necessary to be somewhat selective! And so we turned to entomology, the particular interest of our immediate past president, Professor M.F. Claridge.

The invited speakers were presented with a daunting task – to introduce their own particular research interests, be they in fossils, cells, organisms or populations to the wider audience, in the context of patterns and processes in evolution. Their success was evidenced by the discussion sessions. The palaeontologists persuaded the neobiologists that detailed analysis of the fossil record *can* make a contribution to understanding not only patterns, but also processes, although even the most ardent had to admit that there are many characters used in species recognition that will always evade fossilization. These include some of the most striking images and sounds produced at the meeting, ranging from the quite brilliantly stained plant chromosomes of Ann Kenton and co-workers to the fascinating and apparently specific acoustic signals of certain insects.

We are grateful to the many people who contributed to the success of the regional meeting. They include our hosts, Professor M.F. Claridge, Department of Pure & Applied Biology and Professor M. Brooks, Department of Geology. We also thank Ms. Debbie Harrington for her assistance in preparing this volume.

D.R. Lees & D. Edwards
University of Wales College of Cardiff

Part I

Patterns in the Fossil Record

CHAPTER

1

The inexorable logic of the punctuational paradigm: Hugo de Vries on species selection

STEPHEN JAY GOULD

Introduction 4

Convergence and correlation in the identification of idea-systems 4

Punctuated equilibrium and the punctuational paradigm 6

De Vries's version of the punctuational paradigm 9

Summary and recommendations 15

References 17

Keywords: Punctuated equilibrium – species selection – convergence – Hugo de Vries.

Abstract

The punctuational paradigm has an inexorable logic for its operation within Darwinian theory, once the basic empirical premises of rapid origin by cladogenesis and prolonged stasis thereafter are accepted. The primary implications include: explanation of trends as higher level sorting of species; powering of trends by differential birth, persistence, or directional origin of species; and critique of adaptationism as the motor of trends. Like morphologies, complex systems of ideas can converge as a joint result of empirical veracity (the 'external' and 'adaptationist' component) and logical implication among key notions (the 'internalist' or 'correlational' component). In the first decade of this century, Hugo de Vries developed a full version of the punctuational paradigm; he recognized all the major implications and even coined the name 'species selection' to designate trends by species sorting. But his version has an entirely different basis (origin of species by saltational macromutation) from the punctuated equilibrium of Eldredge and Gould (peripatric speciation properly scaled into geological time). (De Vries's saltationism was anti-Darwinian at the

Evolutionary Patterns and Processes
ISBN 0-12-440895-8

populational level, but he was driven by fierce loyalty to Darwin, who was his personal hero and guru, to reinstate the master's system at the higher level of trends.) These two punctuational systems are convergent, and their remarkable similarity in independent origin illustrates the attractiveness of the idea and its status as a major 'nucleating point' for Darwinian arguments.

INTRODUCTION

This paper presents an idiosyncratic perspective on the 20th anniversary of punctuated equilibrium. (I first introduced the term and concept at the Geological Society of America meeting in Washington, D.C., almost exactly 20 years before presenting a first version of this paper to the Linnean Society meeting at Cardiff in September, 1991. Eldredge and I then published an expanded version of the original talk in 1972.) I say idiosyncratic because the subject matter might, at first glance, seem to denigrate our contribution (even while raising the prestige of the topic). For I have discovered an earlier, quite complete, version of the punctuational evolutionary paradigm – including a formal designation of 'species selection' – in the work of Hugo de Vries, most prominent evolutionary theorist of the early 20th century. Yet, rather than demoting punctuated equilibrium, de Vries's system enhances the mode of thinking common to his version and to our different formulation. For de Vries came to the same logic of evolutionary explanation through a route quite different from punctuated equilibrium. His system is a convergence upon ours, and punctuated equilibrium is not derivatively homologous. Convergences enhance the import of idea-systems, just as they point to a functionally general status for independently evolved morphologies.

CONVERGENCE AND CORRELATION IN THE IDENTIFICATION OF IDEA-SYSTEMS

Evolutionists have devised a powerful concept and terminology in their notion of 'convergence'. As a test for functional response to an external reality, convergence invokes the notion of an 'only way' (or at least a strictly limited set of options) for fitting some construction (a morphology, a set of ideas) to an outer environment – thus providing a second reason for striking similarities between constructions. (In a genealogical science like evolution, complex similarity usually suggests passive retention in historical continuity, or homology. Convergence is, of course, the conceptual opposite of active and independent acquisition in separate lineages, leading to analogy, the primary confounder of phyletic reconstruction.)

Our classical convergences are morphological, for example, the fish-like reptilian ichthyosaurs of Mesozoic seas; the extensive marsupial surrogates which evolved on Australia in striking functional similarity to placentals of similar ecology on other continents; or the old textbook chestnut of similar aerodynamic features in the wings of birds, bats and pterosaurs.

Convergence and homologies have different and opposite essential meanings – the latter as a sign of continuity in history, the former as a mark of functional constraint, a classic example of replication in the general theory of experimentation. Successful replication implies sufficient similarity in conditions of the trial. Since we are testing for limits of functional possibility by using different phyletic lines, the resulting convergence must be imposed by similarity in external shapers of natural selection etc. For example, we yearn to discover some sign of independently evolved life in the universe because all variety on earth offers not a whit of insight into the fundamental issue of how or whether organisms might be constructed differently. (A Martian prokaryotic palaeontology from the planet's early history as an adequately wet and warm body offers real promise for fulfilment of this dream in our lifetime.) For all earthly organisms are replicates of a single experiment in origin, and our stunning biochemical similarities are homologies. We need to know whether an independent origin would yield the same features; then we could begin to talk about functional limits.

Our colleagues in the social sciences may not use the same terminology, but they also search for convergence in ideas and cultural practices – and for the same reason: they wish to discover functional constraint and predictability by studying independent replication in separate cultures. Hence, the endless discussion on whether deep and common elements in the sagas of widely scattered peoples (flood legends, for example) are just preserved and linked memories from a common ancestry (homologies), or convergences independently wrought. And if separately initiated, is the inspiring constraint an externality (actual floods remembered and mythologized) or an expression of common internal mental structures (Jungian archetypes in this case)? And consider what the multiply convergent invention of agriculture (in the Fertile Crescent, the Orient, and the Americas) means for a claim – which I would defend despite my predilection for contingency – that certain tendencies towards complexity in human culture are predictable.

Systems of ideas are also strongly subject to convergence – for the world contains many smart folks, while logic may provide rather fewer basic ways to consider any cardinal issue. So pervasive, indeed, is the phenomenon of joint, independent discovery that our leading sociologist of science, Robert K. Merton (1963), has proposed a revised criterion of fame: ask not how many discoveries a scientist has made, but rather in how many 'multiples' he has taken part.

Our own profession began in the most famous example of a multiple discovery – Wallace's independent derivation, in uncanny detail point for point, of natural selection. The fascination of this example – and of any good convergence – lies in the complexity of coordinated ideas so faithfully formulated twice. We would not be awe-struck by one item independently derived. In fact, we don't even refer to repetition as convergence if the item and cause are simple enough. (When I find another four-leafed clover in a separate field miles away, I may feel lucky but I do not speak of convergence.) But Darwin and Wallace's natural selection is a complex web of interconnected ideas, both constructed in uncannily similar ways. Both received a key inspiration from Malthus. Both devised natural selection to account for anagenesis in lineages, but understood the need for another related principle to explain branching speciation. Both constructed such a second principle and linked it to natural selection in the same way.

The world of ideas is no more homogeneous or isotropic than the morphospace of organisms. 'Nucleating points', where related concepts coagulate into powerful theories, must exist just as *Baupläne* set the basic architectures of large organic groups. The operational definition of these nucleating points must be fashioned in terms of internally forced correlation among constituent parts (the covariances that

Darwin termed 'correlations of growth' for the organic example). Ideas are built to theories in webs of implication, just as organisms are constructed in networks of developmental correlation. All important ideas are systems of complex interactions, not items of singular brilliance.

In summary, two broad criteria, corresponding to structuralist and functionalist approaches in evolutionary morphology, might be applied to systems of ideas to test for their status as substantial 'nucleating points' in a decidedly non-homogenous field. First as an internalist or structuralist criterion, substantial theories are rich webs of interacting ideas held together by strong bonds in the logic of implication, not single items of insight (whatever their brilliance). Second, as an external or functional criterion, substantial theories are (usually) multiply identified, rather than uniquely invented because they are actual configurations of ideas, available for discovery in a limited field, not idiosyncratic effulgences of unique intellects. I call these criteria *correlation* and *convergence*, respectively.

PUNCTUATED EQUILIBRIUM AND THE PUNCTUATIONAL PARADIGM

Punctuated equilibrium is a theory about the proper scaling of ordinary speciation into geological time. Our original account focused upon the conventional allopatric (or peripatric) speciation theory of Mayr, but any of the standard versions will serve equally well (except 'dumb-bell allopatry', which would lead to geological gradualism, but which no major evolutionist seems to grant a substantial relative frequency), because all envisage speciation as occurring in a geological moment compared with the average duration of a species. Speciation is, therefore, effectively unresolvable – it occurs on a bedding plane rather than through a sequence of strata – in almost all geological situations. The opposite side of this coin – stasis in established species during the several million years of their average existence – becomes the major empirical testing point for punctuated equilibrium.

If stasis and punctuation are the bare bones of punctuated equilibrium, then the integration of this non-gradualist empirics with selectionist evolutionary theory poses a large set of issues. I argue in this paper that the broader set – which I call the punctuational paradigm – forms one of the nucleating points for expansive theories discussed in the last section. The paradigm begins with agreement on the basic empirics of punctuation and stasis, but does not require that these results be rendered by punctuated equilibrium. (We will see that de Vries uses a different mechanism, thus making his version of the paradigm convergent upon ours.) It then works through the demands placed by selectionist logic upon systems that break the basic uniformitarian assumption for Darwinian extrapolation from everyday variation to changes at greatest scale. For if species are stable, then insufficient variation exists for producing trends by anagenetic transformation of lineages, and such trends – a manifest feature of life's history – must be produced by a higher order process of sorting among species. The punctuational paradigm is a web of inferences impelled by uniting the empirics of stasis and punctuation with the logic of selectionist theory. Its interrelated concepts are mutually implicating, and the entire 'package' works as a unit on the playing field of theories.

I would specify the following major components of the paradigm:

Basic empirics of punctuational evolution:

1. Species originate rapidly, with their morphological differentia, on a geological scale relative to their later duration. (Operationally, I have defined 'rapid' as less than 1% of later survival, but most origins are effectively instantaneous in geological resolution.)

2. Species are stable during their geological lifespan, which tends to be quite long (5–10 million years for marine invertebrates, less for terrestrial tetrapods). Stasis does not mean absolute immobility or failure to vary, but rather that minor fluctuations lead to no net directional vector through time and generally do not stray beyond the bounds of geographic variability within comparable modern species (see Stanley & Yang, 1987).

Logical implications of integration with selectionist theory:

3. Selection depends upon variability among units of a population subject to scrutiny. Trends represent the major phenomenon of macroevolution, and any selectionist theory must account for them. If species are stable, and constrained to be so, then insufficient variability exists for producing trends by gradualistic anagenesis within phyletic lineages. Usable variation only exists among species within clades – and trends must therefore be produced by a higher level process of sorting among species treated as stable entities.

4. As ordinary selection works by differential births and deaths of organisms within populations, species sorting operates by birth and death biases among species. Trends occur because some species either live longer than others, speciate more frequently, or speciate in a directional manner.

5. The punctuational paradigm almost inevitably leads to a critique of adaptationist thinking on the origin and maintenance of trends, and for two reasons: first, as trends are not driven by ordinary adaptive struggles in normal competition on generational scales, the Darwinian apparatus (while fully operative at its own level) does not extrapolate to the causes of trends. Second, phenotypic shifts in trends may be secondary consequences rather than direct causes, for 'standard' features of palaeontological phenotypes are often unrelated to the primary factor of trends considered as a species-level phenomenon, that is, differential rates of punctuational speciation.

6. These views on trends and adaptation inevitably imply a general commitment to hierarchical selection theory (simultaneous action of selection at several levels, including genes, organisms, demes and species) and to non-uniformity and a denial of extrapolationism from immediate events in modern populations to a fully sufficient account of macroevolution. Since a vision of uniformity was so central to Darwin's own worldview, these implications are the broadest and most challenging notions within the punctuational paradigm.

Eldredge and I did not, by any means, have all these pieces in place when we first published in 1972. We were writing explicitly for palaeontologists, and primarily to make the operational point that the fossil record is not an intractably incomplete distortion of the gradualism that actually rules life's history, but a workably adequate representation of the punctuational pattern that truly dominates the tempo and mode of evolution. We had some inkling of a necessarily revised view of trends (Eldredge & Gould, 1972), but little insight into hierarchy or criticisms of adaptation. With a great deal of help from our friends (Stanley, 1975; Vrba, 1980), we did work through the logic to a complete version of the punctuational paradigm (Gould & Eldredge, 1977; Gould, 1982, 1989).

But punctuated equilibrium became a larger issue within evolutionary theory and, therefore, fell prey to an old saw about the history of new ideas in science. My version comes from von Baer (1866:63), quoting Agassiz, though Huxley is also commonly cited, and I do not doubt that the wry cynicism could be traced to Aristotle himself.

> Deswegen sagt Agassiz, dass wann eine neue Lehre vorgebracht würde, sie drei Stadien durchzumachen habe; zuerst sage man, sie sei nicht wahr, dann, sie sei gegen die Religion, und im dritten Stadium, sie sei längst bekannt gewesen. [Therefore, Agassiz says that when a new doctrine is presented, it must go through three stages. First, people say that it isn't true, then that it is against religion, and, in the third stage, that it has long been known.]

By this criterion, Eldredge and I should have been delighted when a rash of petty and negativistic notes began to appear, announcing the supposed discovery that various past worthies, most notably Darwin himself, had enunciated aspects of punctuated equilibrium (see our commentary in Gould & Eldredge, 1986). These notes all committed the worst error of superficial historiography – the confusion of footnote and side comment with general tenor and central logic. No one as brilliant and as far ranging in his writings as Darwin could have failed to note, or puzzle over, some aspect of the apparatus that later became the punctuational paradigm. But we dishonour Darwin and the power of his theory when we make him an avatar for everything; the logic of natural selection is powerful because it makes distinctions and commitments, and because it rejects a host of potential phenomena (directed variation, for example) as contrary to the central theory. You can find a footnote or two in Darwin accepting this or that occasional case of directional variation, but he is not converted thereby into an anti-selectionist. Similarly, Darwin's acceptance of marked variation in evolutionary rates (who could deny it?) does not refute his gradualism.

Ironically, as these notes appeared like weeds, and as proper historiography plucked them one by one, a lovely primrose stood in the same garden, unnoticed by everyone. The punctuational paradigm did have a genuine precursor, a great scientist who proposed the necessary cmpirics (with a rationale different from punctuated equilibrium) and, by commitment to the principles of Darwinism, worked straight through the logic to a full version of the punctuational paradigm. He even, along the way, recognized and proposed the name 'species selection'. He did not write much about macroevolution and did not give his ideas a prominent push – thus perhaps explaining (but not excusing) the neglect of the profession – but he was the most prominent evolutionary theorist of his time, and he did present the argument in the most widely read treatise from the first decade of the 20th century, a work initially published in English. Shame on us, and especially on me, for missing his convergence and permitting the discussion on precursors to focus upon nonsense and triviality. Shame on us all for reading so little, and for not knowing or remembering the basic history of our profession. The term 'species selection' was not proposed by Stanley in 1975; the basic concept was not first mooted by Eldredge & Gould in 1972. The term belongs to Hugo de Vries, and he advocated the concept, in full knowledge of its broader context, in his major book in English, his Berkeley lectures of 1904, published in 1905.

DE VRIES'S VERSION OF THE PUNCTUATIONAL PARADIGM

The star of Hugo de Vries (1848–1935) burned brightly for a time early in our century. We remember him today primarily for his role among the trio of Mendel's resurrectionists in 1900 – traditionally cited as de Vries, Correns and Tschermak-Seysenegg. We also know that he developed and promoted a saltational theory of speciation based upon his work with *Oenothera lamarckiana*, the evening primrose. We may also be aware that his long, two-volume defence of the theory (*Die Mutationstheorie*, published first in German in 1901–03 and later in English in 1909–10) was surely the most influential evolutionary work of its time.

Beyond these bare bones, the usual modern 'take' on de Vries wallows in negativity (on the false criterion of judging the past by modern perceptions of objective truth). We can all state the great irony that, although Mendelian particulate inheritance eventually secured Darwinism, its first major use in evolutionary theory further derailed selectionism by introducing yet another contender, with de Vries as major promotor – the macromutational theory of saltation. We also know that the empirical basis of de Vries's claim later collapsed with the discovery that 'species-forming' saltations of *Oenothera* arose as oddities of an unusual chromosomal system, not as generalities of Mendelism. (*Oenothera lamarckiana* is a permanent heterozygote, a hybrid with chromosomes of each component linked in rings, and thus segregating together in meiosis. Only half the seeds are viable because both homozygotes are lethal.) Thus, de Vries is often dismissed today as a sidetrack in the history of Darwinism triumphant.

De Vries first found his saltational mutants of *Oenothera* in 1886, growing in a wild field of evening primrose at Hilversum, near Amsterdam. He cultivated them carefully, propagated other seeds in controlled fertilizations by the tens of thousands, and waited for other distinct phenotypes to emerge. In his Berkeley lectures of 1904, he claimed seven new species. De Vries made a sharp distinction between ordinary, small, continuous, Darwinian variability (which reverted to type and could not be parlayed into permanent and substantial evolutionary change) and the abrupt, large-scale discontinuous saltation ('mutation' in his terminology) that made new species all at once. These mutations are the source of evolution, and they confute Darwinian gradualism:

> Species have arisen from one another by a discontinuous, as opposed to a continuous process. Each new unit, forming a fresh step in this process, sharply and completely separates the new form as an independent species .... The new species appears all at once; it originates from the parent species without any visible preparation and without any obvious series of transitional forms.
>
> (De Vries, 1909:3)

As a direct consequence of these views on variability, stasis becomes the fate and condition of all species during their geological tenure. First of all, the conventional continuous variation, always present in populations, is impotent to produce evolutionary change. Efficacious mutational variation, on the other hand, is very rare and only characterizes species for minute fractions of their existence, if ever. In fact, de Vries notes that he could only find such variability in *Oenothera*, and in no other plant – thereby obtaining some estimate (one in a thousand or in tens of thousands) for the proportion of species in a 'mutable period' at any time. Moreover, since new species arise suddenly, novelty is separate and instantaneous, and the stability of a parental phenotype is not thereby threatened.

De Vries speaks of stasis as a well-recognized phenomenon: "The constancy of species is a demonstrated fact: their transmutability is still a matter of theory" (1909:205). "The existence of long intervals of time during which characters remain unaltered is, at any rate in the case of a great many species, a matter of tolerable certainty" (1909:206). He cites several supports that have become prominent again in the debate over punctuated equilibrium: species with isolated and widely separated demes that have not diverged over centuries (1909:206); species that survive unchanged through several geological periods (1905:698); and species that do not alter through substantial fluctuations of ice-age climatic cycles (1905:696).

Moreover, if new species arise in single leaps, then their origin must be non-adaptive, however much their future survival may depend upon a fortuitous match with the environment:

> Specific characters have evolved without any relation to their possible significance in the struggle for life. The facts are contrary to the main principle of the selection theory of Darwin. Moreover, intermediate steps between the endemic species and their parents, in the midst of which they are ordinarily still living, are wanting and therefore must be assumed never to have existed.
>
> (De Vries, 1992:226)

Nothing that I have yet presented should elicit the slightest flurry of special interest in de Vries as a possible punctuationalist. After all, these three themes – sudden origin, stability between saltations, and non-adaptive origin in macroevolution – are standard equipment for nearly any saltationist thinker, and saltationism was a leading alternative, espoused by hundred of evolutionists, during this period. These ideas are, to be sure, 'preadapted' for slotting into the punctuational paradigm – for saltation and stasis by failure to generate the right kind of variation are even more evident than the corresponding notions of punctuated equilibrium (ordinary speciation that scales to geological abruptness, and developmental constraint) as rationales for the empirical requirements of the paradigm.

To promote these claims into a version of the punctuational paradigm, one must unite them with complexities of selectionist logic to develop a macroevolutionary theory of levels for the explanation of trends. Here we encounter a daunting pair of difficulties that made it effectively impossible for any saltationist thinker other than de Vries to take such an argument in this direction.

First, the saltationists were anti-Darwinians. Why should they take their chief set of claims, developed to confute Darwinian gradualism in the first place, and then relink them to selectionist logic? Second – and this is a point far too little appreciated by modern evolutionists – virtually no 19th- or early 20th-century scientist grasped the full subtlety of selectionist logic as an abstract mode of argument that might be applied to various levels. Darwin was a man of enormous genius, and he devoted his life to thinking through all of nature in selectionist terms. We have all imbibed the basis of this style today and, although we still have trouble (or else Williams's 1966 classic would not have been necessary), we do manage. But the selectionist mode of thought is so radical (with respect to previous ways and certainties) and so complex, that virtually no one, even among Darwin's most fervent supporters, really fathomed it. Lyell & Huxley never did; neither did Wallace (who roamed indiscriminantly across levels, trying to find adaptationist explanation at any cost of illogic). In fact, I believe that only two early evolutionists fully grasped the argument: August Weismann and Hugo de Vries. Weismann was a committed Darwinian, so we should not be surprised at his dedication. But why de Vries? He was a brilliant man to be sure. But why should an opponent of Darwin, or at least a scientist working laterally to Darwin's ideas, be almost alone in understanding the selectionist system. We now come to the interesting part of the story – to de Vries's personal, psychological and intellectual ties to Darwin, and to the reasons for his unique blend of saltationism with selectionist logic.

Darwin was de Vries's personal hero, both his guru and his scientific guide. Their primary intellectual tie did not reside in evolutionary work, but in studies of plant physiology that dominated the first 20 years of de Vries's career, and the last decade of Darwin's life as he wrote *Insectivorous Plants* (1875a), *The Movement and Habits of Climbing Plants* (1875b), and *The Power of Movement in Plants* (1880).

Moreover, de Vries directly courted and won Darwin's admiration and mentorship. The two men maintained extensive correspondence (reprinted in Van der Pas, 1970). Darwin devoted much effort to helping de Vries. He sent complimentary copies of his books, and wrote to Asa Gray for seeds, so that de Vries could pursue some experiments on movement in tendrils. De Vries, for his part, repeated and extended several of Darwin's experiments on the physiological basis of movement and insectivory in several species.

In the summer of 1878, de Vries visited England and fulfilled his fondest hope of meeting Darwin. He wrote to his fiancée on 14 August:

> Today I have visited Darwin; I am happy that it happened and I must say that Darwin was so very cordial and friendly.... The conversation was quite easy; they all spoke very slowly and clearly and they gave me the time to speak up.... I feel much more happy these last days. It is such a pleasure to find that somebody is really interested in you and that he cares about what you have discovered.

Finally, Darwin's influence extended far beyond physiology, for he did inspire de Vries to shift into evolutionary and genetic studies. De Vries told an interviewer in 1925 (quoted in Van der Pas, 1970:192): "I was led to the study of heredity by my love for Darwin". De Vries's inspiration was not *On the Origin of Species* (1859) or *The Descent of Man* (1871), but Darwin's extended speculation on inheritance, the 'provisional hypothesis' of Pangenesis, presented as the last chapter of the two-volume *The Variation of Animals and Plants Under Domestication* (1868).

In his last letter to Darwin, written on 15 October 1881, de Vries commented: "For some time I have been studying the causes of the variation of animals and plants, as described in your treatise.... I have always been especially interested in your hypothesis of Pangenesis and have collected a series of facts in favour of it."

De Vries's first book (and his best according to many scientists then and now) revised Darwin's ideas by denying that gemmules (Darwin's name for hereditary particles) could cross cell boundaries. De Vries called his theory, and book, *Intracellular Pangenesis* (1889). Among 19th-century works in the speculative tradition, it certainly came closest to a defendable concept of heredity, and also contained much valuable material on development (more than on inheritance). Incidentally, de Vries, in this book, renamed Darwin's gemmules as 'pangenes', thus honouring Darwin's larger conception. Johanssen later shortened this term to 'gene' and identified the particles with Mendel's factors. Thus – as an interesting footnote unknown to most evolutionists – Darwin, via de Vries, is godfather to the term 'gene'.

I do not believe in psychobiographic speculation as a mode of argument, but you do not have to be Sigmund Freud to recognize that de Vries's personal fealty to Darwin placed him in a most uncomfortable intellectual position. He had, after all, committed an act of scientific parricide in developing the mutation theory as a saltational account of speciation. What mechanism could be more anti-Darwinian at the crucial level most central to Darwin's own concerns (and chosen book title) – the origin of species. There can be little room for selection or adaptation if new species arise in single, fortuitous leaps.

Consider de Vries's difficult situation. He has slain natural selection in its own domain, yet he understands selectionist logic better than any contemporary (except Weismann), and he is fiercely committed, by bonds of personal loyalty, to further the work of his guru and intellectual hero. What is he to do? De Vries found a brilliant exit to his dilemma, one open to him alone (for no one else had both the mix of concepts and the commitment). In short, he jacked up the selectionist argument by a level, applying all its principles to the explanation of macroevolutionary trends. In other words, he developed a convergent version of the punctuational paradigm.

Darwinian logic requires that variation should exhibit three properties, so that it will act as raw material only, thus permitting natural selection to operate as the creative force in evolutionary change. (1) Variation must be copious to supply enough raw material. (2) Variation must be small, so that natural selection can build by accumulation, step by countless step (for if variation produces a new species all at once by an occasional fortuitous leap, then selection has no creative role and can only eliminate the unfit). (3) Variation must be, to use the mineralogist's word, 'isotropic' – that is, occurring in all directions, with no preferred bias towards adaptive states (for if variation is 'prepackaged' in the right direction, then even random mortality will yield the trend, and natural selection can only accelerate a process already driven by inherent variation).

Now the mutation theory annihilates selectionist logic in the origin of species, for mutational variation, confined to certain lineages at highly restricted times, is neither copious nor small. But de Vries reasoned – correctly – that all three Darwinian properties would operate at a higher level, in variation among species acting as raw material for trends within clades. Variation would be copious, for lineages in their mutable periods (like *Oenothera*) generate large numbers of 'elementary species' (he had found seven in just a few years at one location). Relative to the full range of a trend – a transformation that may accumulate hundreds of speciation events, linked end to end – the saltational leap of a single species marks but a small increment. Thus, the 'large-scale' variation that wrecks Darwinism in its own realm (by rendering speciation in a single step) becomes an appropriately small increment at the more ample level of an evolutionary trend. The third requirement of isotropy had been met all along, for de Vriesian saltations occur in all directions, and certainly with no preference for adaptive configurations (the source, after all, of de Vries's original confutation of adaptationism). De Vries defended isotropy as an empirical proposition: "[Mutations] seem to be presented by all characters, to proceed in every direction and to be apparently without limit" (1909:33). He also recognized isotropy as a theoretical requirement: "The mutation theory demands that organisms should exhibit mutability in almost all directions" (1909:204).

De Vries, having unsurped his hero in Darwin's own realm, has now developed the logic to recrown him in the higher kingdom of evolutionary trends – a fitting discharge, perhaps, of the obligations of fealty!

All effective evolutionary variation exists only among the 'elementary species' formed by mutation from a parental stock, not among individual organisms within these species. In principle, therefore, if selection works as an evolutionary force at all, it can only sort elementary species, not organisms within populations. De Vries clearly recognizes the difference between the conventional Darwinian form of organismal selection and his proposal for selection among elementary species:

> The struggle for existence, that is to say the competition for the means of subsistence, may refer to two entirely different things. On the one hand, the struggle takes places between the individuals of one and the same elementary species, on the other between the various species themselves. The former is a struggle between fluctuations, the latter between mutations.
>
> (De Vries, 1909:211–212)

Horticulturists, de Vries noted, recognize the two modes and give them different names (1905:604–605): 'race-breeding' for the limited and relatively ineffective Darwinian sorting based on fluctuating variation among organisms in a population; and 'variety-testing' for sorting among discrete elementary species produced by mutation. In a remarkable passage, de Vries then names this second mode 'species selection' and denies that it can be equated with natural selection:

> The word selection has come to have more than one meaning. Facts have accumulated enormously since the time of Darwin, a more thorough knowledge has brought about distinction, and divisions at a rapidly increasing rate, with which terminology has not kept pace. Selection includes all kinds of choice. ... Selection must, in the first place, make a choice between the elementary species of the same systematic form. *This selection of species or species selection* [my italics] is now in general use in practice where it has received the name of variety testing. This clear and unequivocal term however, can hardly be included under the head of natural selection. The poetic terminology of selection by nature, has already brought about many difficulties that should be avoided in the future. On the other hand, the designation of the process as a natural selection of species complies as closely as possible with existing terminology, and does not seem liable to any misunderstanding. It is a selection *between species*. Opposed to it is the selection *within the species*.
>
> (De Vries, 1905:742–744)

De Vries then develops this concept of species selection as a set of guidelines for a general theory of macroevolution. He argues that sustained evolutionary trends must arise by species selection for two reasons: (1) as stated above, variation among species is the only sort available for an effective process of selection; (2) trends are clearly adaptive and accumulative, but the origin of elementary species by mutation is both non-adaptive and discretely sudden. Mutations, therefore, cannot produce trends by themselves; a 'higher-order' selection upon discrete mutational phenotypes is necessary:

> The differentiating characteristics of elementary species are only very small. How widely distant they are from the beautiful adaptive organizations of orchids, of insectivorous plants and of so many others! Here the difference lies in the accumulation of numerous elementary characters, which all contribute to the same end. Chance must have produced them, and this

> would seem absolutely improbable, even impossible, were it not for Darwin's ingenious theory. ... According to Darwin's view, [variations occur] in all directions, or at least in many. If these include the useful ones, and if this is repeated a number of times, cumulation is possible; if not, there is simply no progression, and the type remains stable through the ages. Natural selection is continually acting as a sieve, throwing out the useless changes and retaining the real improvements. ... Hence the increasing adaptations to the more specialized conditions of life.
>
> (De Vries, 1905:572)

De Vries also recognizes that species selection must include two components, corresponding to birth and death biases in conventional organismic selection. Species selection will favour those lineages that (1) yield more elementary species by mutation (birth bias), and (2) produce phenotypes fortuitously adapted to changing local conditions (persistence bias). De Vries writes (1909:200): "The question is obviously: what share ... is to be ascribed to mutability and what to the natural elimination of elementary species." De Vries also anticipated a central theme in punctuated equilibrium's later version of the paradigm by stressing both the potential importance and unconventional nature of birth biases, for trends powered by differential origin are not so subject to adaptive moulding of phenotypes as those produced by elimination of species that fail in competition:

> It is only in those cases in which variability, that is to say the production of new forms, has been most active that the groups continue for longer periods of time. Incapacity to vary dooms a group to death; only those who can most easily and quickly adapt themselves to changing conditions of life can survive. Species forming variability is therefore not a universal capacity for variation, but only the result of quite special conditions which may often be absent from certain groups. ... It is clear that every branch of the pedigree ... is doomed to extinction as soon as the mutable species in it become extinct from some cause or another. On the other hand it is easy to see that the more numerous the mutable types are, the greater is the species forming capacity of the whole group and, consequently, the greater its prospect of maintenance throughout geological ages.
>
> (De Vries, 1909:660)

In his earlier writings, de Vries (1905, 1909) strongly supported adaptation as a primary result of trends forged by species selection. But he altered this view in later articles, and thereby came to espouse the full range of punctuational critiques against Darwin (by questioning both gradualism and adaptationism). In 1922, de Vries contributed a short, but highly revealing chapter to J.C. Willis's famous critique of adaptation based on the correlation of geographic distribution and geological longevity, *Age and Area* (Willis, 1922).

Of course, de Vries had always opposed adaptationism in the origin of species (1922:224), for selection cannot craft good design if species arise in single steps. But differentia of higher taxa arise by cumulation during species selection, and are therefore usually adaptive.

> It is a curious fact that most of the striking instances of beautiful adaptation to special forms of life are characters of genera and sub-genera, or even of whole families, but not of single species. Climbing plants and tendrils, insectivorous plants, desert types ... submerged water plants, and numerous other instances could be adduced.
>
> (De Vries, 1922:22)

(De Vries saw the distinction between non-adaptation in most differentia of species, and adaptation for the differentia of higher taxa, as virtual proof for the non-selectionist origin of species by saltation and the functional origin of higher taxa by species selection.)

Now, however, inspired by Willis late in his career, de Vries re-assessed even this role for adaptation, recognizing that an a posteriori functional correlation of form and environment (especially for broad characters of higher taxa) does not indicate adaptive fashioning under current circumstances. Using an argument of exaptation (Gould & Vrba, 1982), de Vries recognized and embraced the alternative view that such characters arise for whatever reason (often non-adaptively, though we usually cannot tell one way or the other in retrospect), radiate out randomly by Willis's argument, and then survive in those environments that, by good fortune for the species, favour a set of characters that might have originally evolved for purposes different from current function:

> Everywhere in nature, in geological periods as well as at present, the morphological characters of newly originated types have no special significance in the struggle for life. ... They may afterwards prove to be useful or useless, but this has no influence upon their evolution. Obvious instances of usefulness occur, as a rule, only at much later periods during the wandering of the new forms, when unexpectedly they arrive in environments specially fitted for them. The usual phrase, that species are adapted to their environment, should therefore be read inversely, stating that most species are now found to live under conditions fit for them. The adaptation is not on the side of the species, but on that of the environment. In a popular way we could say that in the long run species choose their best environment.
>
> (De Vries, 1922:226–227)

SUMMARY AND RECOMMENDATIONS

De Vries developed the full range of the punctuational paradigm because he began with the same empirical basis as punctuated equilibrium, and then worked through the same principles of selectionist logic (applied at the higher level of trends) to the same evident conclusions. Yet de Vries's version is convergent upon punctuated equilibrium, not homologous, both for the historical reason that Eldredge and I were ignorant of his formulation (our fault, but with a happy result for illustrating a principle in the history of ideas), and for the much more important logical reason that de Vries developed the necessary empirical basis from an entirely different rationale. The basis is punctuation (at geological scales) for origin, and stasis for the duration, of species. We moved to this result by working out the scaling of allopatric speciation in geological time and by thinking about developmental constraint, stabilizing selection, and evolutionary rates in large and small populations. De Vries developed an even stronger version of punctuation and stasis with a truly saltational theory for the origin of species and a view of variability that contrasted an omnipresent, continuous mode with a rare but effective saltational mode (thus confining change to the rare moments, and leaving species in stasis otherwise).

I should hardly need to add that neither Eldredge nor I accept de Vries's version; we have never been saltationists, and the punctuational paradigm requires no such argument since ordinary allopatric speciation, properly scaled into geological time, yields adequate speed for the origin of species. The paradigm requires punctuation, and is agnostic about the explanation offered for this evident empirical phenomenon. Our aim is not to revive de Vries because he was right, but to explicate his fascinating system because he was so incisive in applying Darwinian logic to the basic pattern of the fossil record. De Vries's effort yielded a powerful set of ideas, highly useful and basically correct in our judgement, even though he chose an invalid rationale for his justified empirical conclusions. (Incidentally, I discovered de Vries's convergence while reading his works *in extenso* for the historical half of my forthcoming book, *The Structure of Evolutionary Theory*.)

From this differing rationale for the same empirics of punctuation and stasis, de Vries correctly applied selectionist logic to the higher level of trends (using species within clades as analogous to individuals within populations at the usual Darwinian level of anagenesis in lineages). He developed all major parts of the complex argument – the concept of hierarchical levels for selection, the driving of sorting among species either by differential origin or persistence, and the critique of adaptation. As an added fillip, he even invented the same name – species selection – that supporters of punctuated equilibrium would later propose anew.

I take enormous personal pleasure in recognizing this punctuational precursor by convergence (amidst my chagrin for not discovering de Vries's arguments sooner). If I may make one explicitly arrogant statement on the 20th anniversary of punctuated equilibrium, the existence of such a complex convergence does show that the punctuational paradigm is 'out there' in some sense amidst the field of interesting and coherent evolutionary positions. It is not an idiosyncratic construction, a figment of the personal imaginations of Eldredge & Gould. The paradigm may not be valid, or even very useful (this is an empirical issue requiring much more research), but if two different research programmes, working 70 years apart, could derive the same structure of explanation by applying the same logic to the same empirical basis (differently defended), then the paradigm has philosophical coherence.

Finally, for all the similarities, we must mark one key difference between de Vriesian saltationism and punctuated equilibrium as bases for the punctuational paradigm. De Vries's multilevelled system set up a conflict of causes between levels, with a unitary process assigned to each and little or no spillover. Selection (on species) ruled at the higher level of trends, but was effectively banished at Darwin's own level of anagenesis within populations, where saltation ruled. De Vries wrote (1905:744–745):

> Having previously dealt with species selection at sufficient length, we may now confine ourselves to the consideration of the intra-specific selection process. In practice it is of secondary importance, and in nature it takes a very subordinate position.

Punctuated equilibrium, on the other hand, has tried to build a fully hierarchical system, with selection acting simultaneously at all levels, albeit in different ways, and with interactions of both positive and negative feedback between levels. Since punctuated equilibrium denies saltation and supports geological scaling of ordinary speciation as a rationale for the empirics of punctuation, Darwinian processes remain operative, and potentially dominant, in Darwin's own realm of immediate change within populations.

De Vries exiled his personal hero from his hero's own domain, and then put him back at a higher level. But imagine the frustration: de Vries's Darwin sits and rules in a world above a glass floor; but he looks through the glass in utter frustration at the world he truly loved and cared about, unable to exert any force, and watching it work by a process that was anathema to him. The Darwin of punctuated equilibrium influences, and may even control, all levels. He cannot build the entire system, as he hoped, by extrapolating up from the floor. The horizontal glass partitions are in place, but they are porous – and information passes, clashes and melds in the most fascinating manner to produce those "endless forms most beautiful and most wonderful". I will not presume intimacy with the psyche of Charles Darwin any more than I can claim similar insight into the complex mind and motivations of Hugo de Vries. Yet, somehow, I rather think that Darwin would prefer ranging through the floors of the edifice of punctuated equilibrium to restricted residence on high in de Vries's building.

REFERENCES

DARWIN, C., 1859. *On the Origin of Species*. London: John Murray.

DARWIN, C., 1868. *The Variation of Animals and Plants Under Domestication*. London: John Murray.

DARWIN, C., 1871. *The Descent of Man and Selection in Relation to Sex*. London: John Murray.

DARWIN, C., 1875a. *Insectivorous Plants*. London: John Murray.

DARWIN, C., 1875b. *The Movement and Habits of Climbing Plants*. London: John Murray.

DARWIN, C., 1880. *The Power of Movement in Plants*. London: John Murray.

DE VRIES, H., 1889. (English translation, 1910). *Intracellular Pangenesis*. Chicago: Open Court Publishing Co.

DE VRIES, H., 1905. *Species and Varieties: Their Origin by Mutation*. Chicago: Open Court Publishing Co.

DE VRIES, H., 1909. *The Mutation Theory*. London: Kegan, Paul, French, Trübner.

DE VRIES, H., 1922. Age and area and the Mutation Theory. In J.C. Willis (Ed.), *Age and Area:* 222–227. Cambridge: Cambridge University Press.

ELDREDGE, N. & GOULD, S.J., 1972. Punctuated equilibria: an alternative to phyletic gradualism. In T.J.M. Schopf (Ed.), *Models in Paleobiology:* 82–115. San Francisco: Freeman, Cooper & Co.

GOULD, S.J., 1982. The meaning of punctuated equilibrium and its role in validating a hierarchical approach to macroevolution. In R. Milkman (Ed.), *Perspectives on Evolution*. Sunderland, Massachusetts: Sinauer Associates.

GOULD, S.J., 1989. *Wonderful Life: The Burgess Shale and the Nature of History*. New York: W.W. Norton.

GOULD, S.J. & ELDREDGE, N. 1977. Punctuated equilibria: the tempo and mode of evolution reconsidered. *Paleobiology, 3:* 115–151.

GOULD, S.J. & ELDREDGE, N. 1986. Punctuated equilibrium at the third stage. *Systematic Zoology, 35:* 143–148.

GOULD, S.J. & VRBA, E.S. 1982. Exaptation – a missing term in the science of form. *Paleobiology, 8:* 4–15.

MERTON, R.K., 1963. Resistance to the systematic study of multiple discoveries in science. *European Journal of Sociology, 4:* 237–282.

STANLEY, S.M., 1975. A theory of evolution above the species level. *Proceedings of the National Academy of Sciences, U.S.A., 72:* 646–650.

STANLEY, S.M. & YANG, S., 1987. Approximate evolutionary stasis for bivalve morphology over millions of years: a multivariate, multilineage study. *Paleobiology, 13:* 113–139.

VAN DER PAS, P.W., 1970. The correspondence of Hugo de Vries and Charles Darwin. *Janus, 57:* 173–213.

VON BAER, K.E., 1866. Über Prof. Nic. Wagner's Entdeckung von Larven, die sich fortpflanzen, Herrn Garren's verwandte und ergänzende Beobachtung und Überhaupt. *Bulletin de l'Académie impériale des Sciences de St Petersbourg, 9:* 63–137.

VRBA, E.S., 1980. Evolution, species and fossils: How does life evolve? *South African Journal of Science, 76:* 61–84.

WILLIAMS, G.C., (1966). *Adaptation and Natural Selection: A Critique of Some Current Evolutionary Thought*. Princeton: Princeton University Press.

WILLIS, J.C., 1922. *Age and Area*. Cambridge: Cambridge University Press.

CHAPTER

2

Making sense of microevolutionary patterns

PETER R. SHELDON

Descriptive bias 20

Reversals 21

A spectrum of patterns in different environments 26

Acknowledgement 29

References 29

Keywords: Microevolution – punctuation – gradualism – stasis – reversals – descriptive bias – environment — Ordovician – trilobites.

Abstract

Three major points are often overlooked in the continuing debate over the relative frequency of gradualistic and punctuated evolutionary patterns: (1) biases exist during the collection and description of fossils that tend to generate an impression of punctuation and stasis, irrespective of the real pattern; (2) reversals can be *expected* to occur in any prolonged trend in a single character; (3) there may be important discrepancies between the nature of evolutionary patterns observed in the short term (over usual biological timescales) and those that emerge over geological timescales.

Unidirectional evolution over geological time intervals is highly unlikely. Even more improbable is prolonged unidirectional evolution at a constant rate, yet gradualistic evolution is still depicted in theoretical diagrams as straight-line change, a very special case. Reversals probably account for the observed inverse relationship between measured rates of evolution and the timespan under consideration, and have many other consequences for evolutionary studies.

According to a recent model, continuous evolution is characteristic of relatively stable environments, whereas stasis tends to prevail in unstable environments. The model predicts a tendency for continuous phyletic evolution offshore, and in the tropics generally, and for stasis in shallow waters (the setting for most of the macrofossil record), and in temperate zones.

Evolutionary Patterns and Processes
ISBN 0-12-440895-8

DESCRIPTIVE BIAS

The initial work that led to some of the points discussed here was a study of *c.* 15 000 Ordovician trilobites from central Wales, which revealed a pattern of broadly parallel, gradualistic evolution in eight lineages (Sheldon, 1987). Over a period of about 2 million years, the mean number of pygidial ribs (a character used widely for species discrimination) increased in all eight benthic lineages, but changes took place at different times in different lineages. Changes in other features (e.g. axial width, glabellar ridges and furrows, pits on trinucleid cephalic fringes) seem also to have followed a pattern closer to gradualism than punctuated equilibrium. The end members of most lineages had previously been assigned to different species and, in one case, to different genera. The 'missing links', however, were no longer missing: intermediate horizons yielded specimens of intermediate morphology. For this reason, and because of temporary trend reversals, taxonomic subdivision of each lineage into successive chronospecies proved impracticable. To have forced specific names on isolated segments of the lineages would also have obscured the pattern of evolution, and led to future taxonomic problems with new collections from stratigraphically intermediate sites.

The sampling resolution was unusually high for a study of macrofossils. Samples were collected from 380 levels with an average stratigraphic thickness of 23 cm. The Teretiusculus Shales, which yielded the trilobites, are remarkably uniform in lithology by comparison with many other deposits, and they accumulated at a very high sediment accumulation rate: *c.* 500 m in total thickness over *c.* 2 million years. Each collected sample represents, *on average*, a timespan that cannot be more than about 900 years (this is the 'minimum acuity' *sensu* Schindel, 1982). In practice, because of inevitable gaps in the continuity of sedimentation, the samples will represent, on average, significantly less time than this.

The apparent success of earlier Linnean nomenclature (with its implications of discrete species) could easily have been misinterpreted as evidence of punctuation and stasis in these trilobites. It therefore seems very likely that gradualistic patterns elsewhere, and in other groups, have been obscured by standard descriptive procedures that encourage an impression of punctuation and stasis, irrespective of the real pattern.

The main descriptive biases, each one of which can be seen in retrospect to have influenced the perception of evolutionary patterns in the Builth trilobites, are as follows:

1. The requirement to apply binominal taxonomy to fossils as well as living organisms. Linnean names, particularly when plotted as vertical bars in stratigraphic range charts, cannot but give an impression of abrupt transition between discrete, static species (Sheldon, 1987).

2. The amalgamation of specimens from different horizons in order to amass enough material for full 'species' description, often a necessary practice among palaeontologists describing macrofossils. Any evolution during the timespan encompassed by the collection of specimens – including net directional change, net stasis (i.e. after reversals), or a change of variance – will be submerged within one large range of variation. In fact, any tendency for a taxonomist to lump together a wide range of forms (whatever the source of that variation) under one morphospecies,

rather than to split them into separate morphospecies, will produce large apparent differences between described 'species'. In the case of the Builth trilobites, the differences that Hughes (1969–79), in his rigorous species descriptions, had previously used to discriminate species were of a similar kind and degree to those differences used to separate other trilobite species. The Builth trilobites do not seem to have been particularly 'oversplit' or 'overlumped' compared with other trilobites and with other macrofossil groups such as bivalves, ammonites and brachiopods, lending support to the claim that this descriptive bias is widespread.

3. A natural reluctance to use terms implying uncertainty or ambiguity (such as *aff.*, *cf.*, or ?) in faunal lists, in case this is mistaken for imprecision or even lack of effort (Sheldon, 1988a). For obvious reasons, palaeontologists, like biologists, would probably prefer to describe, or to record, discrete species wherever possible. Gradually evolving lineages with relatively continuous fossil records are, however, bound to present taxonomic problems.

4. An absence of formal nomenclature to signify small morphological differences between samples from different stratigraphic levels or from different geographic areas. For instance, there is no accepted convention to denote intermediate morphology between subspecies, and the erection of new subspecies is itself discouraged by some journal editors. When referring to, or identifying the Builth trilobites, for example, it proved most practicable to use simply the generic name (or, in the case of the nileids, the family name), and to state in words the character value(s) of an individual trilobite or the mean character value(s) of an assemblage. Using a species name that is unworkable with large sample sizes, or with specimens collected from strata immediately above or below, is a sham.

In addition to these descriptive biases – an influence which could be termed punctuated taxonomy – another bias, long recognized, generates an impression of abrupt change: gaps in the stratigraphic record. The more incomplete the stratigraphic record, the more likely that intermediate forms will have no fossil record, irrespective of the pattern of evolution.

REVERSALS

Another major bias that has hindered understanding of evolutionary patterns is a deep predisposition not to expect reversals, a tendency perhaps more common among palaeontologists than biologists. Reasons for this predisposition are numerous, and include the following:

1. Conscious or subconscious reference to Dollo's Law, a principle generally phrased as 'evolution is irreversible'. However, this idea should only be taken to signify the extreme improbability that a descendent lineage will exhibit exactly the same range of phenotypes as an ancestor, and Dollo's Law should never be applied to just one or a few traits. Even complex traits may occasionally be resurrected a long time after apparent loss: the existence of atavisms suggests that genomes can sometimes maintain a reservoir of latent integrated features that may remain unexpressed for millions of years (for discussion see Sheldon, in press b). Simpson (1953:311) gave a clear discussion of irreversibility and stated "evolution should be reversible, and it is". Likewise, in a far-reaching but neglected paper entitled 'Zig-zag

evolution', Henningsmoen (1964) considered reversals, and, describing various patterns, expressed reservation about unidirectional evolution (1964:346): "It is debatable whether or not some apparently one-way evolutionary trends are truly irreversible". He referred (1964:351) to "the disbelief in reversible evolution and the non-acceptance of repetitive (zig-zag) evolution", positions that were especially prevalent in the first half of this century. Judging from the frequency with which I have experienced a similar reaction when presenting trilobite data showing frequent reversals, this bias is still current, albeit to a lesser extent.

2. The almost universal textbook depiction of phyletic evolution as unidirectional change. For example, in Stanley's influential book on macroevolution (Stanley, 1979), not one of the many theoretical diagrams contrasting punctuated equilibrium with phyletic gradualism depicts any reversals. Such diagrams create a false dichotomy, as explained below.

3. 'The Roman Road Philosophy'; Occam's Razor; and the Principle of Parsimony. As for archaeologists starting to excavate a Roman Road between two towns, there is always the temptation to imagine that any two points will be connected by a straight line. However, unlike Roman roadbuilders, evolution, of course, has no purpose, and the most simple way of linking successive data points – a straight line – is, in fact, a very special case. A parsimonious approach, so often employed in science elsewhere, as in cladistics, is not appropriate here. A unidirectional trend with varying rate is unlikely enough, and becomes increasingly less likely the longer the timespan being considered; straight-line evolution (i.e. unidirectional evolution at constant rate) is even more improbable, yet this is the most frequent theoretical representation of phyletic change.

4. Untenable notions of inevitable progress in evolution, and the lingering legacy of orthogenesis.

By contrast, we should *expect* to find reversals in any feature of a lineage that maintains a range of variation and undergoes prolonged change. Given the enormous number of traits that are potential candidates for selection at any one time, it is unreasonable to expect that the one feature we choose to plot was consistently the only one favoured by selection, or that it was always linked to every other favoured trait. There is also the possibility (if not probability) that selection on a feature may reverse its sign (+/−). Indeed, for these reasons, I would argue that if one has *not* found reversals, then the lineage has not yet been sampled in sufficient detail.

Straight-line evolution is a very special case, almost like drawing one random walk and expecting evolution to exhibit the pattern of that particular path. I am not aware of any high-resolution study of individual features in a single fossil lineage – that is one represented by many successive samples – in which reversals do not occur. In the case of the Builth trilobites, the lineage that appears at first glance to have undergone the most uniform, unidirectional rate of increase in pygidial ribs is that of the rare *Nobiliasaphus* (see Sheldon, 1987: fig. 4). Yet this impression is almost certainly an artefact; with larger sample sizes and more collecting horizons, a pattern that includes reversals is sure to emerge, as indeed was the case with all the better-represented trilobite lineages. An example showing reversals in just part of one Builth trilobite lineage (*Cnemidopyge*) is given in Fig. 1. Interestingly, Swan (1990), in a computer simulation of evolution by natural selection, found that reversals in individual variables were common over all timescales, despite a consistent environment.

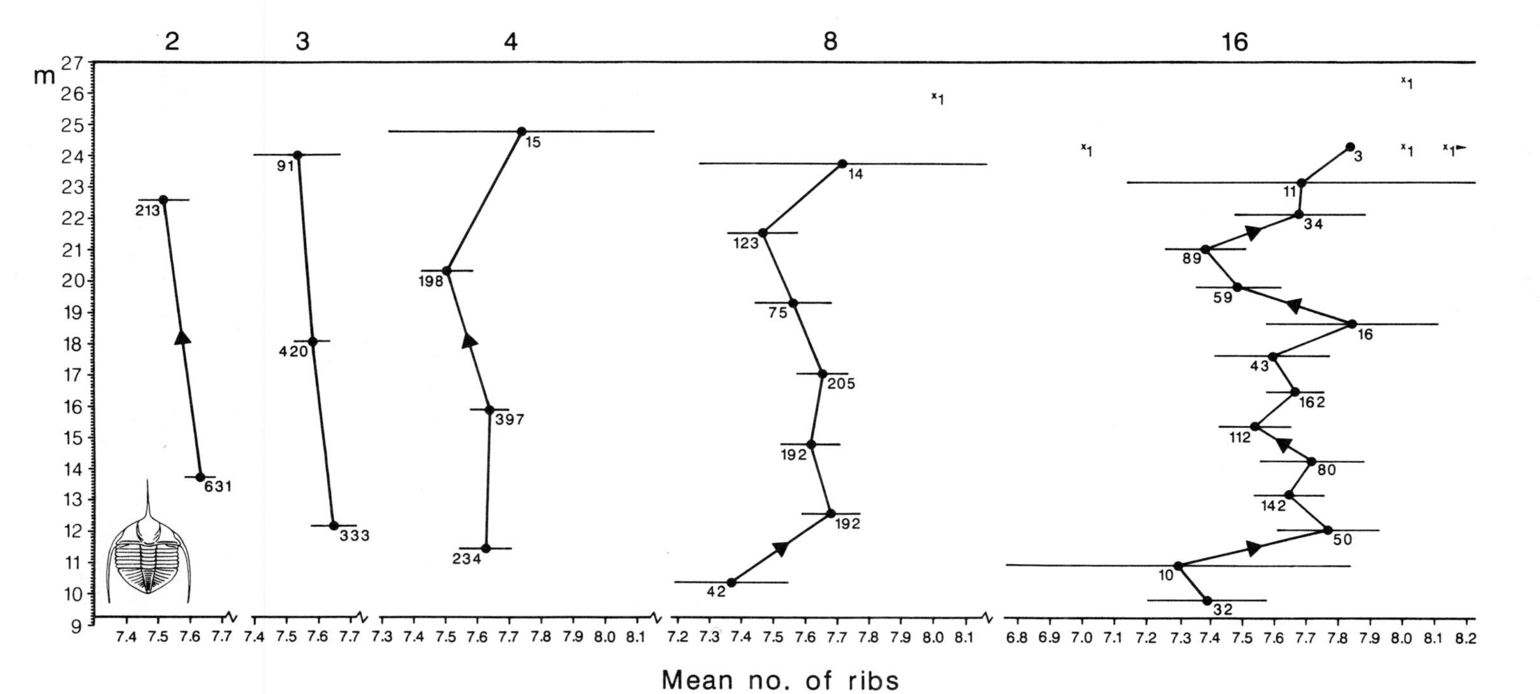

Figure 1 Data for part of the *Cnemidopyge* trilobite lineage (Ordovician) from a single stratigraphic section, showing the effect of sampling strategy, time-averaging and reversals on the perception of evolutionary patterns. The number above each column is the number of subdivisions into which the section has been partitioned. For example, '3' indicates that the section has been split into a lower third, a middle third, and an upper third. Each point represents the mean number of pygidial ribs within that particular subdivision. Beside the mean is its 95% confidence interval and the number of specimens with ribs counted. An arrow indicates a significant difference between two successive means at the 95% confidence level. The stratigraphic thickness of the section analysed is 17.75 m, from 9.28 m to 27.03 m above the base of the exposed sequence. Subdivisions for any given resolution vary slightly in thickness, and the means have been placed at their midpoint. The 17.75 m was originally subdivided into 69 successive collecting localities (of slightly varying thickness), and the finest resolution presented here is 16 subdivisions, for which the locality combinations are: 10–12; 13–18; 19–26; 27–32; 33–37; 38–41; 42–47; 48–52; 53–55; 56–58; 59–63; 64–66; 67–68; 69–72; 73–75 (which had no *Cnemidopyge* specimens with countable ribs); 76–78. The total number of specimens with ribs counted is 844. The data are from the Newmead Lane Stream Section in the Teretiusculus Shales near Builth Wells, Powys, Wales.

For these many reasons, gradualistic evolution should be depicted with reversals in theoretical diagrams, similar to the kind of pattern on the right-hand side of Fig. 2. Taking an analogy from the world of sailing, and without wishing to imply any intentionality to the direction of evolution, one might call these *tacking patterns* produced by *tacking evolution*.

Even if a feature is said to have been in stasis (because it shows the same average value at the beginning and end of the time interval being considered), reversals can be expected to have occurred some time in that period if the feature has maintained a range of variation. For the mean value not to have deviated from the original value over, say, one million years, would be extremely unlikely.

Reversals have many consequences for evolutionary biology and palaeobiology, including the following:

1. Many trends driven by selection may, when resolved in detail, be indistinguishable from random walks (Sheldon, 1987, 1990a). Contrary to the work of Bookstein (e.g. Bookstein, 1988), the fact that patterns may be statistically indistinguishable from random walks does not mean that selection was not involved in determining many successive individual steps within that trend. Runs tests of the kind discussed by Bookstein may enable discrimination of uniform selection pressures on individual traits sustained over geological time periods, but that is all. Bookstein's methods, though mathematically correct, make assumptions concerning the consistency of natural selection that are biologically unrealistic. In any case, the precise nature of any pattern can easily be shown to vary enormously according to the degree of sample aggregation and time-averaging (see Fig. 1, and paragraph 2 below), invalidating the method of runs tests. The impressive, detailed work by Bell (1988 and references therein) on the fossil record of stickleback evolution addresses the complex biological issues that need to be considered when interpreting microevolutionary patterns.

2. Given reversals, most evolutionary patterns will be highly timescale-dependent and sensitive to data amalgamation. Patterns will generally become increasingly dynamic as finer timespans are resolved, as is very clear with the Builth trilobite data (Sheldon, 1990a, and in prep.; see also report in Funnell & Owen, 1990). By merging data from many successive horizons and then breaking them down into increasingly smaller divisions, it is easy to show how perception of patterns is influenced by the stratigraphic interval encompassed by each sample, as illustrated in Fig. 1. As the finest resolution is approached, however, it becomes harder to find significant differences between successive samples because of decreasing sample sizes, and brief intervals of genuine unidirectional change might go undetected.

3. The pervasiveness of reversals explains, at least in part, the general observation (Gingerich, 1983) of an inverse relationship between measured rates of evolution and the timespan under consideration. This effect is closely analogous to the inverse relationship between measured rates of sedimentary deposition and the timespan under consideration (Sadler, 1981). The pervasiveness of reversals probably also explains why evolutionary trends derived from fossils commonly have calculated rates of directional selection so low that they might be expected to be swamped even by genetic drift (Charlesworth, 1984). Short-term rates for such lineages could well have been the same as expected from observations of selection over tens or hundreds of years, but dividing net change in fossil lineages by the total timespan will yield apparent rates that are orders of magnitude lower (see also Levinton, 1988:345).

4. Reversals should not be taken automatically to indicate that the observed pattern reflects ecophenotypic effects or is produced by the migration of ecotypes. The tendency to regard a pattern of reversals as ecophenotypic is rife. For example, Chaline (1990:160), discussing modifications occurring after speciation, writes of "the reversible (ecophenotypic variations) and irreversible morphological variations (phyletic gradualism)". An example in which the reversed migration of a cline is invoked (in theory) to explain a trend reversal, without any consideration of an equally plausible evolutionary reversal is found in Johnson & Colville (1982).

5. Reversals may possibly account for some cases of 'iterative evolution', in which short-lasting 'lineages' showing similar trends are assumed to have budded off repeatedly from a hypothetical unchanging stock (e.g. some of the olenid trilobites discussed by Kaufmann, 1933). Another example in which iterative evolution is invoked for a pattern that might be explained alternatively by rapid phyletic reversals is that of the Jurassic ammonites discussed by Bayer & McGhee (1984, 1985). Their interpretation omits any consideration of the possibility of reversals.

6. Reversals may make it impractical, if not impossible, to subdivide a lineage into successive chronospecies. The more that details of the pattern become resolved, the more difficult lineage subdivision becomes, as is the case with the Builth trilobites.

7. Reversals complicate the theoretical arguments (Fortey, 1985) concerning differentiation between cases of gradualism and punctuated equilibria.

8. Jumps in morphological trends cannot be used to estimate the amount of time missing (i.e. unrepresented) in strata, nor even to indicate where discontinuities might exist, contrary to the method of Brinkmann (1929), who assumed straight-line evolution. As Raup & Crick (1982) showed in their re-analysis of Brinkmann's data for a lineage of the ammonite *Kosmoceras*, there was no close correspondence between known sedimentary breaks and apparent morphological jumps, though morphological jumps were associated with known sedimentary breaks more often than would be expected by chance. The majority of morphological jumps were not at sedimentary breaks, and the majority of sedimentary breaks were not coincident with morphological jumps.

9. If relative ages of strata are based on presumed unidirectional trends in lineages with patchy local records, reversals may produce spurious geological field relationships, such as apparent structural inversion or repetition, and give anomalous species ranges. Reversals may be responsible for several unsuspected miscorrelations and some puzzling correlations.

Reversals in any given pattern of morphological change (other than those resulting from sampling error due to small sample sizes) could be due to one or more of the following possibilities: fluctuating selection pressures, ecophenotypic effects, migration of ecotypes, and genetic drift. Without knowledge of the susceptibility of characters to ecophenotypic variation, or information on geographical variation through time, distinguishing between these possibilities can never be certain; as usual it is a matter of assessing the relative probabilities of competing hypotheses.

An approach to distinguishing ecophenotypic from genetic changes in the fossil record was suggested by a comparison between changes in mean size and in mean number of ribs in two trilobite lineages (Sheldon, 1988c). For *Ogygiocarella* and *Cnemidopyge*, changes in mean rib count are parallel in the long term, but clearly out of phase in the short term, as expected of a heritable variable with only moderate or intermittent selective value in each lineage. By contrast, there is a remarkable degree of short-term parallelism in changes in mean size, probably reflecting a shared ecophenotypic response to some external variable influencing growth, such as temperature and food supply (Sheldon, 1988b). Given sufficiently intense selection pressure for a particular trait, short-term parallel genetic changes can be expected to occur, at least for a while, as, for example, with industrial melanism in moths, spiders and ladybirds (see Lees, 1981 for a review). However, as the vast majority of heritable traits that organisms possess are not preserved in the fossil record, only in very rare cases are we likely to find ourselves plotting traits subject to intense directional selection. In general, therefore, the greater the degree of short-term parallelism in a temporarily reversible trend, the more likely that the short-term changes are due to ecophenotypic effects.

A SPECTRUM OF PATTERNS IN DIFFERENT ENVIRONMENTS

The model of punctuated equilibrium proposed by Eldredge & Gould (1972) has had an immense influence on the way we investigate and interpret evolutionary patterns. Some, for example Levinton (1988), have argued, however, that their version of phyletic gradualism was to some extent a straw man, because they identified one of its tenets as "transformation is even and slow" (Eldredge & Gould, 1972:89). Many earlier authors, such as Simpson (1953), believed that evolution embraced a wide spectrum of rates. Nevertheless, the claim by Eldredge, Gould, Stanley and others that stasis is a major, hitherto largely unrecognized, feature of the fossil record is now indisputable. Stasis, however, remains poorly understood.

The Builth trilobites thrived in a relatively stable, narrowly fluctuating, low energy,

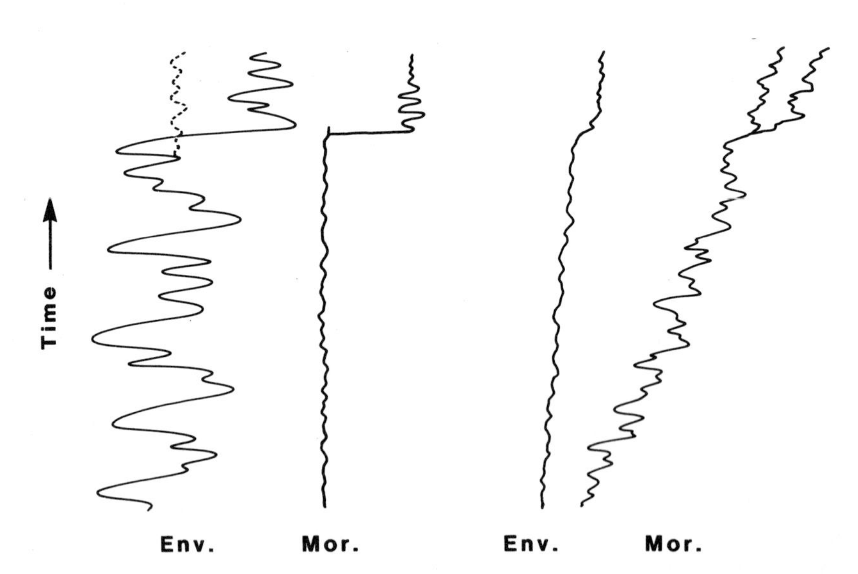

Figure 2 Model illustrating the hypothesis that, over geological timescales, gradual, continuous phyletic evolution is characteristic of narrowly fluctuating, relatively stable environments, whereas stasis tends to prevail in more unstable environments. Env. = some long-term aspect of environment, e.g. sea-level, substrate or mean temperature. Mor. = some aspect of hard-part morphology, such as mean numbers of ribs in molluscs or trilobites. See text for discussion.

dysaerobic environment. They were benthic forms, at least as adults, and probably lived in a silled basin several hundred metres deep (Sheldon, 1987, 1989, in press a). Despite this quiet environmental setting, the Builth trilobites underwent a pattern of long-term phyletic evolution that resembles the gradualism model far more than stasis.

An attempt to embrace evidence from this and other studies, and to bridge the usual timescales of biology and palaeontology, suggested a somewhat counter-intuitive model: continuous phyletic evolution is characteristic of relatively stable environments, whereas stasis tends to prevail in unstable environments (Sheldon, 1990b).

The kind of environmental variables for which the generalization may hold are physical variables studied over geological timescales such as changing sea-level, substrate and climate (e.g. mean temperature). There is overwhelming evidence of morphological stasis lasting millions of years in shallow marine shelly organisms such as bivalves (Stanley & Yang, 1987), bryozoans (Cheetham, 1987; Cheetham & Jackson, 1990), and 'living fossils' such as lingulid brachiopods and some limulids. Cenozoic marine ostracods (Cronin, 1985) and Pleistocene beetles (Coope, 1979, 1990) exhibit the *most* stasis during major high-frequency climatic oscillations. Stasis is certainly more prevalent than would have been predicted from studies of living organisms alone; for example, investigations of natural selection in the wild show that directional selection is more common than stabilizing selection (Endler, 1986).

Perhaps more widely fluctuating environments (seen over *geological* rather than ecological timescales) maintain their own kind of stability within wide reflecting boundaries, and selection soon tends to favour lineages with 'all-purpose' hard-part morphologies that are relatively inert to each environmental twist and turn (Fig. 2). Intuitively, one might perhaps expect a changing environment to lead to changing morphology, and a stable environment to stable morphology. However, over long timescales, the saying "Plus ça change, plus c'est la même chose" may be especially apt, as far as the relationship between physical environment and phyletic evolution is concerned. Whether by selection for habitat tracking (as with the Pleistocene insects described by Coope, 1990) or for widely tolerant morphotypes (eurytopy), such lineages once established might, from their perspective, experience widely fluctuating environments as stable ones until thresholds are reached. In the case of a widely fluctuating environment, such a threshold might occur not only when some environmental variable exceeds wide reflecting boundaries, but also when it *contracts* to become narrowly fluctuating (see Fig. 2).

In quieter, less dynamic environments, perhaps organisms need not be so 'generalist' in the above sense, and finely tuned, less time-averaged adaptations will have longer term value, i.e. short-term fitness benefits are not countered by environmental perturbation over a good many generations. But when a narrowly fluctuating environment does undergo change (as is inevitable), so perhaps must its lineages, particularly if it becomes impossible for them to track (i.e. remain within) a precisely defined environment. Such lineages would be more sensitive (morphologically speaking) to minor environmental nudges, but the nudges themselves would be hard to detect in the geological record.

Perturbations exceeding rather narrow limits of speed and extent might make such lineages prone to cladogenesis or extinction (Sheldon, 1990b). If cladogenesis tends to occur when certain environmental variables exceed 'reflecting boundaries', another factor contributing to high tropical and offshore diversities could be a higher frequency of such events there, albeit events with excursions of far lower amplitude than those elsewhere. For simplicity, only one such event is shown for the narrowly fluctuating environment on the right of Fig. 2.

It is highly significant, and generally consistent with the model in Fig. 2, that the upheaval of the Pleistocene ice age has not (at least, yet) generated many new invertebrate species, and that extinctions occurred mainly at its initiation, except for some late Pleistocene and Recent extinctions for which humans are implicated. As Coope (1990:214) put it, "Those [insects] that survived the first onslaughts of the climatic changes were well fitted to withstand subsequent ones." Note too that, in the model, the amplitude of morphological changes is greater for organisms in a narrowly fluctuating environment. This might be linked to the production of a greater range of variation in a relatively stable environment, under relaxed selection pressures (for examples see Williamson, 1987). Preliminary evidence from the Builth trilobites suggests that variation increased when conditions were most narrowly fluctuating (using sediments and faunal composition as environmental indicators). Presumably lineages that produce a wider range of variation 'when the going is good' are more likely to survive severe new environmental perturbations. This would be an important mechanism for 'lineage selection' (i.e. species selection).

The model predicts a tendency for more continuous phyletic evolution offshore, and in the tropics generally, and for more stasis (and occasional punctuations) in shallow waters and in temperate zones (Sheldon, 1990b). Shallow-water sediments provide the vast majority of the macrofossil record, so if this model is valid as a general tendency, it is not surprising that many macrofossil lineages show stasis and occasional punctuated change. This may help explain why Linnean taxonomy often seems to work rather well for macrofossils as well as for living organisms.

Recent support for the model has come from Parsons (1991a, b; see also Chapter 9) who argues that it is consistent with the evolutionary genetics of populations under varying degrees of environmental stress. Maximum evolutionary rates are expected in habitats characterized by narrowly fluctuating (but not invariant) environments implying moderate stress, moderate genetic variability of ecologically important traits, and not unduly restrictive metabolic costs.

Presumably planktonic lineages, such as planktonic foraminiferans, can often, in effect remain within a narrowly fluctuating environment over long timespans; simply by passive drifting with currents over huge distances some individuals are likely to find themselves in precisely the most suitable environment by chance; those that do not, perish. It is not surprising, therefore, that they often show gradualism despite global climatic oscillations (e.g. Malmgren & Kennett, 1981).

The model is bound to have many exceptions. An immensely complex interplay of factors such as effective population size, dispersal strategy, behavioural complexity, coloniality, to name but a few, will determine the details of patterns in individual cases. Also, individual taxa are likely to exhibit different patterns at different times, and different morphological characters in the same lineage may evolve at different rates (Sheldon, 1990a). One exception to the model in Fig. 2 would appear to be that stasis is prevalent in some benthic deep-sea micro-organisms, such as some bathyal benthic foraminiferans, and ostracods (e.g. Whatley, 1985). However, such environments may not be so narrowly fluctuating as hitherto supposed, and particular life-history strategies (discussed below) may exert very strong influences on patterns. Wilson & Hessler (1987) considered many aspects of deep-sea environments and summarized the little that is known about speciation in the deep sea. They concluded that the basic natural history of most deep-sea species needs to be far better understood before many of the conflicting hypotheses about speciation and evolutionary patterns can be resolved. The fossil record of macro-organisms living at depths much greater than the continental shelf edge is, alas, extremely poor.

There may prove to be a loose association between *K*-strategists and gradualism, and *r*-strategists and stasis (Sheldon, 1989, 1990b), again with a spectrum of intermediate strategies and patterns, and many exceptions. Where the *r*–*K* dichotomy can be applied, *K*-strategists, on ecological timescales, are associated with fairly constant, predictable environments, whilst *r*-strategists are characteristic of variable and unpredictable environments (for discussion see Begon, Harper & Townsend, 1990). But in the same way that 'one man's punctuated equilibrium may be another's evolutionary gradualism', depending on the observational timescale to which the investigator is accustomed (Jones, 1981), so an ecologist's 'unpredictable environment' might be highly predictable over geological timescales. In a similar way, a species that an ecologist reports as very susceptible to local extinction may be highly resistant to *global* extinction (it is likely to be extinction-resistant on a global scale in order for us to observe it becoming extinct frequently on a local scale!).

McKinney (1985) suggested that punctuations in shallow benthic settings, reported as common by Johnson (1982), may be artefacts due to a relatively incomplete stratigraphic record compared with offshore, pelagic environments. However, on the above model, the incompleteness of the record *may only be exaggerating the greater evolutionary abruptness that is there anyway*. The depositional conditions that affect stratigraphic completeness might themselves affect original evolutionary patterns (Sheldon, 1990b). We would expect to find more cases of genuinely gradualistic evolution in relatively complete sections with uniform lithology, and more stasis (and occasional punctuated change) in less complete sections from dynamic environments (e.g. Jurassic and Pleistocene shallow shelf seas around Britain). We could test this by following lineages through sections of varying completeness.

These hypotheses remain to be tested; we are not yet in a position to assess accurately the relative frequency of particular evolutionary patterns and the domain of their expected settings. It is, however, clear that punctuated equilibrium and phyletic gradualism (as usually characterized) are ends of a wide spectrum of possible evolutionary patterns, and that there is no longer any justification for polarization into two opposing and exclusive camps.

ACKNOWLEDGEMENT

This paper has benefited from lively discussion with Peter Skelton.

REFERENCES

BAYER, U. & McGHEE, G.R., 1984. Iterative evolution of Middle Jurassic ammonite faunas. *Lethaia, 17:* 1–16.

BAYER, U. & McGHEE, G.R., 1985. Evolution in marginal epicontinental basins: the role of phylogenetic and ecological factors. In U. Bayer & A. Seilacher (Eds), *Sedimentary and Evolutionary Cycles:* 164–220. Berlin: Springer-Verlag.

BEGON, M., HARPER, J.L. & TOWNSEND, C.R., 1990. *Ecology: Individuals, Populations and Communities.* Cambridge, Massachusetts: Blackwell Scientific.

BELL, M.A., 1988. Stickleback fishes: bridging the gap between population biology and paleobiology. *Trends in Ecology and Evolution, 3:* 320–325.

BOOKSTEIN, F.L., 1988. Random walk and the biometrics of morphological characters. *Evolutionary Biology, 23:* 369–398.

BRINKMANN, R., 1929. Statistisch-biostratigraphische Untersuchungen an mitteljurassischen Ammoniten uber Artbegriff und Stammesenwicklung. *Abhandlungen der Gesellschaft der Wissenschaften zu Gottingen, Mathematisch-Physikalische Klasse,* Neue Folge, Bd. 13, Teil 3.

CHALINE, J., 1990. *Paleontology of Vertebrates.* New York: Springer-Verlag.

CHARLESWORTH, B., 1984. The cost of phenotypic evolution. *Paleobiology, 10:* 319–327.

CHEETHAM, A.H., 1987. Tempo of evolution in a Neogene bryozoan: are trends in single morphological characters misleading? *Paleobiology, 13:* 286–296.

CHEETHAM, A.H. & JACKSON, J.B.C., 1990. Evolutionary significance of morphospecies: a test with cheilostome bryozoa. *Science, 248:* 579–583.

COOPE, G.R., 1979. Late Cenozoic fossil Coleoptera: evolution, biogeography, and ecology. *Annual Review of Ecology and Systematics, 10:* 247–267.

COOPE, G.R., 1990. The invasion of Northern Europe during the Pleistocene by Mediterranean species of Coleoptera. In F. Di Castri, A.J. Hansen & M. Debussche (Eds), *Biological Invasions in Europe and the Mediterranean Basin:* 203–215. Dordrecht: Kluwer Academic Publishers.

CRONIN, T.M., 1985. Speciation and stasis in marine Ostracoda: climatic modulation of evolution. *Science, 227:* 60–63.

ELDREDGE, N. & GOULD, S.J., 1972. Punctuated equilibria: an alternative to phyletic gradualism. In T.J.M. Schopf (Ed.), *Models in Paleobiology:* 82–115. San Francisco: Freeman, Cooper & Co.

ENDLER, J.A., 1986. *Natural Selection in the Wild.* Princeton: Princeton University Press.

FORTEY, R.A., 1985. Gradualism and punctuated equilibria as competing and complementary theories. In J.C.W. Cope & P.W. Skelton (Eds), Evolutionary case histories from the fossil record. *Special Papers in Palaeontology, 33:* 17–28.

FUNNELL, B.M. & OWEN, A.W., 1990. Palaeocomputing: keyboard to the past. *Journal of the Geological Society, London, 147:* 393–395.

GINGERICH, P.D., 1983. Rates of evolution: effects of time and temporal scaling. *Science, 222:* 159–161.

HENNINGSMOEN, G., 1964. Zig-zag evolution. *Norsk Geologisk Tiddskrift, 44:* 341–352.

HUGHES, C.P., 1969-79. The Ordovician trilobite faunas of the Builth-Llandrindod inlier, central Wales. *Bulletin of the British Museum (Natural History)* (Geology), 18 (1969): 39–103, *20*(1971): 115–182, *32*(1979): 109–181.

JOHNSON, J.G., 1982. Occurrence of phyletic gradualism and punctuated equilibria through geologic time. *Journal of Paleontology, 56:* 1329–1331.

JOHNSON, M.E. & COLVILLE, V.R., 1982. Regional integration of evidence for evolution in the Silurian *Pentamerus-Pentameroides* lineage. *Lethaia, 15:* 41–54.

JONES, J.S., 1981. An uncensored page of fossil history. *Nature, 293:* 427–428.

KAUFMANN, R., 1933. Variationsstatistische Untersuchungen uber die 'Artabwandlung' und 'Artumbildung' an der oberkambrischen Trilobitengattung *Olenus* Dalm. *Abhandlungen-Geologisch Palaeontologisches Institut Universität Greifswald, 10:* 1–54.

LEES, D.R., 1981. Industrial melanism: genetic adaptation of animals to air pollution. In J.A. Bishop & L.M. Cook (Eds), *Genetic Consequences of Man Made Change:* 129–176. London: Academic Press.

LEVINTON, J., 1988. *Genetics, Paleontology, and Macroevolution.* Cambridge: Cambridge University Press.

McKINNEY, M.L., 1985. Distinguishing patterns of evolution from patterns of deposition. *Journal of Paleontology, 59:* 561–567.

MALMGREN, B.A. & KENNETT, J.P., 1981. Phyletic gradualism in a Late Cenozoic planktonic foraminiferal lineage; DSDP Site 284, southwest Pacific. *Paleobiology, 7:* 230–240.

PARSONS, P.A., 1991a. Stress and evolution. *Nature, 351:* 356–357.

PARSONS, P.A., 1991b. Evolutionary rates: stress and species boundaries. *Annual Review of Ecology and Systematics, 22:* 1–16.

RAUP, D.M. & CRICK, R.E., 1982. *Kosmoceras*: evolutionary jumps and sedimentary breaks. *Paleobiology, 8:* 90–100.

SADLER, P.M., 1981. Sediment accumulation rates and the completeness of stratigraphic sections. *Journal of Geology, 89:* 569–584.

SCHINDEL, D.E., 1982. Resolution analysis: a new approach to the gaps in the stratigraphic record. *Paleobiology 8:* 340–353.

SHELDON, P.R., 1987. Parallel gradualistic evolution of Ordovician trilobites. *Nature, 330:* 561–563.

SHELDON, P.R., 1988a. Making the most of the evolution diaries. *New Scientist, 117:* 52–54.

SHELDON, P.R., 1988b. Trilobite size-frequency distributions, recognition of instars, and phyletic size changes. *Lethaia, 21:* 293–306.

SHELDON, P.R., 1988c. An approach to distinguishing ecophenotypic from genetic changes in the fossil record. *Geological Society of America Abstracts with Programs, 20:* A46–47.

SHELDON, P.R., 1989. Environmental setting, community structure and ecological strategy of gradually evolving Ordovician trilobites. *Geological Society of America Abstracts with Programs, 21:* A253.

SHELDON, P.R., 1990a. Microevolution and the fossil record. In D.E.G. Briggs & P.R. Crowther (Eds), *Palaeobiology: a Synthesis:* 106–110. Oxford: Blackwell Scientific.

SHELDON, P.R., 1990b. Shaking up evolutionary patterns. *Nature, 345:* 772.

SHELDON, P.R., 1992. Descriptive bias, reversals, and a counter-intuitive model of evolution in different environments. *Geographical Society of America Abstracts with Programs 24:* A138–139.

SHELDON, P.R., in press a. Evolutionary and palaeobiological insights from Ordovician trilobites of central Wales. *Proceedings of the Geologists' Association.*

SHELDON, P.R., in press b. The evolution of form. In P.W. Skelton (Ed.), *Evolution – a Biological and Palaeontological Approach.* London: Addison Wesley.

SIMPSON, G.G., 1953. *The Major Features of Evolution.* New York: Columbia University Press.

STANLEY, S.M., 1979. *Macroevolution – Pattern and Process.* San Francisco: Freeman.

STANLEY, S.M. & YANG, X., 1987. Approximate evolutionary stasis for bivalve morphology over millions of years: a multivariate, multilineage study. *Paleobiology, 13:* 113–139.

SWAN, A.H.R., 1990. A computer simulation of evolution by natural selection. *Journal of the Geological Society, London, 147:* 223–228.

WHATLEY, R.C., 1985. Evolution of the ostracods *Bradleya* and *Poseidonamicus,* in the deep-sea Cainozoic of the south-west Pacific. In J.C.W. Cope & P.W. Skelton (Eds), *Evolutionary Case Histories From the Fossil Record. Special Papers in Palaeontology, 33:* 103–116.

WILLIAMSON, P.G., 1987. Selection or constraint?: a proposal on the mechanism of stasis. In K.S.W. Campbell & M.F. Day (Eds), *Rates of Evolution:* 129–142. London: Allen & Unwin.

WILSON, G.D.F. & HESSLER, R.R., 1987. Speciation in the deep sea. *Annual Review of Ecology and Systematics, 18:* 185–267.

CHAPTER

3

How big was the Cambrian evolutionary explosion? A taxonomic and morphological comparison of Cambrian and Recent arthropods

DEREK E.G. BRIGGS, RICHARD A. FORTEY & MATTHEW A. WILLS

Introduction	34
Recognizing *Baupläne*	34
The database	35
Cladistic analysis	35
Quantifying disparity	41
Conclusions	43
Acknowledgements	43
References	44

Keywords: Disparity – *Baupläne* – morphological variety – Cambrian explosion – arthropods – problematica – evolution.

Abstract

Claims that the Cambrian 'explosion' resulted in a greater variety of fundamental bodyplans (disparity) than has existed before or since have fuelled speculation on the nature of this most significant of adaptive radiations. Here disparity among Cambrian and Recent arthropods is assessed using both taxonomic and morphological criteria. A database is presented for those Cambrian taxa preserved in sufficient detail to allow consistent coding of attributes, together with representatives of most modern arthropod classes or subclasses. Disparity amongst Cambrian arthropods does not appear to be markedly greater than among those of today, by either of the measures considered.

Evolutionary Patterns and Processes
ISBN 0-12-440895-8

INTRODUCTION

The Cambrian radiation coincided with the appearance in the fossil record of many of the metazoan phyla we recognize today. Its magnitude and 'explosive' nature proved enigmatic for Darwin in formulating his ideas on natural selection, and explanations were sought in terms of clandestine evolution or missing tracts of the fossil record. In spite of the dramatic events near the Precambrian/Cambrian boundary the overriding trend, allowing for perturbations caused by mass extinctions, has traditionally been viewed as one of increasing complexity and diversity through the Phanerozoic.

Work on the most exquisitely preserved of Cambrian fossils (Conway Morris, 1989a, b), most notably the soft-bodied occurrences in the Middle Cambrian Burgess Shale of British Columbia, has revealed numerous forms that cannot be accommodated readily within established taxonomic groups (Briggs & Conway Morris, 1986) – perhaps as many as 23 among the arthropods, for example (Whittington, 1985). It has been argued that these problematic animals represent new bodyplans of equivalent importance or rank to modern classes or even phyla, and that the 'disparity' or variety of *Baupläne* was much greater immediately following the Cambrian radiation than it has been at any time since (Valentine & Erwin, 1987; Gould, 1989). This prompted Gould (1989) to reinterpret the pattern of metazoan evolution as an inversion of the traditional 'cone of increasing diversity' and encouraged speculation on the possibility that special genetic mechanisms were operating in the earliest Phanerozoic. To test this underlying assertion requires the development of quantitative methods of assessing disparity.

The concept of disparity has not been precisely defined, and there is a risk of conflating two issues: how to characterize *Baupläne*, and how to describe and quantify the differences between taxa. In an assessment of morphological disparity in the Cambrian (Briggs *et al.*, 1992) arthropods were used as a proxy for the Metazoa as a whole. Not only are they the most important Cambrian group in terms of numbers of taxa and individuals (Conway Morris, 1986), but they also possess a large range of fossilizable characters that can be used in various forms of analysis.

RECOGNIZING *BAUPLÄNE*

Existing higher taxa can often be defined on the basis of relatively few characters. Within the Arthropoda, the four major divisions (Crustacea, Uniramia, Chelicerata and extinct Trilobita) are clearly diagnosed by arrangements of head appendages and tagmatization. The gaps separating them have made their classification relatively simple, but at the same time concealed their interrelationships. The Cambrian problematica, on the other hand, form no obvious groups, and blur the borders between the established classes. It is therefore necessary to determine whether any 'natural' higher taxa emerge when both Cambrian and Recent forms are considered together. Cladistics provides the most suitable tool for recognizing such groups.

Gould (1991) denied that claims for greater Cambrian disparity rest simply on the assignment of bizarre forms to higher taxa, but continued to assert that some attributes are far more significant in evaluating disparity than others, and should accordingly receive greater weight. He focused on those characters used to define existing higher taxa, a choice justified with reference to 'architectural depth' and temporal conservatism. But how are we to recognize such characters a priori in an assemblage of fossil animals, most of which have no subsequent history? Why should individual fossil arthropods be identified as separate *Baupläne* simply on the basis of variation on a set of themes familiar in modern examples, namely patterns of tagmosis and appendage morphology? An assessment of cladistic sister-group relationships provides a better means for determining where major branching events have probably occurred, and what character transitions or *de novo* features identify them. It seems that cladistic logic can provide at least one approach to the problem of assessing the degree to which claims of disparity can be invested in characters that define *Baupläne* (Gould, 1991).

The database

In our analysis (Briggs *et al.*, 1992) 25 Cambrian and 21 Recent arthropods were scored for 134 morphological attributes (Tables 1 and 2). One representative of each major class or subclass of extant arthropod was included to provide a sample of the range of disparity amongst living arthropods. Standard textbook examples were selected as far as possible to avoid using unusual forms which might bias the analyses. Discrete binary or multistate characters were coded for the presence/absence or number of particular features to produce a database which could be analysed both cladistically and phenetically.

Cladistic analysis

The data were analysed with PAUP 3.0 (Swofford, 1990), using Wagner parsimony. *Aysheaia* was designated as the outgroup, based on its possession of many primitive arthropodan characters, and general agreement on its status as member of a sister group (Onychophora) to the rest of the arthropod phylum. A heuristic search using 200 random additions of taxa, and implementing TBR branch swapping was made, producing six equally parsimonious, 402-step trees. These differed only in details of relatively low-level topology, and a single solution is therefore presented here (Fig. 1).

The number of high-level *Baupläne* can be appraised by a consideration of the topology of the trees. Only one high-level clade, *Branchiocaris* and *Marrella*, comprises extinct animals alone, but both are highly plesiomorphic. It is difficult to support a model of greater diversity at the highest taxonomic level in the Cambrian than the Recent based on this topology.

Some other general observations emerge from the cladogram. The uniramians occupy a primitive position and do not show a close relationship to any of the crustaceans. The current diagnosis of Crustacea eliminates a range of 'crustaceanomorphs' which are interspersed on the cladogram in a way that suggests a close relationship to the living forms. The bona fide chelicerates and the trilobites appear somewhat isolated clades within a larger 'arachnomorph' clade which includes many Burgess problematica. The inclusion of the soft-bodied fossils reveals a large number of taxa with intermediate combinations of character states, making the distinction between the extant major groups less clear.

Table 1 Character codes for Cambrian and Recent arthropods used in constructing Table 2. Binary characters score 1 when the structure or condition pertains and 0 where it does not, except where stated otherwise.

1.	Trunk annulated (0 – the condition in *Aysheaia*), trunk clearly segmented (1), trunk showing no clear correspondence between segments and limbs (2).
2.	Exoskeleton calcium phosphate.
3.	Exoskeleton calcium carbonate.
4.	Body flexure lateral and dorsoventral (0), dorsoventral only (1).
5.	Post cephalic articulation without overlap (0), with overlapping pleurae (1), with articulating half-rings (2).
6.	Cephalic shield absent (0 – the condition in *Aysheaia*, uniramians and some crustaceans), present entire (1), present with tergites (2).
7.	Cephalic shield bivalved.
8.	Trilobation present.
9.	Pleura absent (0), pleura present (1), pleura grooved (2).
10.	Pygidium present.
11.	Doublure present.
12.	Labrum primitively absent (0 – the condition in *Aysheaia*), labrum present (1), labrum lost (2).
13.	Labrum attached (0), detached (1).
14.	Eyes free or pedunculate.
15.	Eyes on dorsal cuticle of head.
16.	Median eye present.
17	Ecdysial sutures absent (0), marginal (1), dorsal (2).
18.	Genal spines present.
19.	Marginal rim present.
20.	Number of cephalic limbs; 1 (0), 2 (1), 3 (2), 4 (3), 5 (4), 6 (5).
21.	Cephalic gnathobases present.
22.	First appendage primitively lacking outer ramus (0 – the condition in uniramians and *Aysheaia*), first cephalic appendage with outer ramus (1), outer ramus lost (2).
23.	First appendage antenniform reduced.
24.	First appendage chelicera.
25.	First appendage 'Great Appendage' with three large proximal podomeres and distal spines.
26.	First appendage 'Great Appendage' terminating in claw.
27.	First appendage n-1 podomere bar-like, with flagellar extension.
28.	First appendage n-2 podomere bar-like, with flagellar extension.
29.	Second appendage secondarily lost.
30.	Second appendage primitively lacking outer ramus (0 – the condition in *Aysheaia*), first cephalic appendage with outer ramus (1), outer ramus lost (2).
31.	Second appendage antenniform.
32.	Third appendage mandibulate, biting with a gnathobasic surface.
33.	Third appendage mandibulate, biting with its tips (uniramians).
34.	Third appendage with gnathal lobe.
35.	Fourth appendage independent (0), medially coalesced (1), fused as a gnathochilarium (2).
36.	Fifth appendage independent (0), medially coalesced (1), fused as a labium (2).
37.	Gut diverticula present.
38.	Number of trunk tagmata (excluding telson); 1 (0), 2 (1), 3 (2).
39.	Number of trunk somites; 25+ (0), 16–24 (1), 15 (2), 12–14 (3), 11 (4), less than 11 (5).

Table 1 *Continued*

40. Trunk limbs not diminishing in size significantly (0), diminishing significantly (1).
41. Trunk gnathobases absent (2); on some (1); on all (0).
42. Inner ramus of trunk appendages reduced or absent.
43. Inner ramus of trunk appendages lobate (0 – the condition in *Aysheaia*) segmented stenopodous (1), foliaceous (2).
44. Inner ramus of trunk appendages lacking spines (0), spinose (1). (Spines not articulated, bristles articulated.)
45. Outer ramus of trunk appendages primitively absent (0), outer ramus present (1), outer ramus lost (2).
46. Outer ramus of trunk appendages unfilamentous.
47. Outer ramus of trunk appendages unsegmented.
48. Outer ramus of trunk appendages lamellate.
49. Distal lobe on outer ramus absent (0), present (1).
50. Anus terminal (0), anus ventral in terminal division of body (1), anus ventral and between somites (2).
51. Anus ventral in pygidium.
52. Post-ventral plate absent (0), present (1).
53. Terminal extension styliform and articulated at the base.
54. Terminal extension paddle-shaped and articulated at the base.
55. Appendages on posteriormost trunk segment incorporated into tail fan.
56. Movable telson appendages present.
57. Appendages on posteriormost tagma (excluding telson) present on all somites (0), present on some (1), absent (2).
58. Terminal extension fringed with setae.
59. Cuticle with tubercles.
60. Articulating rostral spine.
61. Exopod multisegmented, each podomere with rigid, long seta.
62. Abdominal combs.
63. Fusion of adjoining somites to form diplosegments.

Characters unique to single taxa (autapomorphies), not included in Table 2.

Agnostus

64. Second appendage as in *Agnostus*.
65. Outer ramus of trunk appendages as in *Agnostus*.
66. Exoskeleton articulating as in *Agnostus*.
67. Pliable sheath supporting hypostome (or labrum).

Androctonus

68. Pectines.
69. Chelate pedipalps.
70. Aculeus or terminal sting (also homologous to the terminal spine of *Limulus*).

Argulus

71. First appendage triangular, two-segment, plate-like, hooked antennule.
72. Sucking, disc-shaped fourth appendages (first maxillae).
73. Piercing proboscis.
74. Pre-oral spine.
75. Bifurcating abdomen.
76. Exopods on first two thoracopods bearing medially-directed flabellae.

Aysheaia

77. Oral papillae.
78. Cuticle soft, and lacking sclerotized plates.

Table 1 *Continued*

Branchiocaris

79. Second appendage composed of many short, broad podomeres, terminating in a bulbous spine or ?chela.
80. Non-articulating telson processes as flat blades or spines.

Bredocaris

81. Dorsal organ.

Burgessia

82. Outer ramus of cephalic appendages flagellate.
83. Prominent anterior lobes of the gut.
84. Posteriormost appendages unjointed and sickle-shaped.

Calanus

85. Intercoxal plates.
86. First thoracic appendages maxillipeds.

Canadaspis

87. Line of short, peripheral telson spines.
88. Cephalic spine (contrasts with rostral spine).

Cypridina

89. All-enveloping carapace.
90. Third trunk limb lacking true joints – long vermiform organ with long bristles at its distal end.
91. Broad, lamelliform caudal rami bearing several claws.

Derocheilocaris

92 Claw-like caudal rami.
93. Cephalic constriction.
94. Limbs of first thoracic segment modified as maxillipeds but not incorporated into cephalon.

Habelia

95. Terminal extension styliform with single articulation and large serrations.

Homarus

96. Chelate, dimorphic pereiopods.
97. Fused cephalothorax.
98. Epistome between the bases of thc first and third cephalic appendages.
99. Inarticulate rostral spine.
100. First three thoracic appendages specialized as maxillipeds.
101. Antennal scale.

Leanchoilia

102. Dorsally-curved snout on carapace.

Lepas

103. Capitulum.
104. Calcium carbonate plates developed external to the cuticle.
105. Development of plates independent of segmentation.
106. First appendage pedunculate, with associated cement gland.
107. Setose cirri.

Limulus

108. Fused opisthosomal shield.
109. Spines articulating with abdominal somites.

Table 1 *Continued*

Lithobius
110. First trunk appendages poison claws.

Marrella
111. Second appendage composed of single, long, unsetose proximal podomere and five distal setose podomeres.
112. Wedge-shaped cephalic shield with prominent lateral and dorsal arm-like extensions.

Martinssonia
113. Spinose telson projection bifurcating.

Molaria
114. Terminal extension with numerous joints.

Naraoia
115. '*Naraoia*-type' configuration of cephalic and pygidial shield.

Nebalia
116. Bivalved carapace lacking a true hinge.

Odaraia
117. Carapace tubular.
118. Three-fluked tail.

Olenoides
119. Eye ridges.
120. Trilobite caudal cerci.
121. Calcitic eye.

Pauropus
122. First antenna bifurcating.

Periplaneta
123. Mesothoracic wings.
124. Metathoracic wings.

Perspicaris
125. Four spine rows on caudal rami.
126. Bulbous anterior projection of the cephalon.

Sanctacaris
127. First appendage raptorial biramous.
128. Sixth appendage with an outer, antenna-like ramus, and short projections fringed with radiating spines or setae.

Sarotrocercus
129. First appendage composed of many short, broad podomeres, terminating in small chela.

Smynthurus
130. Collophore.

Speleonectes
131. Anterior part of cephalon bearing a small pair of rod-like frontal processes, each with a stout spine, distinctly smaller than the main branch.
132. First maxilliped modified as huge grasping appendage.

Waptia
133. Segmented caudal furca.

Yohoia
134. Subrectangular cephalic shield.

Table 2 Data matrix of character states for 21 Recent and 25 Cambrian arthropods (see Table 1).

Taxon/Char.	1	2	3	4	5	6	7	8	9	10	11	12	13	14	15	16	17	18	19	20	21	22	23	24	25	26	27	28	29	30	31	32	33	34	35	36	37	38	39	40	41	42	43	44	45	46	47	48	49	50	51	52	53	54	55	56	57	58	59	60	61	62	63
Acerentomon	1	0	0	0	0	0	0	0	0	0	0	1	0	0	0	0	0	0	0	4	0	0	1	0	0	0	0	0	1	?	?	0	1	0	0	2	0	1	3	0	2	0	1	0	0	0	0	0	0	0	0	0	0	0	0	0	1	0	0	0	0	0	0
Aglaspis	1	1	0	0	1	2	0	1	2	0	1	1	0	0	1	0	0	1	1	3	1	2	0	?	0	0	0	0	0	2	0	0	0	0	0	0	0	0	4	0	0	0	1	0	2	?	?	?	?	2	0	1	1	0	0	0	0	0	1	0	0	0	0
Agnostus	1	0	1	1	0	2	0	1	2	1	1	1	1	0	0	1	1	1	1	3	1	2	0	0	0	0	0	0	0	2	0	0	0	0	0	0	1	1	5	0	0	0	1	1	1	1	0	0	0	0	1	0	0	0	0	0	0	0	0	0	1	0	0
Alalacomenaeus	1	0	0	1	1	2	0	0	0	0	1	1	0	1	0	1	0	0	0	3	1	2	0	0	1	1	1	1	0	1	0	0	0	0	0	0	0	0	4	0	0	0	1	1	1	0	1	0	0	1	0	0	0	1	0	0	0	1	0	0	0	0	0
Androctonus	1	0	0	1	0	2	0	0	0	0	0	2	?	0	0	1	0	0	0	5	1	2	0	1	0	0	0	0	0	2	0	0	0	0	0	0	1	1	3	0	2	0	1	0	2	?	?	?	?	2	0	0	1	0	0	0	2	0	0	0	0	0	0
Argulus	1	0	0	1	0	1	0	0	0	0	0	2	?	1	0	0	0	0	0	4	1	2	0	0	0	0	0	0	0	1	1	1	0	0	0	0	0	1	5	0	2	0	1	0	1	1	0	0	0	0	0	0	0	0	0	0	2	0	0	0	0	0	0
Artemia	1	0	0	0	0	0	0	0	1	0	0	1	1	1	0	1	0	0	0	4	1	2	1	0	0	0	0	0	0	2	1	1	0	0	0	0	1	1	1	0	0	0	2	0	1	1	1	0	0	0	0	0	0	0	0	1	2	0	0	0	0	0	0
Aysheaia	0	0	0	0	0	0	0	0	0	0	0	0	0	0	0	0	0	0	0	0	0	0	0	0	0	0	0	0	0	0	0	0	0	0	0	0	0	0	3	0	2	0	0	1	0	0	0	0	0	0	0	0	0	0	0	1	0	0	1	0	0	0	0
Branchiocaris	2	0	0	0	0	0	1	0	0	0	0	1	1	0	0	0	0	0	0	1	0	2	0	0	0	0	0	0	0	2	0	0	0	0	0	0	0	0	0	1	2	1	?	?	1	1	0	1	0	0	0	0	0	0	0	0	0	0	0	0	0	0	0
Bredocaris	1	0	0	1	0	1	0	0	0	0	0	1	0	1	0	0	0	0	0	4	1	2	0	0	0	0	0	0	0	1	0	1	0	0	0	0	?	1	5	0	0	0	1	1	1	1	1	0	0	0	0	0	0	0	0	1	2	0	0	0	1	0	0
Burgessia	1	0	0	0	0	1	0	0	0	0	0	2	?	0	0	0	0	0	1	3	0	2	0	0	0	0	0	0	0	1	0	0	0	0	0	0	1	0	5	0	2	0	1	0	1	1	1	1	0	2	0	0	1	0	0	0	1	0	0	0	0	0	0
Calanus	1	0	0	0	0	1	0	0	0	0	0	1	0	0	0	1	0	0	0	4	1	2	0	0	0	0	0	0	0	1	1	1	0	0	0	0	0	1	5	0	1	0	1	0	1	1	0	0	0	0	0	0	0	0	0	1	2	0	0	0	0	0	0
Campodea	1	0	0	0	0	0	0	0	0	0	0	1	0	0	0	0	0	0	0	4	0	0	0	0	0	0	0	0	1	?	?	0	1	0	0	2	0	1	3	0	2	0	1	0	0	0	0	0	0	0	0	0	0	0	0	1	0	0	0	0	0	0	0
Canadaspis	1	0	0	1	0	0	1	0	0	0	0	1	1	1	0	0	0	0	0	3	1	2	1	0	0	0	0	0	0	2	1	1	0	0	0	0	0	1	2	0	0	0	1	1	1	1	1	1	0	0	0	0	0	0	0	0	2	0	0	0	0	0	0
Cypridina	1	0	1	1	0	0	1	0	0	0	1	1	1	1	0	0	0	0	0	3	1	2	0	0	0	0	0	0	0	1	1	1	0	0	0	0	0	0	5	0	1	0	1	1	1	1	1	0	0	0	0	0	0	0	0	1	1	0	0	0	0	0	0
Derocheilocaris	1	0	0	0	0	0	0	0	0	0	0	1	1	0	0	1	0	0	0	4	1	2	0	0	0	0	0	0	0	1	1	1	0	0	0	0	0	1	5	0	1	0	1	0	1	1	0	0	0	0	0	0	0	0	0	1	2	0	0	0	0	0	0
Emeraldella	1	0	0	0	1	2	0	0	0	0	1	1	0	0	0	0	0	0	0	5	1	2	0	0	0	0	0	0	0	2	0	0	0	0	0	0	0	1	3	0	0	0	1	1	1	0	1	0	1	2	0	1	1	0	0	0	1	0	0	0	0	0	0
Habelia	1	0	0	1	1	2	0	1	1	0	0	?	?	0	0	0	0	0	1	2	?	2	0	0	0	0	0	0	0	1	0	0	0	0	0	0	0	0	3	0	?	0	1	1	1	0	1	0	0	1	0	0	1	0	0	0	0	0	1	0	0	0	0
Homarus	1	0	1	1	1	1	0	0	0	1	1	1	1	0	0	0	0	0	0	4	1	1	0	0	0	0	0	0	0	1	1	1	0	0	0	0	1	1	3	0	1	0	1	0	1	1	0	0	0	1	0	0	0	1	1	0	1	0	0	0	0	0	0
Julus	1	1	0	0	2	0	0	0	1	0	0	1	0	0	1	0	0	0	0	3	0	0	0	0	0	0	0	0	1	?	?	0	1	1	2	0	0	0	0	0	2	0	1	0	0	0	0	0	0	0	0	0	0	0	0	0	0	0	0	0	0	0	1
Leanchoilia	1	0	0	1	1	2	0	1	0	0	0	1	0	0	0	0	0	0	0	2	0	2	0	0	1	1	1	1	0	1	0	0	0	0	0	0	0	0	4	0	2	0	1	0	1	0	1	0	0	2	0	0	1	0	0	0	0	1	0	0	0	0	0
Lepas	2	0	1	1	0	0	1	0	0	0	0	1	1	0	0	1	0	0	0	4	1	2	0	0	0	0	0	0	1	?	?	1	?	?	0	0	0	0	5	0	2	0	1	0	1	1	0	0	0	0	0	0	0	0	0	0	0	0	0	0	0	0	0
Lepisma	1	0	0	0	0	0	0	0	0	0	1	1	0	0	1	1	0	0	0	4	0	0	0	0	0	0	0	0	1	?	?	0	1	0	0	2	0	1	3	0	2	0	1	0	0	0	0	0	0	0	0	0	0	0	0	1	1	0	0	0	0	0	0
Lightiella	1	0	0	0	1	2	0	0	1	0	1	1	0	0	0	0	0	0	0	4	1	2	0	0	0	0	0	0	0	1	1	1	0	0	0	0	0	1	1	0	0	0	1	0	1	1	0	0	0	0	0	0	0	0	0	1	2	0	0	0	0	1	0
Limulus	1	0	0	1	0	2	0	1	0	0	1	1	1	0	1	1	1	1	1	5	1	2	0	1	0	0	0	0	0	2	0	0	0	0	0	0	0	0	5	0	2	1	?	?	1	1	1	1	0	2	0	0	1	0	0	0	0	0	0	0	0	0	0
Lithobius	1	0	0	0	0	0	0	0	0	0	0	1	0	0	1	0	0	0	0	4	0	0	0	0	0	0	0	0	1	?	?	0	1	1	1	1	0	0	1	0	2	0	1	0	0	0	0	0	0	0	0	0	0	0	0	0	1	0	0	0	0	0	0
Marrella	1	0	0	1	0	1	0	0	0	0	1	1	0	0	0	0	0	0	0	1	0	2	0	0	0	0	0	0	0	2	0	0	0	0	0	0	0	0	0	1	2	0	1	0	1	0	0	0	0	0	0	0	0	0	0	0	0	0	0	0	0	0	0
Martinssonia	1	0	0	1	0	1	0	0	0	0	0	1	0	0	0	0	0	0	0	4	1	2	0	0	0	0	0	0	0	1	0	0	0	0	0	0	0	1	5	0	1	0	1	0	1	1	1	0	0	1	0	0	0	0	0	0	2	0	0	0	1	0	0
Molaria	1	0	0	0	1	2	0	1	1	0	1	1	0	0	0	0	0	0	1	3	1	2	0	0	0	0	0	0	0	1	0	0	0	0	0	0	1	0	5	0	0	0	1	1	1	0	1	0	0	2	0	0	1	0	0	0	1	0	0	0	0	0	0
Naraoia	1	0	0	1	1	2	0	1	0	1	1	1	0	0	0	0	0	0	1	3	1	2	0	0	0	0	0	0	0	1	0	0	0	0	0	0	1	0	1	0	0	0	1	1	1	0	1	0	1	0	1	0	0	0	0	0	0	0	0	0	0	0	0
Nebalia	1	0	0	1	1	0	1	0	0	0	0	1	1	1	0	0	0	0	0	4	1	1	0	0	0	0	0	0	0	2	1	1	0	0	0	0	0	1	2	0	1	0	2	0	1	1	1	0	0	0	0	0	0	0	0	1	1	0	0	1	0	0	0
Odaraia	2	0	0	0	0	0	1	0	0	0	0	1	1	1	0	1	0	0	0	4	0	2	1	0	0	0	0	0	0	2	1	1	0	0	0	0	0	1	0	1	0	0	1	0	1	1	1	1	0	0	0	0	0	0	0	0	0	0	0	0	0	0	0
Olenoides	1	0	1	1	2	2	0	1	2	1	1	1	0	0	1	0	2	1	1	3	1	2	0	0	0	0	0	0	0	1	0	0	0	0	0	0	1	1	3	0	0	0	1	1	1	0	1	0	1	0	1	0	0	0	0	0	0	0	1	0	0	0	0
Pauropus	1	0	0	0	0	0	0	0	0	0	0	1	0	0	0	0	0	0	0	3	0	0	0	0	0	0	0	0	1	?	?	0	1	1	2	0	0	0	4	0	2	0	1	0	0	0	0	0	0	0	0	0	0	0	0	0	0	0	0	0	0	0	1
Periplaneta	1	0	0	0	0	0	0	0	0	0	0	1	0	0	1	1	0	0	0	4	0	0	0	0	0	0	0	0	1	?	?	0	1	0	0	2	1	1	3	0	2	0	1	0	0	0	0	0	0	0	0	0	0	0	0	0	2	0	0	0	0	0	0
Perspicaris	1	0	0	1	0	0	1	0	0	0	0	1	1	1	0	0	0	0	0	?	?	2	1	0	0	0	0	0	0	?	1	1	0	0	0	0	0	1	2	0	?	1	?	?	1	1	1	1	0	0	0	0	0	0	0	1	2	0	0	0	0	0	0
Plenocaris	1	0	0	0	0	0	1	0	0	0	1	1	1	0	0	0	0	0	0	?	0	2	0	0	0	0	0	0	?	?	?	?	?	?	?	?	0	1	3	0	?	0	1	0	?	?	?	?	?	0	0	0	0	0	0	1	2	0	0	1	0	0	0
Sanctacaris	1	0	0	0	1	2	0	1	1	0	1	1	0	1	0	0	0	0	1	5	1	1	0	0	0	0	0	0	0	1	0	0	0	0	0	0	0	1	4	0	0	0	1	1	1	1	1	1	0	2	0	0	0	1	0	0	1	1	0	0	0	0	0
Sarotrocercus	1	0	0	0	1	2	0	1	1	0	1	1	0	1	0	0	0	0	0	1	0	2	0	0	0	0	0	0	0	1	0	0	0	0	0	0	0	0	5	0	2	1	?	?	1	1	1	1	0	2	0	0	1	0	0	0	?	0	0	0	0	0	0
Scutigerella	1	0	0	0	0	0	0	0	0	0	0	1	0	0	0	0	0	0	0	4	0	0	0	0	0	0	0	0	1	?	?	0	1	1	1	2	0	0	3	0	2	0	1	0	0	0	0	0	0	0	0	0	0	0	0	0	0	0	0	0	0	0	0
Sidneyia	1	0	0	0	1	2	0	0	0	0	1	?	?	1	0	0	1	0	0	0	0	2	0	0	0	0	0	0	0	1	0	0	0	0	0	0	0	2	3	0	0	0	1	1	1	0	1	0	0	1	0	0	0	1	1	0	1	0	0	0	0	0	0
Skara	1	0	0	0	0	2	0	0	0	0	1	1	0	0	0	0	0	0	0	4	1	2	0	0	0	0	0	0	0	1	0	1	0	0	0	0	0	1	4	0	0	0	1	0	1	1	0	0	0	0	0	0	0	0	0	1	2	0	0	0	1	1	0
Smynthurus	1	0	0	0	0	0	0	0	0	0	0	1	0	0	1	1	0	0	0	4	0	0	0	0	0	0	0	0	1	?	?	0	1	0	0	2	0	1	5	0	2	0	1	0	0	0	0	0	0	0	0	0	0	0	0	0	1	0	0	0	0	0	0
Speleonectes	1	0	0	0	1	2	0	0	0	0	1	1	0	0	0	0	0	0	0	4	1	1	0	0	0	0	0	0	0	1	0	1	0	0	0	0	0	1	0	1	1	0	1	0	1	0	0	0	0	0	0	0	0	0	0	1	0	0	0	0	0	0	0
Waptia	1	0	0	0	0	0	1	0	0	0	0	1	1	1	0	0	0	0	0	4	?	2	1	0	0	0	0	0	0	?	1	0	0	0	0	0	0	2	2	0	?	0	1	0	1	0	1	0	0	0	0	0	0	0	0	1	2	0	0	0	0	0	0
Yoboia	1	0	0	0	1	2	0	1	2	0	1	1	0	1	0	0	0	1	0	3	0	2	0	0	1	0	0	0	0	2	0	0	0	0	0	0	0	1	3	0	2	1	?	?	1	1	1	1	0	2	0	0	0	1	0	0	2	0	0	0	0	0	0

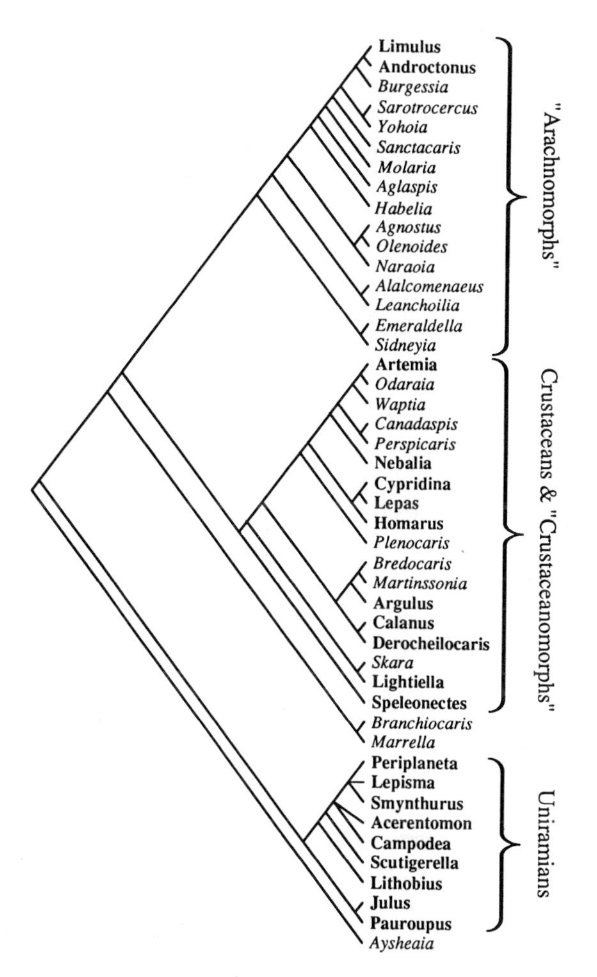

Figure 1 One of six equally parsimonious, 402-step cladograms produced from the data set. The overall consistency index is 0.267, excluding cladistically uninformative characters. Recent arthropods are shown in bold type; Cambrian arthropods are in italics. (After Briggs *et al.*, 1992)

QUANTIFYING DISPARITY

The cladistic approach may help to identify *Baupläne*, but it considers the arrangement of taxa in a hierarchy of nested sets rather than measuring the morphological distance between them. Ideally such distances are measured using morphometric methods which either rely on identifying arrays of point homologies (Bookstein *et al.*, 1985; Foote, 1991a), or comparing the form of entire homologous structures (Foote, 1991b). The range of morphology amongst the arthropods under consideration in our study made this impossible. Instead we applied a phenetic approach to the same database as that used for the cladistic analysis, without standardizing the variables (Briggs *et al.*, 1992). This obviously gives greater weight to ordered, multistate characters (as in the cladistic interpretation), but avoids the undesirable effects caused by the expression of characters with different proportions of states in standard deviation units.

The phenetic approach uses principal component analysis (PCA) based on a covariance matrix, which assists appreciation of the distribution of arthropods in morphological hyperspace. A plot of taxa on their character axes is rotated and referred to a new set of orthogonal dimensions, explaining successively smaller fractions of the total variance. A simple scatterplot of the taxa on the first three components gives a visual impression of their distribution with reference to these abstracted dimensions (Fig. 2). However, points which apparently lie close together on these axes may be far apart in reality, when later components are taken into account. For this reason a minimum-length spanning tree (MST) based on the first 19 components (explaining 90% of the total variance) has been superimposed on the plot. This also assists in visualizing those taxa which are peripherally situated in morphospace.

Three 'groups' of taxa, linked at one point only by the MST, fall in a sequence from the bottom left to the top right of the plot: the uniramians, the crustaceans and 'crustaceanomorphs', and the 'arachnomorphs', i.e. trilobites, chelicerates, and a large number of Cambrian problematica. Cambrian and Recent taxa are interspersed to a great extent. This grouping echoes the larger clades identified by the cladistic analysis.

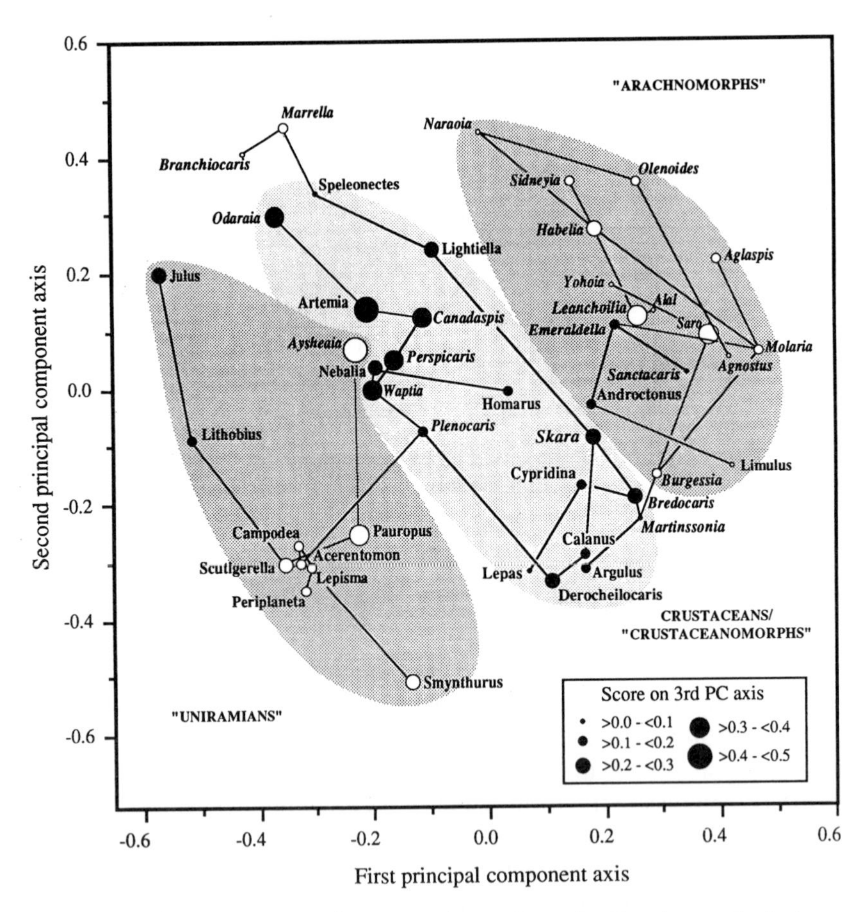

Figure 2 Scatterplot of Cambrian and Recent arthropods on the first three principal components, which together account for 49% of the global variance. The symbols indicate the score of each taxon on the third principal component axis. Open circles indicate negative values. The taxa have also been joined by a Euclidean minimal-length spanning tree, based on their scores on the first 19 PCs (which encompass over 90% of global variance). Recent arthropods are shown in bold type; Cambrian arthropods are in italics. (After Briggs *et al.*, 1992)

Table 3 Cambrian disparity.

Taxa	n	Distance from centroid*	Morphospace occupied†
Recent arthropods	21	0.574	0.0103/0.401
Cambrian arthropods	25	0.586	0.0115/0.420

*Mean Euclidean distance from the centroid of multidimensional attribute space.
†Volume of morphospace occupied by different subsets of taxa calculated as the product of the variances and ranges of the scores in each of the first 19 dimensions of the PCA plot that encompasses over 90% of the total variance.
After Briggs *et al.*, 1992.

A 'volume' disparity measure was calculated for Cambrian and Recent taxa in two ways, as the product of the ranges, and of the variances, of the scores of each subset of animals on each of the first 19 components, which has then been expressed for convenience as the 19th root. Neither estimate of hypervolume (Table 3) suggests that there is markedly more disparity in the Cambrian sample than the Recent, and certainly there is no reason to suppose that the outlying taxa are predominantly Cambrian.

CONCLUSIONS

Disparity is not an easy concept to quantify, but our attempt goes some way toward testing assertions about the nature of the Cambrian 'explosion'. We found no evidence for vastly greater disparity or variety of *Baupläne* in arthropods of the Cambrian than those of the Recent, either in terms of numbers of bodyplans we can recognize, or as expressed phenetically by attribute space occupied. We therefore consider it likely that rates of evolution and levels of morphological 'experimentation' in the Cambrian 'explosion' can be explained in terms of traditionally modelled genetic mechanisms, rather than requiring processes unique to this period of history (Gould, 1989). The recognition of taxa as problematic acknowledges only that we do not yet understand their relationships.

ACKNOWLEDGEMENTS

We acknowledge, with thanks, comments and discussion on the quantification of disparity and on phylogenetic analysis from G.A. Boxshall, E. CoBabe, S. Conway Morris, A. Dennis, S.J. Gould, M. Foote, P.L. Forey, C.B. Moncrieff, J.M.V. Rayner, J.J. Sepkoski, P.H.A. Sneath, A. Swan, J.T. Temple, and M. Wilkinson. This does not, of course, necessarily imply full agreement with the methods used. J.W. Archie generously supplied software, M.A.W. was funded by a University of Bristol Postgraduate Scholarship.

REFERENCES

BOOKSTEIN, F.L., CHERNOFF, B., ELDER, R.L., HUMPHRIES, J.M., SMITH, G.R. & STRAUSS, R.E., 1985. Morphometrics in evolutionary biology. Special publication 15 of the Academy of Natural Sciences of Philadelphia.

BRIGGS, D.E.G. & CONWAY MORRIS, S., 1986. Problematica from the Middle Cambrian Burgess Shale of British Columbia. In A. Hoffman & H. Nitecki (Eds), *Problematic Fossil Taxa*. Oxford: Clarendon Press.

BRIGGS, D.E.G., FORTEY, R.A. & WILLS, M.A., 1992. Morphological disparity in the Cambrian. *Science., 256:* 1670–1673.

CONWAY MORRIS, S., 1986. The community structure of the Middle Cambrian phyllopod bed (Burgess Shale). *Palaeontology, 29:* 423–467.

CONWAY MORRIS, S., 1989a. Burgess Shale faunas and the Cambrian explosion. *Science, 246:* 339–346.

CONWAY MORRIS, S., 1989b. The persistence of Burgess Shale-type faunas: implications for the evolution of deeper-water faunas. *Transactions of the Royal Society of Edinburgh: Earth Sciences, 80:* 271–283.

FOOTE, M., 1991a. Morphological and taxonomic diversity in a clade's history: the blastoid record and stochastic simulations. *Contributions from the Museum of Paleontology, University of Michigan, 28:* 101–140.

FOOTE, M., 1991b. Morphologic patterns of diversification: examples from trilobites. *Palaeontology, 34:* 461–485.

GOULD, S.J., 1989. *Wonderful Life. The Burgess Shale and the Nature of History*. London: Hutchinson Radius.

GOULD, S.J., 1991. The disparity of the Burgess Shale arthropod fauna and the limits of cladistic analysis: why we must strive to quantify morphospace. *Paleobiology, 17:* 411–423.

SWOFFORD, D.L., 1990. PAUP: Phylogenetic Analysis Using Parsimony, Version 3.0. Computer program distributed by the Illinois Natural History Survey, Champaign, Illinois.

VALENTINE, J.W. & ERWIN, D.H., 1987. Interpreting great developmental experiments: the fossil record. In R.A. Raff & E.C. Raff (Eds), *Development as an Evolutionary Process*. New York: Liss.

WHITTINGTON, H.B., 1985. *The Burgess Shale*. New Haven: Yale University Press.

CHAPTER

4

Adaptive radiation: definition and diagnostic tests

PETER W. SKELTON

Introduction 46

Meaning and usage 46

Adaptive and non-adaptive radiations 48

Adaptive radiation 48

Defining and testing for radiation 48

Non-adaptive radiation 49

Diagnostic tests for adaptive radiations 51

Comparative cladistics 51

Stratigraphical range analysis 53

Microevolution and adaptive radiations 54

Conclusion 56

Acknowledgements 56

References 56

Keywords: Adaptive radiation – cladistics – evolutionary terminology – historical testing – preadaptation – radiation – speciation – stratigraphical ranges.

Abstract

A strict definition of adaptive radiation is offered, to avoid the threat of vacuousness arising from over-generalized usage: 'an episode of significantly sustained excess of cladogenesis over extinction, with adaptive divergence cued by the appearance of some form of ecological stimulus'. The effective cue may be either environmental (physical or biotic) or intrinsic (preadaptive) in any given example, although both factors are necessary.

A significant trend for increasing species numbers in a clade (radiation) may be recognized by rejection of the null hypothesis of a random walk in diversity levels. Non-adaptive radiations may result from successive vicariance events and/or as a consequence of species selection.

Evolutionary Patterns and Processes
ISBN 0-12-440895-8

Two diagnostic tests for adaptive radiation have been used: cladistic comparison and stratigraphical range analysis. Cladistic comparison tests for elevated diversity in a clade possessing a key evolutionary innovation, relative to its sister group, with corroboration sought from other, analogous instances. But the method lacks a statistical test, and cannot determine if the innovation was itself the cue. Stratigraphical range analysis plots the changing diversities of sister taxa, and thus allows testing both for radiation (as above) and the nature of the cue (co-incident with the inception of radiation).

There are many possible microevolutionary paths to adaptive radiation. In particular, pre-speciational and post-speciational modes may be recognized, according to whether resource-based adaptive divergence precedes, or follows speciation.

INTRODUCTION

One of the most familiar icons of the neo-Darwinian literature is that of the adaptive radiation. The concept of taxonomic proliferation driven by some new opportunity for multiple adaptive divergence has been applied at all categorical levels, whether to explain the species of Darwin's finches in the Galápagos Islands, or the exponential diversification of metazoan higher taxa through the early Cambrian. Yet the very familiarity of the term has fostered uncritical application of it to almost any history of increase in numbers of taxa, so threatening its explanatory power. This paper is an attempt to rescue it from thus drowning in a sea of sloppy usage.

MEANING AND USAGE

The concept of adaptive radiation combines two components – the adaptive divergence of populations and some cue, or change of circumstances, serving as a trigger to the process. The first is derived from Darwin's own views on the effects of competition between individuals (Darwin, 1859). His one, famous figure in *On the Origin of Species*, depicting an hypothetical phylogenetic scheme with repeated branching of some of the lineages, was intended precisely to show how natural selection might initiate adaptive divergence between varieties within species, and, in time, transform such divisions into species and higher taxa. The relationship between adaptive divergence and speciation is now considered to be not as straightforward as Darwin envisaged (e.g. see Coyne, 1992), and some models of speciation, in particular, dispense with prior adaptive divergence (as discussed later). Nevertheless, some linkage between the two processes is a logical necessity in the context of adaptive radiation.

Darwin did not regard the occurrence of adaptive divergence as especially episodic in the longer term, believing deficiencies in the fossil record to give an exaggerated picture of biotic replacement. In this respect he was paying the dues of his allegiance to Lyellian gradualism, which, he believed, refuted the catastrophic (and anti-evolutionary) ideas of earlier geologists. Subsequent stratigraphical palaeontology has, of course, confirmed a strongly episodic overprint on the background pattern of biotic turnover (Sepkoski, 1986), while modern evolutionary theory is no longer embarrassed by the phenomenon of catastrophic extinction and repopulation. Indeed the latter is now seen as a major opportunity for adaptive radiation by the survivors.

Four kinds of cue – the second component of adaptive radiations – are thus available to today's neo-Darwinian theorist (Simpson, 1949, 1983):

1. Chance colonization of new or previously unoccupied habitats (e.g. the classic examples of Darwin's finches and the Hawaiian honeycreepers).
2. Following a mass extinction, the invasion, by the survivors, of adaptive zones previously denied them by incumbent competitors and/or predators, as noted above (e.g. the Tertiary radiations of mammals following the terminal Cretaceous mass extinction).
3. The appearance of new biotic stimuli to co-evolutionary divergence afforded by other radiations (e.g. the radiations of flower-pollinating insects in tandem with that of the angiosperms).
4. The evolution of features that prove to be preadaptive for the penetration of previously inaccessible adaptive zones (e.g. the evolution of the shelled egg in reptiles and the consequent invasion of terrestrial habitats).

The first three factors are extrinsic and the fourth, intrinsic. Eventual radiation relies on a combination of both kinds of factor: only suitably preadapted organisms (4) can radiate when a given environmental opportunity (1, 2, 3) arises; conversely, a preadaptation will only serve as such in a receptive environment. Nevertheless, one of these factors can usually be identified as having been the effective cue for radiation in any given instance, and the other(s) will thus have been prior 'enabling circumstances' (Skelton, 1991).

That such radiations have occurred, triggered by the kinds of causes listed above, would probably be disputed by few. The real problem, as stated earlier, has been the tendency for the term adaptive radiation to be applied uncritically to almost any history of taxonomic increase. If used in so generalized a sense as merely to refer to any combination of increased taxonomic diversity and eventual adaptive scope within a clade, then it is indeed difficult to imagine an example of taxonomic increase that would not thus qualify: an exception would require the unlikely state of affairs whereby all the species concerned shared identical niches, with no consequent broadening of the clade's ecological repertoire. Used thus, the expression risks becoming vacuous as an explanation. A sample of definitions of adaptive radiation from recent texts will serve to illustrate how far the rot has gone in this respect.

> Evolutionary divergence of members of a single phyletic line into a series of different niches or adaptive zones.
>
> (Mayr, 1970:413)

> The phenomena and process of different adaptations arising in descendants of related ancestry.
>
> (Simpson, 1983:227)

> Evolutionary divergence of members of a single phylogenetic line into a variety of different adaptive forms; usually with reference to diversification in the use of resources or habitats.
>
> (Futuyma, 1986:550)

> Adaptive radiation is the rapid proliferation of new taxa from a single ancestral group.
>
> (Stanley, 1979:65)

Only the last definition specifies the episodic nature of the phenomenon (implying causation by the appearance of a particular cue), but that definition is itself deficient in its tacit implication that all such instances are adaptively driven.

What is needed, then, is a more restrictive definition of adaptive radiation, which in turn begs the question of what might constitute a 'non-adaptive radiation' (Gittenberger, 1991). This consideration in turn requires that means of testing be established for distinguishing between adaptive and non-adaptive radiations. Such are the objectives of this paper.

ADAPTIVE AND NON-ADAPTIVE RADIATIONS

Adaptive radiation

Adaptive radiation itself presents the least of the definitional problems, although, as discussed later, recognition of the phenomenon can be problematical. Expressing the concept as outlined in the previous section in slightly more formal terms, it may be defined as an episode of significantly sustained excess of cladogenesis over extinction (as explained below), with adaptive divergence cued by the appearance of some form of ecological stimulus. Notwithstanding the inadequacy of the other definitions cited earlier (and others in a similar vein), it is usually evident, from the examples which are given, that something more like the definition stated above is actually intended. The form of the cue, be it a key evolutionary innovation or an environmental opportunity of some kind, is generally specified in each case (as for the examples mentioned in the previous section), and the exponential nature of the initial phase of diversification often remarked upon (e.g. Stanley, 1979, 1990). These, and other such examples are the adaptive radiations we know and love (and teach).

What then is a non-adaptive radiation? Before answering that, it must be decided what constitutes an evolutionary radiation in the first place, so that the constituency within which the adaptive and non-adaptive versions are to be distinguished can be delimited.

Defining and testing for radiation

The number of species existing at any one time in a clade is a budgetary consequence of earlier speciations and extinctions. Taxonomic diversity rises, stays the same, or falls as the net rate of speciation respectively exceeds, equals or falls below that of extinction. Raup *et al.* (1973) established a model for generating stochastic phylogenies in which the probabilities of speciation and extinction were set equal (after an initial start-up phase) and clade diversities were allowed to fluctuate randomly (Fig. 1). Such a model provides a null hypothesis against which the dynamics of real clades may be tested for deviation from random patterns of change. Given that, from any one time interval to the next, the diversity of a clade can rise or fall with equal probability according to the null hypothesis, the sequence of rises and falls of a real clade over a series of many intervals can readily be tested against the binomial probability distribution of random walks with the corresponding number of steps (Skelton *et al.*, 1990). Should a real clade show a more or less sustained trend of growth that reaches the selected critical region of the probability distribution (e.g. $P = 0.05$; one-tailed test) then an "episode of significantly sustained excess of cladogenesis over extinction" (above) may be recognized, and it is to such a pattern that I propose to restrict the term *radiation*.

Stanley *et al.* (1981) have criticized the particular model used by Raup *et al.* (1973) as having unrealistic clade sizes and probability values to serve as a null hypothesis

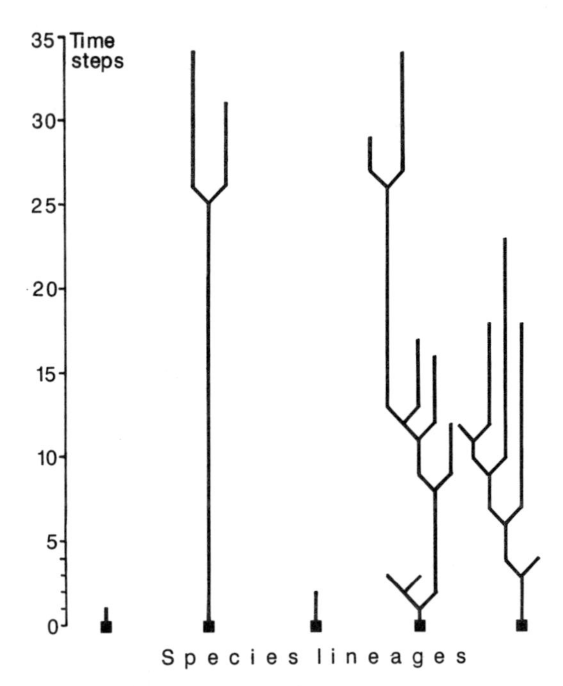

Figure 1 Five stochastic phylogenies produced after the manner of Raup *et al.* (1973). At each successive step in time (left) each lineage could split or become extinct ($P = 0.1$, for each), or continue ($P = 0.8$).

for the real phylogenetic patterns investigated by them. Nevertheless, there is no problem in the present context because only the sequence of diversity gains and losses, not their magnitude, is of interest. Moreover, Skelton *et al.* (1990) recommend including those steps in which diversity has remained the same (for which there is a finite probability in the null hypothesis) with the steps for diversity decrease, so in fact biasing the test against rejection of the null hypothesis.

Although two of the five lineages shown (at the right) in Fig. 1 give rise to relatively diverse clades, over a number of intervals, neither does so with significant persistence, and so both fail to qualify as radiations, as defined here. Such stochastic clade expansions can only be illustrated by hypothetical examples (as in Fig. 1), for the conventional statistical reason that we may only fail to reject the null hypothesis for a given real example, not confirm it. Here I am concerned only to show how radiations might be identified, by rejection of this null hypothesis.

Non-adaptive radiation

There are two main processes which might conceivably yield non-adaptive radiations:

1. *Pure vicariance radiation* (Fig. 2). A sequence of vicariance events might lead to allopatric biological speciation, though with only negligible adaptive divergence between the vicariant species. The result would be a proliferation either of sibling species, or of species of closely similar character (Fig. 2), such as the pairs of cognate species of benthic molluscs either side of the Panama Isthmus, produced by the completion of the latter in the Pliocene (Vermeij, 1978).

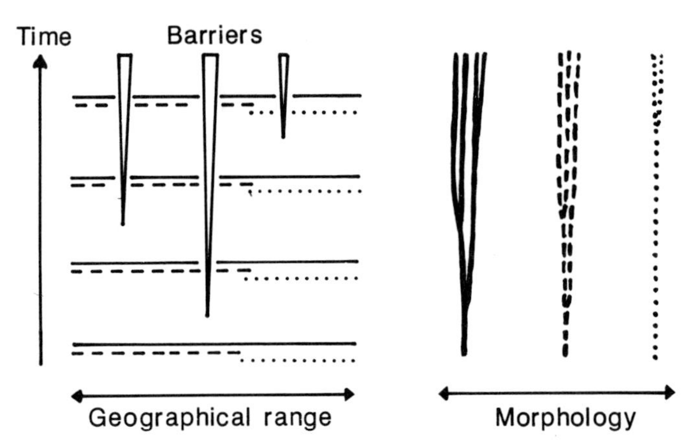

Figure 2 Radiation by pure vicariance. Left: three species with different geographical ranges (indicated respectively by solid, dashed and dotted lines) become progressively split up by the successive appearances of barriers (vertical wedges). Right: the resulting successions of allopatric speciations may show negligible adaptive (morphological) divergence. (N.B. This example would need to continue, to achieve significance.)

2. *Radiation through species selection.* The idea that entire species, as well as individuals, may be subject to selection has a venerable, if unorthodox pedigree (see Chapter 1). Over the last decade or so, however, the concept has been greatly refined (Vrba & Gould, 1986): it is now restricted to differential speciation or extinction caused by any irreducible (though necessarily emergent) property of species populations that shows some degree of cladogenetic heritability. Perhaps the clearest illustration is afforded by selective extinction according to geographical range, as with Tertiary neogastropod species of the American Gulf Coast (Hansen, 1980). Geographical range is itself an irreducible property of a species population although of course it is emergent from (i.e. an effect of) the mode of dispersal of individuals. Insofar as species beget daughter species which tend to acquire similarly sized geographical ranges, because of similarities in the mode of dispersal of the descendent individuals, there is scope for the differential growth of clades through such species selection.

Despite scepticism in some quarters as to its evolutionary importance, there is a growing acceptance that species selection is at least theoretically valid (Maynard Smith, 1989). Its importance in the present context is that, because it has to do with the differential growth of clades directly contingent upon the characteristics of species populations, rather than the adaptive attributes of individuals, it could be responsible for some non-adaptive radiations.

An important form of species selection is that operating through the emergent effects on speciation rate of the pattern of reproduction. As Paterson (Chapter 13) stresses, the emergence of reproductive isolation between sexually reproducing organisms can be merely an incidental effect of changes in their specific mate recognition systems (SMRSs). The appearance of a characteristic that enhanced the narrowness of focus of the SMRS in a species might thus accidentally increase the propensity to speciate. If radiation ensued, it could hardly be said to have been cued by any form of ecological stimulus involving access to new resources, and so would be a case of non-adaptive radiation. Likewise, factors affecting the dispersal of gametes, propagules or adults might thereby influence proneness to speciation. Taylor & Larwood (1990), for example, attribute the spectacular initial radiation of ovicellate (brooding) cheilostome bryozoans, from mid-Cretaceous times, to the effect of the appearance of the brooding habit on the propensity to speciate, rather than to the entry of a new adaptive zone.

Because of the involvement of patterns of dispersal in many likely cases of radiation through species selection, vicariance frequently also plays an influential rôle here, too. Concerning the numerous sibling, and closely similar species of plethodontid salamanders, Larson (1989:593) observed:

> Lack of long-distance dispersal produces the geographic fragmentation observed among them; the long-term histories of these populations appear to feature gradual expansion of dense populations into geographically contiguous and ecologically favourable regions, with geographical fragmentation being imposed subsequently by climatic changes ... speciation could be achieved gradually in geographically isolated populations by selective alteration of the mate recognition system.

A similar kind of history is inferred by Gittenberger (1991) for the non-adaptive radiation of the pulmonate gastropod genus *Albinaria* in eastern Crete, which shows a mosaic distribution of ecologically equivalent species.

Such instances of non-adaptive radiation can perhaps only be satisfactorily demonstrated among living taxa, between which the relative lack of adaptive divergence may be directly monitored. But, rather than attempting to show how to identify these, it is the aim of this paper to show how adaptive radiations may be confirmed as such.

Finally, before leaving this survey of kinds of radiation, it is necessary to establish a definitional dustbin term of compound radiations. Many cited examples of 'adaptive radiation' cannot be satisfactorily attributed to any one predominant cue, but seem instead to be the compounded product of many separate episodes of radiation (both adaptive and non-adaptive). The bivalves, for example, are frequently held up as exemplars of adaptive radiation. Yet, as discussed by Skelton *et al.* (1990), their overall radiation is a complex amalgam of many independent radiations occurring at different times and among different clades.

DIAGNOSTIC TESTS FOR ADAPTIVE RADIATIONS

Two kinds of test for adaptive radiation are available, according to the nature of the data to be analysed. These are comparative cladistics and stratigraphical range analysis.

Comparative cladistics

The method of comparative cladistics has been extensively developed by Lauder (1981), Liem (1990, 1991), Jensen (1990) and others. The last named author sets out, and illustrates the method with admirable clarity. The starting point is the proposal of a key evolutionary innovation (KEI: Liem, 1974), which, on the basis of morphological analysis, is hypothesized to have been preadaptive for new ecological opportunities. Adaptive radiation is thus (retrospectively) predicted to have ensued (Liem, 1990).

The test of the hypothesis then consists of two steps (Jensen, 1990:174): "In the first step, a group possessing a proposed key evolutionary novelty is compared to its sister group to assess its properties (e.g. species richness, diversity, ecological breadth). The second step investigates independent cases where the novelty has arisen to determine whether the pattern seen in step 1 is maintained" (Fig. 3).

The KEI used by Jensen to illustrate the test is that of the evolution of an independent pharyngeal jaw apparatus in the cichlid fishes, operationally decoupled from the mandibular and premaxillary jaw mechanism (Liem, 1974). It had been hypothesized that this change allowed a much wider repertoire of specialized feeding mechanisms than was available with the more primitive coupled jaw system found in generalized percoids. In fact, the cichlid arrangement was found to characterize a somewhat broader group, the Labroidei (Jensen, 1990). The latter are a monophyletic grouping comprising some 1800 species. Noting that no other perciform taxon which might be a sister group to the Labroidei was likely to contain such a large number of species, Jensen (1990) concluded that the first step in the test had been satisfied.

Finding a duplicate natural experiment of this type for the second step is always likely to be problematical, but Jensen felt that the Beloniformes could serve the purpose. One monophyletic grouping within these, the Exocoetoidea, is characterized by synapomorphies concerned with a pharyngeal jaw system, analogous to those in the Labroidei. Again, the taxon with the KEI, the Exocoetoidea, was found to have a much greater species diversity (at least 130 species) than its sister group, the Scomberesocoidea (with 36 species). This second step was thus judged to provide corroboration of the hypothesized rôle of the KEI.

There are, however, some problems with this approach. The requirement for duplicated examples of the postulated KEI in comparable but unrelated taxa means that satisfactory subjects for analysis may be uncommon. More serious is the lack, so far, of any statistical means for testing the results. Thus, can the difference in species diversity between the Exocoetoidea and the Scomberesocoidea, cited above, be considered significant? The problem here is that the lack of a real time-frame for these phylogenetic patterns makes it difficult to propose a suitable null hypothesis: how long have the taxa had for those differences in diversity to arise and how consistent was the trend through time (i.e. could the differences be due to chance alone)? Moreover, even though the KEI may have licensed the eventual radiations both of the Labroidei and of the Exocoetoidea, it remains to be established whether each taxon did in fact radiate immediately from the time of first appearance of the

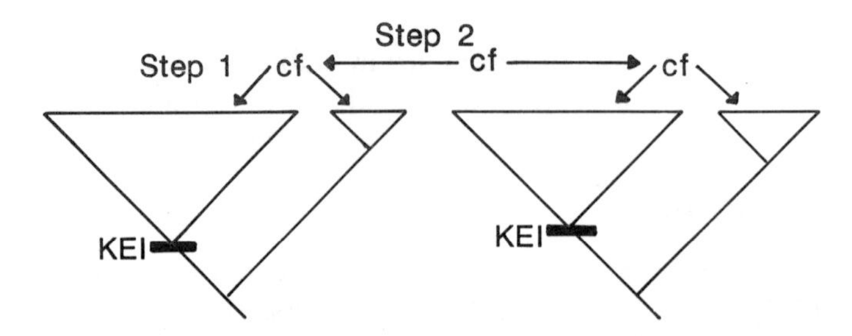

Figure 3 Synopsis of test for radiation due to key evolutionary innovation (KEI) using cladistic comparison, based on Jensen (1990). Step 1 involves comparison of the species diversity of a clade possessing a synapomorphic KEI with that of its sister taxon. Step 2 repeats the comparison in another group in which the KEI appears. Further explanation in text.

KEI, or whether later environmental opportunities triggered their radiations. In other words, were these adaptive radiations ultimately cued by the morphogenetic change itself or by environmental changes (Skelton, 1991)? In the absence of adequate palaeontological data, some of these questions of timing might be tackled using molecular distance data.

Stratigraphical range analysis

Stratigraphical range analysis permits testing both for the presence of an adaptive radiation and of the postulated cue. Skelton (1985) tested the hypothesis that an evolutionary change in the structure of the ligament in an early species of the extinct rudist bivalves was the preadaptive cue for an adaptive radiation of rudists of uncoiled tubular shell form in the Cretaceous. Changes in the generic diversity of the clade of uncoiled rudists were plotted for successive stratigraphical intervals (stages), and compared with those for the remaining co-eval spirogyrate rudists, with the primitive form of ligament (Fig. 4).

The trend for increasing diversity of the uncoiled rudists was found to be significantly persistent. That this radiation was not matched by any such pattern amongst the spirogyrate rudists was taken to imply that the cause of the radiation was intrinsic to the clade of uncoiled rudists, and not attributable to some environmental change affecting all rudists. The radiation was indeed found to be

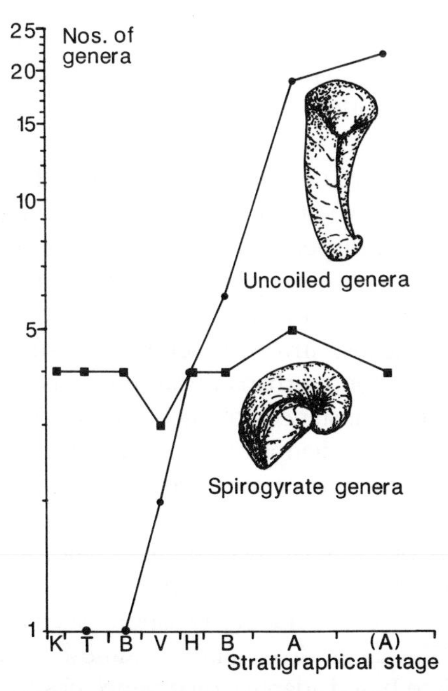

Figure 4 Semilogarithmic plot of changes in numbers of genera of uncoiled and spirogyrate rudist bivalves, stage by stage through the Upper Jurassic (K,T) and Lower Cretaceous (B–(A)), modified from Skelton (1985) with additional information from Skelton (1991) and Masse (1988: *Lovetchenia* Masse). The stages (representing a total of some 58 million years) are as follows: K, Kimmeridgian; T, Tithonian; B, Berriasian; V, Valanginian; H, Hauterivian; B, Barremian; A, Aptian; and (A) Albian. Further explanation in text.

approximately exponential over most of the interval studied (hence the almost straight ascending line of the semilogarithmic plot of Fig. 4); and the near co-incidence of its inception with the first appearance of the modified ligament was as (retrospectively) predicted by the hypothesis of adaptive radiation. That hypothesis was therefore judged to be corroborated by the test data.

Some later episodes of radiation within the clade of uncoiled rudists, in contrast, were judged to have been environmentally cued (Skelton, 1991). The evidence for this conclusion was that the polyphyletic radiations in question, all within a particular adaptive zone, were found to be synchronous. Moreover, features preadaptive for occupation of the new adaptive zone were already present in much earlier species.

Hence, where an adequate fossil record is available, the analysis of taxonomic pattern may permit identification of the likely nature of the cue for an adaptive radiation, as well as confirmation of the latter.

MICROEVOLUTION AND ADAPTIVE RADIATIONS

Much more difficult than the identification of an adaptive radiation, and of the likely nature of its principal cue, is the interpretation in detail of the microevolutionary processes involved. Evidence exists for a wide spectrum of models of speciation, as papers in this volume, and in others (e.g. Otte & Endler, 1989) make clear: indeed it seems only reasonable that a diversity of mechanisms should exist in nature, given the huge variation among organisms in their patterns of reproduction, dispersal, adaptive commitment and other such influences on population structure and gene flow. Correspondingly, there are many theoretically possible ways of linking adaptive divergence, speciation and radiation. Figure 5 is a crude attempt to illustrate at least some of these possible pathways.

Various possible instigating circumstances may arise under the headings shown in ellipses. The effects they may have are indicated in the boxes. Thus, through chance colonization of a new habitat ('dispersal'), new, uncontested resources may become available ('new resources'). Selective divergence from the parent population ('resource-based adaptive divergence') leads to niche spearation. From thence, speciation may ensue, via a number of means, involving either post-mating isolation ('genetic isolation') or pre-mating isolation, perhaps because of changes in the SMRS ('sexual adaptive divergence'). Multiple instances of such speciation (as on newly colonized islands) could then generate a radiation. In this case, the adaptive radiation would have been cued ultimately by the original chance dispersal, and resource-based adaptive divergence would have preceded the speciations.

However, the latter part of the sequence might be the other way around: chance dispersal and consequent geographical separation may lead to speciation directly via sexual adaptive divergence and pre-mating isolation. Such seems to have been the case with some of the Hawaiian drosophilids (Kaneshiro & Boake, 1987). Only subsequently might resource-based adaptive divergence ensue, on secondary sympatry. The resulting separation of niches could assist survivorship, such that radiation ensued from the excess of speciation over extinction. Here, adaptive radiation, ultimately cued by the secondary sympatry, would have involved resource-based adaptive divergence following speciation. It is conceivable that something like this occurred in the cichlids of Lake Victoria. Greenwood (1981:72) posed the possibility that:

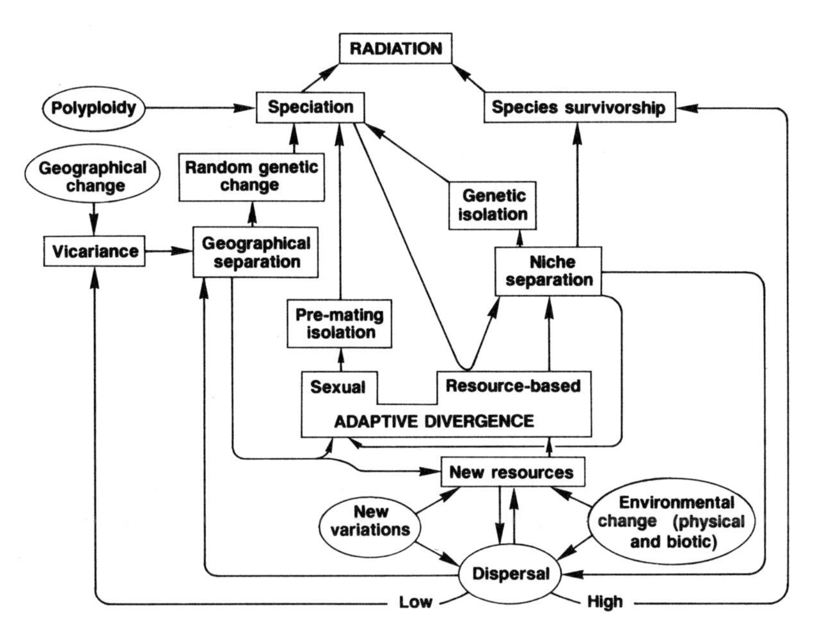

Figure 5 Diagram showing some of the theoretically possible microevolutionary relationships involved in evolutionary radiations. Ellipses indicate the means by which the process may be triggered. Possible effects are indicated in the boxes. Adaptive divergence is shown as having two facets – adaptation in relation to sex and that in relation to physical resources. Further explanation in text.

> these species are the products of speciation events that did not involve the evolution of any change in anatomical or physiological characters associated with feeding habits, but simply the evolution, in different peripheral lakes, of only specific reproductive barriers between daughter species derived from segments of the same parental species.

The different feeding specializations of the species in the main lake (Greenwood, 1974) would thus be largely the products of post-speciational divergence in secondary sympatry.

It is therefore possible, in theory at least, to distinguish between pre-speciational, and post-speciational adaptive radiation, insofar as resource-based adaptive divergence is concerned. In practice, however, it would probably be difficult if not impossible to distinguish between the two in most cases, and for this reason it is pragmatic to regard both as comprising kinds of adaptive radiation (rather than restricting the term to the classic Darwinian pre-speciational form).

Other routes to adaptive radiation on Fig. 5 include those involving the opening up of new resources either by the appearance of preadaptive traits ('new variations'), or as a consequence of, say, a climatic change or a mass extinction ('environmental change ... '). The routes to non-adaptive radiations may also be picked out, such as that via vicariance, resulting from the formation of barriers ('geographical change'), and leading to speciation without necessarily involving resource-based adaptive divergence. However, it is not my intention to review the possibilities comprehensively. Rather, Fig. 5 is intended only to illustrate the wide range of microevolutionary possibilities that may be encompassed within the scope of radiations in general, and of adaptive radiations in particular.

CONCLUSION

When strictly defined, adaptive radiation can be a testable proposition with explanatory power, and not the vacuous concept that generalized definitions imply. The definition offered here is 'an episode of significantly sustained excess of cladogenesis over extinction, with adaptive divergence cued by the appearance of some form of ecological stimulus' (p. 48). The cue may be intrinsic (preadaptation) or extrinsic (access to new niches through chance dispersal or through physical or biotic change in the environment). Both factors are involved in any given adaptive radiation, though with one serving as the cue and the other(s) as enabling circumstances. The identification of examples and of their cues, however, is an onerous task.

Radiation may be recognized as a more or less persistent trend for increasing numbers of species in a clade, that deviates significantly from a random walk. Non-adaptive radiations may result from pure vicariance, or through species selection, with negligible adaptive divergence of the species. The two influences commonly also interact.

Testing for adaptive radiation requires both the establishment of a link between a particular cue and the ensuing adaptive repertoire of a clade, and the identification of an associated radiation (as above). Cladistic comparison can corroborate the probable link between a key evolutionary innovation and differential diversity, but (so far) lacks a means of statistical testing and cannot detect if the innovation itself was indeed the eventual cue. Stratigraphical range analysis offers the possibilities both of a statistical test for radiation and for linking that with a hypothesized cue.

The possible microevolutionary pathways to adaptive radiation are many and complex, and in most cases may be intractable. In some cases, however, there may be grounds for distinguishing between pre-speciational, and post-speciational adaptive radiations, in which adaptive divergence (in relation to resources) has tended, respectively, to precede or follow speciation.

ACKNOWLEDGEMENTS

Peter Sheldon offered helpful advice on the recipe for the spaghettigram in Fig. 5, and John Taylor (Open University) drew it, though I alone am to blame for the outcome. Janet Dryden processed the typescript at the Open University. The Linnean Society, and the organizers are gratefully acknowledged for their invitation to the meeting at Cardiff.

REFERENCES

COYNE, J.A., 1992. Genetics and speciation. *Nature, 355:* 511–515.

DARWIN, C.R., 1859. *On the Origin of Species by Means of Natural Selection*, London: John Murray.

FUTUYMA, D.J., 1986. *Evolutionary Biology*, 2nd edition. Sunderland, Massachusetts: Sinauer Associates.

GITTENBERGER, E., 1991. What about non-adaptive radiation? *Biological Journal of the Linnean Society, 43:* 263–272.

GREENWOOD, P.H., 1974. Cichlid fishes of Lake Victoria, East Africa: the biology and evolution of a species flock. *Bulletin of the British Museum (Natural History)*, Zoology series, Supplement 6. London: Trustees of the British Museum (Natural History).

GREENWOOD, P.H., 1981. Species-flocks and explosive evolution. In P.L. Forey (Ed.), *The Evolving Biosphere:* 61–74. London: British Museum (Natural History) and University of Cambridge Press.

HANSEN, T.A., 1980. Influence of larval dispersal and geographic distribution on species longevity in neogastropods. *Paleobiology, 6:* 193–207.

JENSEN, J.S., 1990. Plausibility and testability: assessing the consequences of evolutionary innovation. In M.H. Nitecki (Ed.), *Evolutionary Innovations:* 171–190. Chicago: University of Chicago Press.

KANESHIRO, K.Y. & BOAKE, C.R.B., 1987. Sexual selection and speciation: issues raised by Hawaiian drosophilids. *Trends in Ecology and Evolution, 2:* 158–167.

LARSON, A., 1989. The relationship between speciation and morphological evolution. In D. Otte & J.A. Endler (Eds), *Speciation and its Consequences:* 579–598. Sunderland, Massachusetts: Sinauer Associates.

LAUDER, G.V., 1981. Form and function: structural analysis in evolutionary morphology. *Paleobiology, 7:* 430–442.

LIEM, K.F., 1974. Evolutionary strategies and morphological innovations: Cichlid pharyngeal jaws. *Systematic Zoology, 22:* 425–441.

LIEM, K.F., 1990. Key evolutionary innovations, differential diversity, and symecomorphosis. In M.H. Nitecki (Ed.), *Evolutionary Innovations:* 147–170. Chicago: University of Chicago Press.

LIEM, K.F., 1991. A functional approach to the development of the head of teleosts: implications on constructional morphology and constraints. In N. Schmidt-Kittler & K. Vogel (Eds), *Constructional Morphology and Evolution:* 231–249. Berlin: Springer-Verlag.

MASSE, J.-P., 1988. Importance relative, chronologie et signification phylogenetique des modifications morphologiques et anatomiques chez les Requieniidae (Rudistes) due Crétacé inférieur. In M. Sladić-Trifunović (Ed.), *First International Conference on Rudists, October 1988*, Abstracts: 15. Belgrade: Serbian Geological Society.

MAYNARD SMITH, J., 1989. *Evolutionary Genetics*. Oxford: Oxford University Press.

MAYR, E., 1970. *Populations, Species, and Evolution*. Cambridge, Massachusetts: Belknap Press of Harvard University Press.

OTTE, D. & ENDLER, J.A., (Eds) 1989. *Speciation and its Consequences*. Sunderland, Massachusetts: Sinauer Associates.

RAUP, D.M., GOULD, S.J., SCHOPF, T.J.M. & SIMBERLOFF, D.J., 1973. Stochastic models of phylogeny and evolution of diversity. *Journal of Geology, 81:* 525–542.

SEPKOSKI, J.J. Jr., 1986. Phanerozoic overview of mass extinctions. In D.M. Raup & D. Jablonski (Eds), *Patterns and Processes in the History of Life, Dahlem Konferenzen 1986*. Berlin: Springer-Verlag.

SIMPSON, G.G., 1949. *The Meaning of Evolution*. New Haven: Yale University Press.

SIMPSON, G.G., 1983. *Fossils and the History of Life*. New York: W.H. Freeman & Co. (Scientific American Books).

SKELTON, P.W., 1985. Preadaptation and evolutionary innovation in rudist bivalves. *Special Papers in Palaeontology, 33:* 159–173.

SKELTON, P.W., 1991. Morphogenetic versus environmental cues for adaptive radiations. In N. Schmidt-Kittler & K. Vogel (Eds), *Constructional Morphology and Evolution:* 375–388. Berlin: Springer-Verlag.

SKELTON, P.W., CRAME, J.A., MORRIS, N.J. & HARPER, E.M., 1990. Adaptive divergence and taxonomic radiation in post-Palaeozoic bivalves. In P.D. Taylor & G.P. Larwood (Eds), *Major Evolutionary Radiations, Systematics Association Special Volume, 42:* 91–117. Oxford: Clarendon Press.

STANLEY, S.M., 1979. *Macroevolution, Pattern and Process*. San Francisco: Freeman.

STANLEY, S.M., 1990. Adaptive radiation and macroevolution. In P.D. Taylor & G.P. Larwood (Eds), *Major Evolutionary Innovations, Systematics Association Special Volume, 42:* 1–16. Oxford: Clarendon Press.

STANLEY, S.M., SIGNOR, P.W. III, LIDGARD, S. & KARR, A.F., 1981. Natural clades differ from 'random clades': simulations and analyses. *Paleobiology, 7:* 115–127.

TAYLOR, P.D. & LARWOOD, G.P., 1990. Major evolutionary radiations in the Bryozoa. In P.D. Taylor

& G.P. Larwood (Eds), *Major Evolutionary Radiations, Systematics Association Special Volume, 42:* 209–233. Oxford: Clarendon Press.

VERMEIJ, G.J., 1978. *Biogeography and Adaptation. Patterns of Marine Life*. Cambridge, Massachusetts: Harvard University Press.

VRBA, E.S. & GOULD, S.J., 1986. The hierarchical expansion of sorting and selection: sorting and selection cannot be equated. *Paleobiology, 12:* 217–228.

CHAPTER

5

Depth habitats and the Palaeocene radiation of the planktonic foraminifera monitored using oxygen and carbon isotopes

RICHARD M. CORFIELD

Introduction 60

The depth habitats of fossil planktonic foraminifera 60

The evolution of the planktonic foraminifera after the K/T boundary 63

Acknowledgements 68

References 69

Keywords: Planktonic foraminifera – depth habitats – Palaeocene evolution.

Abstract

The planktonic foraminifera offer significant advantages to scientists wishing to study patterns of evolution in relation to environmental change in the fossil record. Planktonic foraminifera in cores recovered by the Ocean Drilling Program are abundant, generally well-preserved and, using modern drilling technology, are available in stratigraphically long and complete sections. Over and above this, oxygen and carbon isotope measurements of their carbonate shells makes available uniquely detailed palaeoenvironmental information. Oxygen isotope measurements on appropriate foraminifera can provide information concerning the temperature variability at several levels in the water column, while carbon isotope measurements of foraminiferal shells are offering new insights into the relationships between ocean productivity and ventilation, atmospheric pCO_2 and global temperature change.

The radiation of the planktonic foraminifera after the Cretaceous/Tertiary boundary may be related to profound changes in the surface water productivity of the world ocean. Intervals of maximum species turnover are concentrated at times of systematic carbon isotope change.

Evolutionary Patterns and Processes
ISBN 0-12-440895-8

INTRODUCTION

There can be little doubt that the planktonic foraminifera have much to offer the study of evolution in the fossil record. Specific advantages derive from the large numbers of these minute organisms recovered from the deep sea through the activities of the Ocean Drilling Program, and the fact that due to modern drilling technology (e.g. the advent of the hydraulic piston corer) long and stratigraphically continuous successions can be recovered.

In addition, and importantly, their simple morphology lends itself to morphometric treatment and hence the analysis of phylogenetic rates of evolution (Malmgren & Kennett, 1981; Malmgren, Berggren & Lohmann, 1983; Corfield & Granlund, 1988) while simple biostratigraphic information (benefiting from the technical advantages outlined above) yield detailed taxonomic turnover data frequently on absolute timescales (e.g. Wei & Kennett, 1983; Corfield, 1987; Corfield & Shackleton, 1988).

Another significant advantage is the fact that $\delta^{13}C$ and $\delta^{18}O$ measurements of planktonic foraminiferal shells provide uniquely detailed palaeoenvironmental information which potentially add another dimension to the interpretation of evolutionary patterns and processes within their fossil record.

In this contribution I examine two specific examples of the way in which stable isotope data can add a new dimension to the study of planktonic foraminiferal evolution. First, I examine the $\delta^{13}C$ and $\delta^{18}O$ evidence for the depth stratification of fossil planktonic foraminifera. Secondly, I examine the environmental changes which may have influenced the radiation of the planktonic foraminifera in the aftermath of the extinctions at the Cretaceous/Tertiary (K/T) boundary.

THE DEPTH HABITATS OF FOSSIL PLANKTONIC FORAMINIFERA

One of the most striking features of the evolution of the planktonic foraminifera is the similarity between the range of phenotypes evolved in the four major radiations of the group (Cifelli, 1969; Norris, 1991). The first radiation of the planktonic foraminifera occurred in the early Cretaceous or late Jurassic from globigerine root-stock (Grigelis & Gorbatchik, 1980; Banner & Desai, 1988) and culminated in pre-Cenomanian/Turonian boundary time in a spectrum of morphologies running from globigerine (e.g. *Hedbergella delrioensis*) to globorotaliid (e.g. *Rotalipora cushmani*). The second radiation of the group (in the late Cretaceous) again from globigerine ancestors (e.g. *Whiteinella archaeocretacea*) resulted in a broad spectrum of morphologies in the Maastrichtian from globigerine (e.g. *Rugoglobigerina scotti*) to globorotaliid (e.g. *Abathomphalus mayaroensis*). The third radiation of the group occurred in the aftermath of the K/T boundary. The globigerine ancestors in this case were forms such as *Hedbergella monmouthensis* and *Globonamalia archaeocompressa* and resulted in the late Palaeocene in globigerine forms such as *Subbotina velascoensis* and the globorotaliid-like *Morozovella velascoensis*. The fourth and youngest radiation of the planktonic foraminifera started after an interval of reduced diversity in the Oligocene and resulted in our diverse modern day fauna with globigerine forms such as *Orbulina universa* and *Globigerina ruber* and globorotaliid forms such as *Globorotalia tumida* and *Globorotalia truncatulinoides*.

In a sequence of papers over the last two decades, several authors (Douglas &

Savin, 1978; Caron & Homewood, 1982; Hart, 1980) have suggested that keeled forms of the planktonic foraminifera have always occupied deepest water habitats in the several radiations of the group. These papers have been valuable in developing models by which to interpret the patterns of planktonic foraminiferal evolution but have now become an axiom of planktonic foraminiferal micropalaeontology based only on the uniformitarian assumption that fossil morphotypes stratify in the same depth order as modern forms. $\delta^{13}C$ and $\delta^{18}O$ measurements of the shells of fossil planktonic foraminifera allow this hypothesis to be rigorously tested. The controls on the distribution of both $\delta^{13}C$ and $\delta^{18}O$ result in vertical gradients in the world ocean (Fig. 1) that allow different depth dwellers to be ranked in terms of depth habitat. Figures 2 and 3 illustrate a sequence of histograms (derived from Corfield & Cartlidge, 1991) which display the relationship between $\delta^{13}C$ and $\delta^{18}O$ and foraminiferal morphology from a variety of time-slices through the Cenozoic and in the late Mesozoic. Species are ranked from the globigerine-like morphotype (on the left) through to the globorotaliid-like morphology (on the right). Clearly this system makes no assertions about specific taxonomic relationships, nor does it imply that the range between globigerine-like and globorotaliid-like forms is the same in the different time-slices.

The data show that the hypothesis that keeled forms are the deepest water dwellers applies only in the Holocene and latest Neogene and probably in the late Cretaceous. It may well be that keeled species were relatively deep water dwelling (e.g. near the thermocline) taxa during much of the Neogene and the Cretaceous, while, even

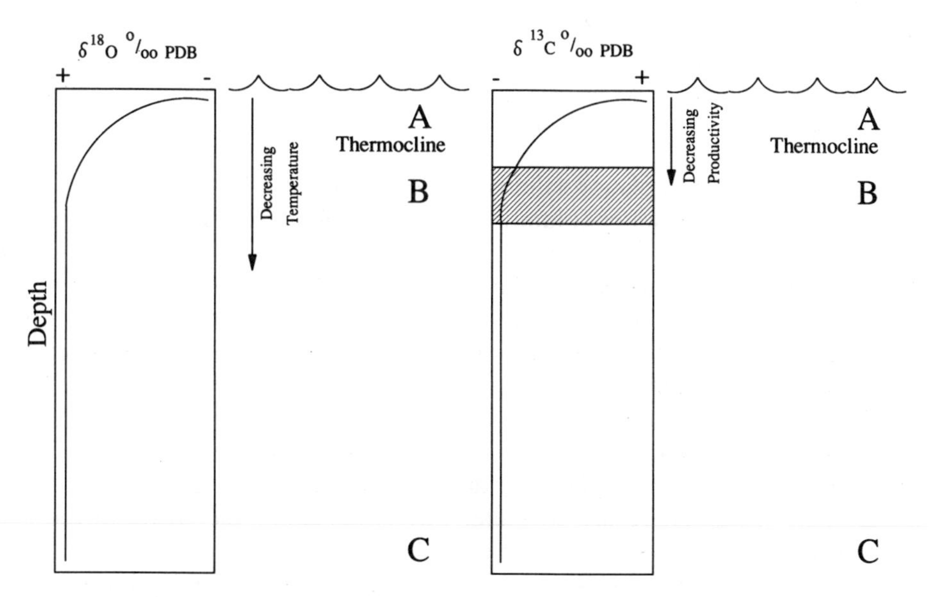

Figure 1 Vertical gradients of $\delta^{13}C$ and $\delta^{18}O$ in the ocean. The vertical gradient of $\delta^{18}O$ is predominantly controlled by declining temperature with increasing depth. The vertical gradient of $\delta^{13}C$ is controlled by the removal of ^{12}C in surface waters due to photosynthesis and its return at depth by oxidation. A = Surface-dwelling planktonic foraminifera (e.g. the genus *Morozovella* in the Palaeocene) with most negative $\delta^{18}O$ and most positive $\delta^{13}C$ values; B = a deeper dwelling planktonic foraminifera (e.g. the genus *Subbotina* in the Palaeocene) with relatively more positive $\delta^{18}O$ values and more negative $\delta^{13}C$ values; C = benthic foraminifera with the most positive $\delta^{18}O$ values and the most negative $\delta^{13}C$ values of the three.

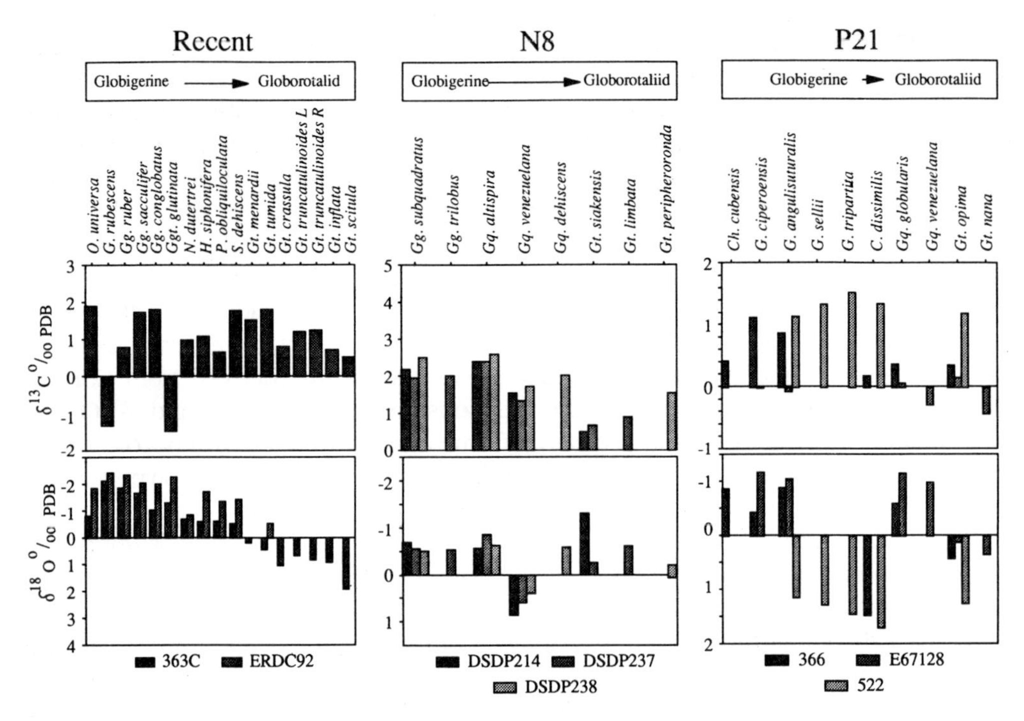

Figure 2 Evolution of planktonic foraminiferal depth habitats during the Neogene and late Palaeogene monitored using the δ^{13}C and δ^{18}O composition of the shells. Genera and their abbreviations used here and in Fig. 3 are: *A.*, *Acarinina*; *C.*, *Catapsydrax*; *Ch.*, *Chiloguembelina*; *G.*, *Globigerina*; *Gg.*, *Globigerinoides*; *Ggt.*, *Globigerinita*; *Gq. Globoquadrina*; *Gt.*, *Globorotalia*; *H.*, *Hastigerina*; *M.*, *Morozovella*; *N.*, *Neogloboquadrina*; *O.*, *Orbulina*; *P.*, *Pulleniatina*; *S.*, *Subbotina*; *Sp.*, *Sphaeroidinella*. Except for Recent, ages in this figure and Fig. 3 are indicated by planktonic foraminiferal zones. (Redrawn from Corfield & Cartlidge, 1991).

during the Palaeogene, the planorotalitids (keeled compressed globorotaliids) appear to have lived deep in the water-column. However, keeled taxa have rarely been the deepest dwelling taxa, this niche having been filled by genera such as *Subbotina*, *Catapsydrax* and *Globoquadrina*. This finding has implications for the interpretation of the functional morphology of the planktonic foraminifera. Rudwick (1964) developed the paradigm of functional morphology which asserts that the function of an organic component may be derived from examination of its structure. If this applies to the planktonic foraminifera, and the repeated evolution of a relatively narrow range of morphologies suggests that the shape of the shell has some adaptive advantage, then how can we explain the different depths occupied at different times by similar shell-shape? There are two alternatives: (1) either the shape of the shell is irrelevant in evolutionary terms (i.e. it is 'transparent' to functional considerations) or (2) some factor other than gross morphology has controlled the depth distribution of planktonic foraminifera during their history.

It is unclear as yet how to distinguish between the two possibilities. Intuitively, the former suggestion seems unlikely. It is known that planktonic foraminifera are influenced during ontogeny by environmental considerations (for example, the occurrence of kummerform individuals, see Kennett, 1976) and this suggests that it is unlikely that the shell of the planktonic foraminiferan would be unaffected by its depth habitat. In addition, there is a strong case from the deep-sea fossil record for

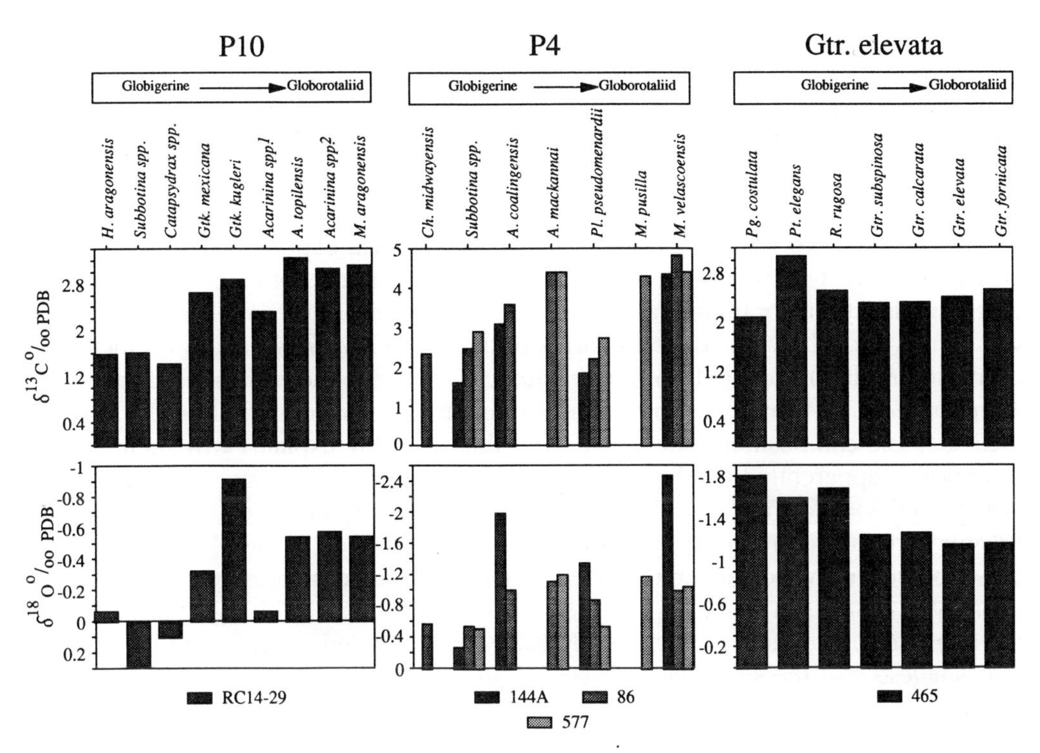

Figure 3 Evolution of planktonic foraminiferal depth habitats during the Palaeogene monitored using the $\delta^{13}C$ and $\delta^{18}O$ composition of the shells. See Fig. 2 for abbreviations. (Redrawn from Corfield & Cartlidge, 1991).

the selection of planktonic foraminiferal morphologies during times of environmental change (e.g. Malmgren *et al.*, 1983; Corfield & Shackleton, 1988) which again suggests some dependence on environmental variability.

This suggests that the second possibility is the more likely, with any shell-shape/depth habitat interrelationship being overridden by some other, more important factor. It is known that many species of planktonic foraminifera co-exist with photosynthetic symbionts and it is probable that this has been a common condition throughout the history of the group. Photosymbionts are of necessity limited to the photic zone and therefore so are the planktonic foraminifera with which they exist. Hence, one explanation for the clear surface-dwelling habitat of, for example Palaeogene globorotaliid-like forms, is that they lived in symbiosis with photosymbionts. Note however that morozovellids are not considered to have been spinose (Boersma & Premoli-Silva, 1991) and hence any physical support of photosymbionts would not have been in an analogous manner to modern forms.

THE EVOLUTION OF THE PLANKTONIC FORAMINIFERA AFTER THE K/T BOUNDARY

Corfield (1987) and Corfield & Shackleton (1988) developed a model linking the evolution of the Palaeocene and early Eocene planktonic foraminifera to changes in surface water productivity and, therefore, by implication with changes in the intensity of the subsurface oxygen minimum zone. They suggested that the increase in $\delta^{13}C$ measured in marine carbonates in the early and early late Palaeocene represents a profound increase in productivity which drove an increase in the rate of taxonomic turnover of the planktonic foraminifera, resulting in a relatively rapid proliferation of *Morozovella* species and a rapid trend toward acquisition of angulo-conical globorotaliid-like morphology. They also suggested that the $\delta^{13}C$ decline from the late Palaeocene into the early Eocene is the geochemical expression of a decline in ocean productivity which again drove an increase in the rate of turnover of surface-dwelling planktonic foraminiferal species, as forms typical of the late Palaeocene gave way to new morphotypes (typically forms of the genus *Acarinina* and a distinct suite of morozovellids), adapted to lower surface-water productivity conditions. Both *Morozovella* and *Acarinina* were surface water dwellers as identified on the basis of their isotopic composition (see above) and this probably explains why their rate of evolution is apparently linked to changes in the $\delta^{13}C$ of surface waters. Since carbon isotope fractionation is controlled by photosynthesis and an increase in the rate of organic carbon production by photosynthesis will be limited to the photic zone, then those organisms which inhabit surface waters will correspondingly respond most strongly to changes in adaptive pressures related to productivity change. Furthermore, the increase in productivity in the early Palaeocene implied by the rapid increase in $\delta^{13}C$ suggests that the subsurface oxygen minimum layer would have intensified and expanded as the increased flux of particulate organic carbon to the sea bed was oxidized. Corfield & Shackleton (1988) suggested that this might have resulted in the establishment of two distinct habitats (a surface water habitat and an oxygen minimum zone habitat) into which two major clades of the early Tertiary planktonic foraminifera evolved; on the one hand *Morozovella* and later *Acarinina* into the surface water habitat, and on the other hand *Subbotina* and *Planorotalites* into the deeper, oxygen-deficient habitat. The radiation of the genus *Acarinina* during the $\delta^{13}C$ decline from the late Palaeocene into the early Eocene is probably not related to separation of ancestral populations along a gradient (parapatry) as suggested for *Morozovella* and *Subbotina* during the $\delta^{13}C$ increase in the early Palaeocene. It is more likely to have been the result of changing selective pressures as productivity decreased in the early Eocene causing the extinction of the Palaeocene morozovellids and the radiation of *Acarinina*.

At the time this model was developed, detailed carbon isotope gradient data were unavailable from DSDP Site 577 on Shatsky Rise (North Pacific) and a suite of DSDP Leg 74 sites on the Walvis Ridge (South Atlantic). These sites provided the faunal data on which their hypothesis was developed. Here I re-examine their model in the light of additional $\delta^{13}C$ and $\delta^{18}O$ data (Corfield & Cartlidge, 1992). Corfield & Shackleton (1988) assessed evolutionary turnover in the Palaeocene planktonic foraminifera in 1 million year increments of modelled time. Here I have calculated per taxon rates of evolution (Sepkoski, 1978; Wei & Kennett, 1986) using half million year increments of modelled time. The rate of turnover (r_t) is calculated by summing the per taxon rate of origination (r_o) and the per taxon rate of extinction r_e (Lasker, 1978; Wei & Kennett, 1986) (Figs 4–6).

Figure 4 illustrates an estimate of 'global' turnover within the planktonic foraminifera using the method of Corfield & Shackleton (1988). The two broad peaks in evolutionary activity are marked A and B. Also shown are $\delta^{13}C$ data from the most intensively studied site for this time interval to date; DSDP Site 577 (data from Corfield &

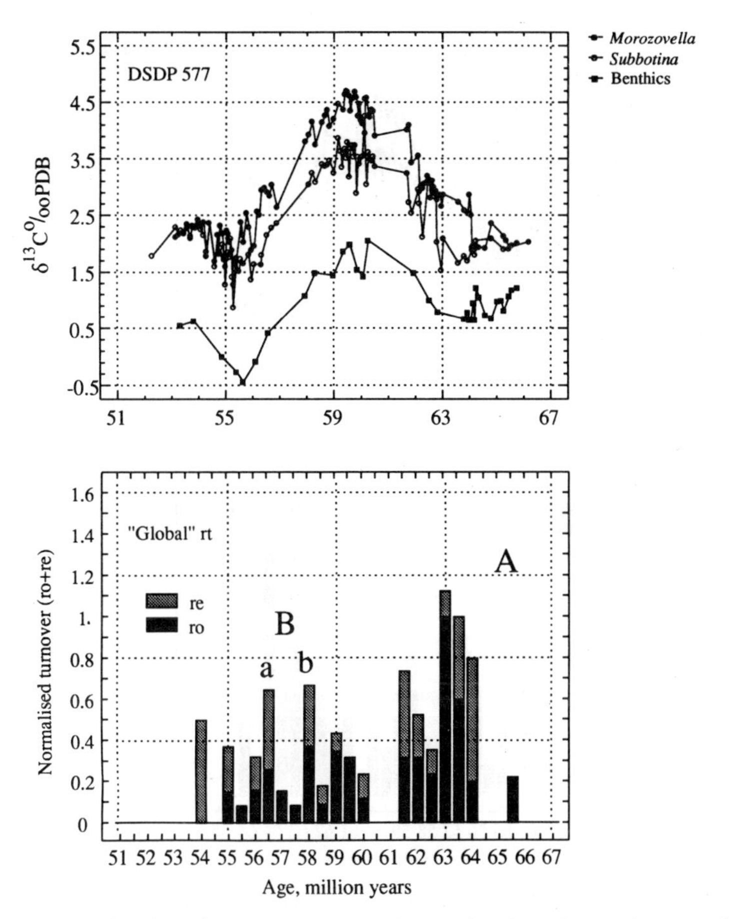

Figure 4 Estimated global evolutionary turnover (normalized to diversity) using the method of Corfield & Shackleton (1988) together with the detailed $\delta^{13}C$ stratigraphy for three depth groupings of foraminifera from the Leg 86 sites on the Shatsky Rise, North Pacific.

Cartlidge, 1992). The $\delta^{13}C$ data are from a surface-dwelling planktonic species (*Morozovella*), a deeper-water planktonic species (*Subbotina*) and the benthic foraminiferal genera *Nuttallides* and *Cibicidoides*. The differences between these curves indicate changes in the vertical carbon isotope gradient ($\Delta\delta^{13}C$) over this time interval.

Peak A corresponds broadly to the increase in $\delta^{13}C$ between 64 and 60 million years Palaeocene and peak B to the decline in $\delta^{13}C$ between 59 and 55 million years as initially documented by Corfield & Shackleton (1988). It is also clear that the initial, large peak in turnover (peak A) is associated with a particularly rapid increase in the carbon isotope gradient at 64 million years. This may indicate that as well as the suggested broad association of heightened turnover with changes in $\delta^{13}C$ and the vertical carbon isotope gradient, rapid changes in $\Delta\delta^{13}C$ are correlated with especially rapid evolutionary turnover of the planktonic foraminifera.

Figure 5 illustrates turnover from DSDP Site 577 only, together with the $\delta^{13}C$ record from DSDP 577 already illustrated in Fig. 4. During the Palaeocene, this site was located near 14°N. The conspicuous steepening in the $\delta^{13}C$ carbon isotope gradient ($\Delta\delta^{13}C$) at approximately 64 million years ago is marked X. The co-incidence of

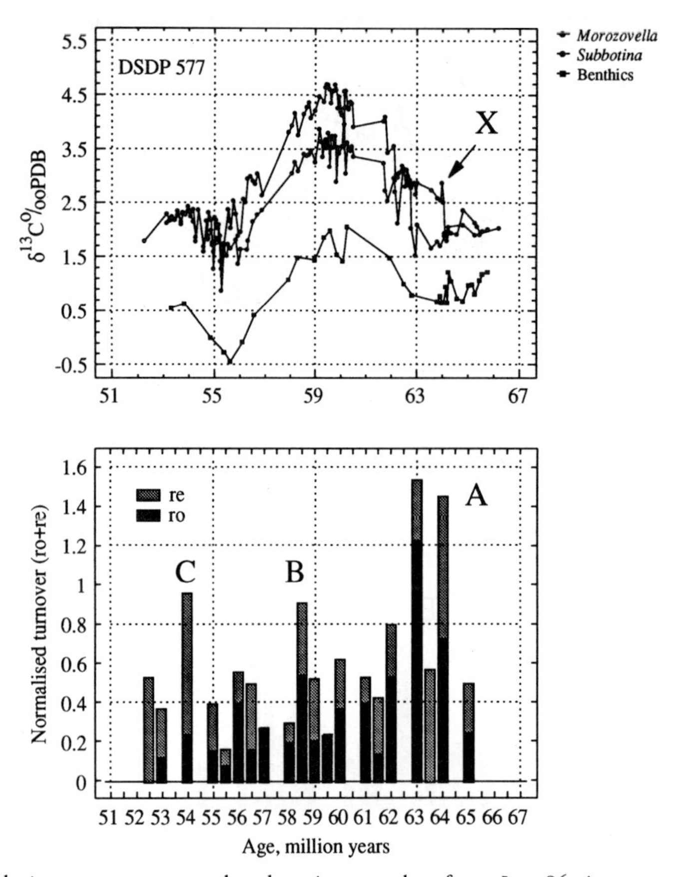

Figure 5 Evolutionary turnover and carbon isotope data from Leg 86 sites.

peak A with the increase in $\delta^{13}C$ and $\Delta\delta^{13}C$ is again apparent even though the comparison is now of data from only a single location. As Corfield & Granlund (1988) and Corfield & Shackleton (1988) have pointed out, the major contributor to peak A is the rapid evolution of the genus *Morozovella* at this time. Rates of evolution in the remainder of the Palaeocene and the Early Eocene at this site were lower with the exception of peaks marked B and C.

Figure 6 illustrates the rate of planktonic foraminiferal evolution in DSDP 527 from the Walvis Ridge area in the South Atlantic. This site occupied approximately the same palaeolatitude during Palaeocene and Early Eocene times as during the present day. $\delta^{13}C$ data (again from *Morozovella*, *Subbotina* and *Nuttallides* and *Cibicidoides*) are compiled from DSDP 527 and the closely adjacent DSDP Sites 525 and 528. Taxonomic turnover data are compiled from DSDP 527. These data show that the Early Palaeocene peak in turnover (peak A) occurs approximately synchronously with the peak in DSDP 577. As at DSDP 577, the $\delta^{13}C$ data from the Leg 74 sites also show a rapid increase (within the limits of the sampling resolution) in the vertical $\delta^{13}C$ gradient at this time (marked X).

After a lull in evolutionary activity between 61 and 59.5 million years ago rates of turnover again show a general increase (marked B). In particular there are conspicuous peaks in turnover centred on 58 and 56.5 million years ago. These are of interest because they are close to the Palaeocene/Eocene boundary interval which has

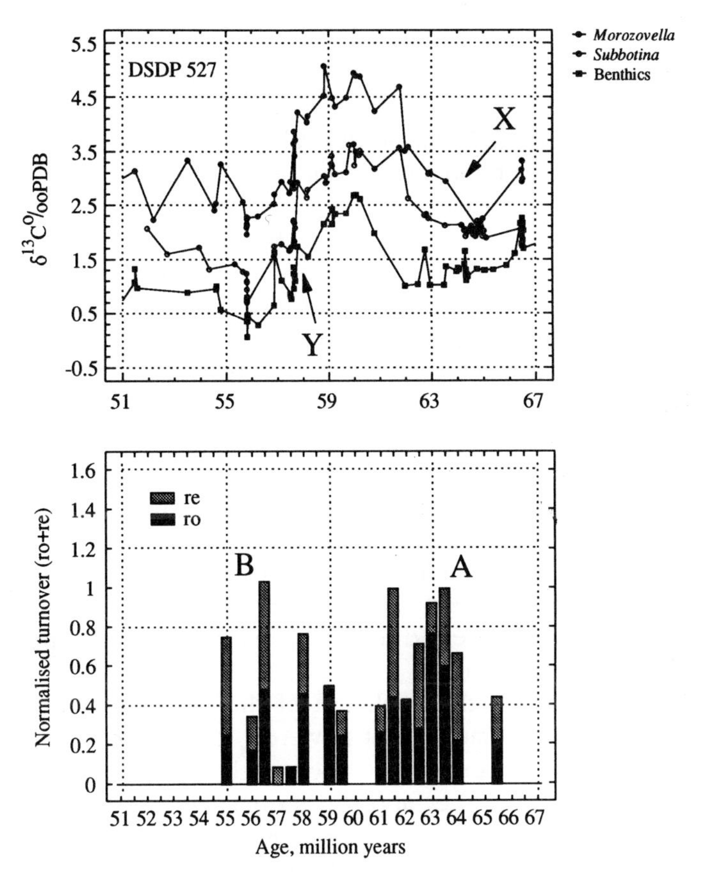

Figure 6 Evolutionary turnover and carbon isotope gradients data from Leg 74 sites.

been identified (Kennett & Stott, 1991) as an interval of significant biotic and palaeoceanographic change. Previous evolutionary studies (Tjalsma & Lohmann, 1983; Thomas, 1990, 1991) have associated this boundary interval with an interval of profound extinction within benthic foraminifera. In addition recent work on samples obtained from an apparently complete Palaeocene/Eocene boundary section drilled during Leg 113 of the ODP in the Weddell Sea (Stott *et al.*, 1990; Kennett & Stott, 1990, 1991) has spawned considerable interest in the isotopic events at that boundary because of their discovery of a marked negative shift in both $\delta^{13}C$ and $\delta^{18}O$ of very short duration. Subsequently this excursion has been discovered in other sites, e.g. ODP 738 in the Antarctic Indian Ocean (Barrera & Keller, 1991) and ODP 762C in the mid-latitude Indian Ocean (Kennett, 1991). Also, and significantly, new work on the Leg 74 sites (Thomas, 1991; Corfield & Cartlidge, 1992) indicate that this profound excursion is also present in these Walvis Ridge sites and occurs at 57.4 million years ago. This event is marked Y on Fig. 6.

If I am correct in linking one of these peaks with the Palaeocene/Eocene boundary excursion of Kennett & Stott (1991) then it is important to explain why there is no corresponding peak in DSDP 577. The most likely explanation is that due to a hiatus in this site in the interval 58–56.5 million years ago the Kennett & Stott $\delta^{13}C$ excursion is missing. In addition, due to the method of calculating taxonomic turnover employed by Corfield & Shackleton (1988) and used here, gaps in the succession will necessarily include error in the accurate positioning of first occurrence or last appearance estimates. The larger the gap, the larger the potential error.

It is also worth noting that the background rate of turnover was greater in DSDP Site 577 than in DSDP Site 527 during the Palaeocene and Early Eocene. This may have been because warmer temperatures and greater vertical stratification provided more niches for potential species to evolve into following the extinctions at the Cretaceous/Tertiary boundary. Alternatively, the higher variability of turnover between adjacent time-slices in DSDP Site 527 may have been because a larger contribution to the record at this temperate site was due to the extension and contraction of the areal extent of species through time.

From these limitations it follows that a better approximation of turnover is derived from the global estimates of species ranges shown in Fig. 4 as originally suggested by Corfield (1987) and Corfield & Shackleton (1988). In addition to the peaks marked A and B, this figure shows two peaks in activity near to the Palaeocene/Eocene boundary marked a and b. As I suggested for the record from DSDP Site 527, it may be that one of these peaks is associated with the $\delta^{13}C$ and $\delta^{18}O$ excursion of Kennett & Stott (1991) at the Palaeocene/Eocene boundary. Clearly a greater number of sites must be examined to show clearly whether the major threshold of the Palaeocene/Eocene boundary affected the evolution of the planktonic foraminifera as well as the benthic foraminifera.

It has been suggested that the evolution of the planktonic foraminifera is not influenced by environmental factors but powered 'internally', resulting in apparent independence of evolution from environmental change (Van Valen, 1973). These arguments, at least in as far as they apply to the planktonic foraminifera, have been cogently refuted in a recent paper (Pearson, 1992) who argued that any change in the rate of real-time extinction precludes the type of survivorship analysis on which Van Valen's 'law' is based. As is now clear, the history of the planktonic foraminifera is characterized by changes in real-time turnover, including of course, the rate of extinction. In particular, the profound variations in the rate of turnover of the planktonic foraminifera during the Palaeocene and Early Eocene, suggest that survivorship curves which attempt to summarize our picture of evolutionary activity during this interval are at best, suspect.

I have argued for a general relationship between changes in the productivity of the Palaeocene oceans and the rate of taxonomic evolution of the planktonic foraminifera. Others have suggested a close link between rates of evolution of the planktonic foraminifera and temperature change through geological time (e.g. Stehli, Douglas and Kafescioglu, 1972; Jenkins & Shackleton, 1979; Thunell, 1981). The likelihood is, of course, that as we refine our understanding of environmental change during geological time, that these and perhaps other factors, acting singly or in combination, will be shown to have influenced the evolution of the planktonic foraminifera throughout their history.

ACKNOWLEDGEMENTS

As always, many thanks to Julie Cartlidge for operating the PRISM mass spectrometer used for the generation of much of the isotope data referred to in this contribution. Thanks also to Dick Norris for stimulating discussion and his careful review of this paper. This is a contribution to IGCP 308 'Events around the Palaeocene/Eocene boundary'.

REFERENCES

BANNER, F.T. & DESAI, D., 1988. A review and revision of the Jurassic–Early Cretaceous Globigerinina with especial reference to the Aptian assemblages of Speeton (North Yorkshire, England). *Journal of Micropalaeontology, 7(2):* 143–185.

BARRERA, E. & KELLER, G., 1991. Late Paleocene to Early Eocene climatic and oceanographic events in the Antarctic Indian Ocean. *Geological Society of America Abstracts 23(5):* A 179.

BOERSMA, A. & PREMOLI-SILVA, I., 1991. Distribution of Paleogene planktonic foraminifera – analogies with the Recent? *Palaeogeography, Palaeoclimatology, Palaeoecology, 83:* 29–48.

CARON, M. & HOMEWOOD, P., 1982. Evolution of early planktonic foraminifers. *Marine Micropaleontology, 7:* 453–462.

CIFELLI, R., 1969. Radiation of Cenozoic planktonic foraminifera. *Systematic Zoology, 18:* 154–168.

CORFIELD, R.M., 1987. 'The environmental control of the evolution of Palaeocene and Early Eocene planktonic foraminifera'. Unpublished Ph.D. thesis, University of Cambridge, U.K.

CORFIELD, R.M. & CARTLIDGE, J.E., 1991. Isotopic evidence for the depth stratification of fossil and Recent Globigerinina. *Historical Biology, 5:* 37–63.

CORFIELD, R.M. & CARTLIDGE, J.E., 1992. Oceanographic and climatic implications of the Palaeocene carbon isotope maximum. *Terra Nova 4:* 443–455.

CORFIELD, R.M. & GRANLUND, A.H., 1988. Speciation and structural evolution in the Palaeocene *Morozovella* lineage. *Journal of Micropalaeontology, 7(1):* 59–72.

CORFIELD, R.M. & SHACKLETON, N.J., 1988. Productivity change as a control on planktonic foraminiferal evolution after the Cretaceous/Tertiary boundary. *Historical Biology, 1:* 323–343.

DOUGLAS, R.G. & SAVIN, S.M., 1978. Oxygen isotopic evidence for the depth stratification of Tertiary and Cretaceous planktonic foraminifera. *Marine Micropaleontology, 3:* 175–196.

GRIGELIS, A., & GORBATCHIK, T., 1980. Morphology and taxonomy of Jurassic and early Cretaceous representatives of the superfamily Globigerinacea (Favusellidae). *Journal of Foraminiferal Research, 10(3):* 180–190.

HART, M.B., 1980. A water depth model for the evolution of the planktonic foraminifera. *Nature, 286:* 252–254.

JENKINS, D.G. & SHACKLETON, N.J., 1979. Parallel changes in species diversity and palaeotemperatures in the lower Miocene. *Nature, 278:* 50–51.

KENNETT, J.P., 1976. Phenotypic variation in some Recent and Late Cenozoic planktonic foraminifera. In R. Hedley & C.G. Adams (Eds), *Foraminifera: 2.* London: Academic Press.

KENNETT, J.P., 1991. Paleoceanographic changes near the Paleocene/Eocene boundary in mid-latitude site ODP Site 762C, Exmouth Plateau, NW Australia. *Geological Society of America Abstracts, 23(5):* A 338.

KENNETT, J.P. & STOTT L.D., 1990. Proteus and Proto-Oceanus: Ancestral paleogene oceans as revealed from Antarctic stable isotope results; ODP Leg 113. *Proceedings of the Ocean Drilling Program Scientific Results, 113:* 865–880.

KENNETT, J.P. & STOTT L.D., 1991. Abrupt deep sea warming, palaeoceanographic changes and benthic extinctions at the end of the Paleocene. *Nature, 353:* 225–229.

LASKER, H.R., 1978. The measurement of taxonomic evolution: Preservational consequences. *Paleobiology, 4:* 135–149.

MALMGREN, B.A. & KENNETT, J.P., 1981. Phyletic gradualism in a late Cenozoic planktonic foraminiferal lineage: DSDP Site 284, Southwest Pacific. *Paleobiology, 7(2):* 230–240.

MALMGREN, B.A., BERGGREN, W.A. & LOHMANN, G.P., 1983. Evidence for punctuated gradualism in the late Neogene *Globorotalia tumida* lineage of planktonic foraminifera. *Paleobiology, 9(4):* 377–389.

NORRIS, R.D., 1991. Biased extinction and evolutionary trends. *Paleobiology, 17(4):* 388–399.

PEARSON, P.N., 1992. Survivorship analysis of fossil taxa when real time extinction rates vary: the Paleogene planktonic foraminifera. *Paleobiology, 18,(2):* 115–131.
RUDWICK, M.J., 1964. The inference of function from structure in fossils. *British Journal for the Philosophy of Science, 15:* 27–40.
SEPKOSKI, J.J. JR., 1978. A kinetic model of Phanerozoic taxonomic diversity I. Analysis of marine orders. *Paleobiology, 4(3):* 223–251.
STEHLI, F.G., DOUGLAS, R.G., & KAFESCIOGLU, I.A., 1972. Models for the evolution of planktonic foraminifera. In T.J.M. Schopf (Ed.), *Models in Paleobiology*. San Francisco: Freeman Cooper.
STOTT, L.D., KENNETT J.P. SHACKLETON, N.J. & CORFIELD, R.M., 1990. The evolution of Antarctic surface waters during the Paleogene: Inferences from the stable isotopic composition of planktonic foraminifers, *Proceedings of the Ocean Drilling Program Scientific Results, 113:* 849–863.
THOMAS, E., 1990. Late Cretaceous–early Eocene mass extinctions in the deep sea. In *Global Catastrophes in Earth History. Geological Society of America Special Publications, 247:* 481–495.
THOMAS, E., 1991. The latest Paleocene mass extinction of deep sea benthic foraminifera: Result of global climate change. *Geological Society of America Abstracts, 23:* A141.
THUNELL, R.C., 1981. Cenozoic palaeotemperature changes and planktonic foraminiferal speciation. *Nature, 289:* 670–672.
TJALSMA, R.C. & LOHMANN, G.P., 1983. Paleocene–Eocene bathyal and abyssal benthic foraminifera from the Atlantic Ocean. *Micropaleontology Special Publication, 4:* 1–90.
VAN VALEN, L., 1973. A new evolutionary law. *Evolutionary Theory: 1:* 1–30.
WEI, K-Y. & KENNETT, J.P., 1983. Nonconstant rates of extinction in Neogene planktonic foraminifera. *Nature, 305(5931):* 218–220.
WEI, K-Y., & KENNETT, J.P., 1986. Taxonomic evolution of Neogene planktonic foraminifera and paleoceanographic relations. *Paleoceanography, 1(1):* 67–84.

CHAPTER

6

Patterns of evolution in Quaternary mammal lineages

ADRIAN. M. LISTER

Introduction **72**

Case histories **74**
Moose 74
Mammoths 78
Water voles 81
Red deer 82

Discussion **87**
Rates of change 87
Evolutionary trends 88

References **89**

Keywords: Evolutionary rates – gradualism – punctuated equilibrium – speciation – island dwarfing – *Mammuthus* – *Alces* – *Cervus* – *Arvicola*.

Abstract

Quaternary mammals provide finely stratified fossil sequences of great potential for tracing evolutionary patterns, forming a link between the fine-scale evolution of modern populations, and coarser palaeontological trends of the more distant past. In the moose lineage (*Alces*), relative antler length decreased directionally across four levels spanning 2 million years. In mammoths (*Mammuthus*), directional dental change is observed from 2 to 0.4 million years BP, followed by 0.4 million years of 'dynamic stasis'. These species-level transitions occurred with (as yet) no evidence of cladogenetic speciation. Across 20 levels in 300 000 years, subspecies-level dental change in water voles (*Arvicola*) is largely directional but with an inversion possibly due to population movement. The dwarfing of red deer (*Cervus elaphus*) on Jersey took 6000 years or less, corresponding to allopatric, geologically 'punctuated' evolution, but without (as yet) evidence of reproductive speciation. These examples invite a pluralistic view of evolutionary modes. In addition, credible adaptive explanations are possible for some morphological shifts among Quaternary mammals, based on our knowledge of related modern forms.

Evolutionary Patterns and Processes
ISBN 0-12-440895-8

INTRODUCTION

The origin of species, and of the traits which distinguish them, continues to be central issues in evolutionary biology, and still poses many unanswered questions. Modern genetics, to its credit, can provide a plausible mechanism for almost any long-term pattern of evolution that a palaeontologist can posit – gradual or 'punctuated', speciation via allopatric isolates or over a broad range, and so on. But by the same token it is difficult to predict which of these modes may actually have predominated in practice and in different circumstances; in this crucial area we benefit greatly from the direct window on the past which palaeontology provides.

The Quaternary, the most recent geological period, spans the time from *c.* 1.7 million years ago to the present day. Geologically, this is a relatively short period, but as such covers precisely the temporal orders of magnitude at which we may hope to observe the evolution of species and species-level characters. There are several reasons why the study of Quaternary mammals is particularly inviting in this respect: (1) There are thousands of easily exposed deposits world-wide, many of which yield abundant mammalian remains. (2) Recent advances in stratigraphy, and especially in absolute dating techniques (e.g. electron spin resonance, uranium series, palaeomagnetism: Smart & Frances, 1991), mean that many of these samples can be placed with some accuracy within the Quaternary sequence, providing relatively fine stratigraphic spacing, at least in comparison with most of the terrestrial record. (3) Intensive palaeoenvironmental work by oxygen-isotope analysis of deep-sea cores (Shackleton & Opdyke, 1973), pollen studies of terrestrial sediments (West, 1977), and other techniques, mean that the sequence of climatic and vegetational changes through the Quaternary is becoming increasingly well-understood, more so than for any earlier period. The varied, sometimes drastic nature of these changes provides a rich backdrop against which to view the responses of the mammalian species. (4) Many of the fossil species are still extant or have close living relatives, providing a known reference point of adaptation, variability, and so on, against which to compare their antecedents. In many ways, research on Quaternary mammals forms a link between neontological studies of modern populations, and the coarser palaeontological patterns of the more distant past.

To gain a picture of the relative frequency of different evolutionary modes, and their incidence in different taxonomic groups and ecological situations, it is essential to examine as many 'case histories' as possible. In the present paper, four such examples are described, in order of increasing stratigraphic resolution. In the terrestrial Quaternary record, long fossiliferous sequences in single locations or formations (like the early Tertiary mammals of the Bighorn Basin, see Gingerich, 1980; or protists in oceanic cores, see Malmgren, Berggren & Lohman, 1983) are uncommon, and we must piece together time-series from a series of separate localities of different ages. All samples included in this study have some independent evidence of dating; this is summarized in Table 1. The taxa examined here are all Eurasian, and most of the samples are from the European region. This is not to deny the potential evolutionary importance of variation across the broader geographical range of the taxa, but research has to start somewhere, and the European Quaternary provides a unique wealth of dated, intensively studied fossil localities and samples. Assessment of the adaptive significance of characters has also largely been omitted

Table 1 Samples of Eurasian Quaternary fossils scored in this study.

Site	Approximate age (million years BP)	Dating methods	Collections	References
Khapry/Liventsovka, Russia	2.0–2.5	f, m	GIN	Baigusheva, 1971; Dubrovo, 1989; Vangengeim & Pevsner, 1991
Senèze, France	1.6–2.0	f, m	UL	Bout, 1970; Thouveny & Bonifay, 1984
E. Runton, U.K.	1.7	f, l, m, p	NHM, BGS	Gibbard *et al.*, 1990; Lister, in press a
Upper Valdarno, Italy	1.6–1.7	f, l	IGF	Azzaroli *et al.*, 1988; Masini, personal communication
Siniaya Balka, Russia	1.2–1.0	f, m	PIM	Dubrovo, 1964; Vangengeim *et al.* 1991
Mosbach III, Germany	0.6–0.5	f, l, m	NM, HLD	Brüning, 1978; Igel, 1984
Süssenborn, Germany	0.6–0.5	f, l, m	IQW	e, 1969; Steinmüller, 1972
Homersfield, U.K.	0.5–0.3	1	NCM	Coxon, 1982; Stuart, 1982
Tourville, France	0.3	f, l, t	GCE	Carpentier & Lautridou, 1986
Ilford, U.K.	0.2	a, f, l	NHM, BGS	Stuart, 1976; Sutcliffe, 1976; Bowen *et al.*, 1989
La Cotte, Jersey	0.24–0.15	l, t	JM	Callow & Cornford, 1986
Balderton, U.K.	0.14	e, f, i, l	UMZC	Brandon & Sumbler, 1991; Lister & Brandon, 1991
Ehringsdorf, Germany	0.2–0.12	f, l	IQW	Kahlke, 1975
Taubach, Germany	0.12	f, l	IQW	Kahlke, 1976
Belle Hougue, Jersey	0.12	a, u	JM	Keen *et al.*, 1981; Lister, 1989
Předmostí, Czechoslovakia	0.026	i, r	AIB	Musil, 1968
Kostenki I, Russia	0.014	i, l, r	ZIN	Klein, 1969

Dating methods: a, amino acid epimerization; e, electron spin resonance; f, fauna; i, industry; l, lithostratigraphy; m, palaeomagnetism; p, pollen; r, radiocarbon; t, thermoluminescence; u, uranium series.
Location of collections: AIB, Anthropos Institute, Brno; BGS, British Geological Survey, Keyworth; GCE, G. Carpentier, Elbeuf; GIN, Geological Institute, Moscow; HLD, Hessisches Landesmuseum, Darmstadt; IGF, Institute of Geology, Florence; IQW, Institut für Quartärpaläontologie, Weimar; JM, Jersey Museum; NCM, Castle Museum, Norwich; NHM, Natural History Museum, London; NM, Naturhistorisches Museum, Mainz; PIM, Palaeontological Institute, Moscow; UL, University of Lyon; UMZC, University Museum of Zoology, Cambridge; ZIN, Zoological Institute, St. Petersburg.

(with the exception of a brief discussion of moose antlers, as an example), in order to concentrate on the pattern of change.

Two of the lineages here described (moose and mammoth) have conventionally been divided taxonomically into a series of chronospecies. In making use of these species names, I am regarding them merely as markers for the fossil samples (or populations from which they derived), without prejudice to the evolutionary pattern which is the subject of investigation. Thus, I am a priori neither asserting nor denying the possible 'individuality' of species as evolutionary players (Ghiselin, 1974; Vrba & Eldredge, 1984).

CASE HISTORIES

Moose

The moose, *Alces alces* (L.), also known in Europe as the elk, is the largest living species of deer. The early Quaternary evolution of the genus occurred in Eurasia, where a series of chronospecies has been described: *A. gallicus* Azzaroli, from the late Pliocene to early Pleistocene of Europe (*c.* 2.0–1.5 million years BP); *A. latifrons* (Johnson), from the early Middle Pleistocene of northern Eurasia (*c.* 0.7–0.5 million years BP) (and Middle to Upper Pleistocene of North America), and *A. alces* from the Upper Pleistocene of northern Europe, Asia and North America (0.1 million years ago to present). Details of the dating and taxonomy of these forms are given by Lister (in press a,b). As a working hypothesis, the three nominal chronospecies are regarded as an approximately direct line of descent, for several reasons. First, they share a derived suite of dental and skeletal features peculiar among deer, indicating close relatedness. Second, they follow each other in time and replace each other in geographical distribution. Third, no other presumptive ancestors are currently known for either of the two later species.

Through this sequence of fossils, significant morphological changes occurred, to a degree at least as great as differences typically separating deer species today. Body size increased dramatically from *A. gallicus* to *A. latifrons*, and then reduced somewhat to *A. alces* (Mauser, 1990; Lister, in press b); skull architecture became modified between *A. latifrons* and *A. alces*, in ways interpreted as adaptive to a shift into boreal forest (Flerov, 1952; Sher, 1987a); and the relative length of the antler beam became reduced in both steps of the sequence. It is the last-mentioned feature which will be discussed here.

The basic structure of the moose antler is seen throughout the sequence: a horizontal, unbranched, laterally directed beam, terminating in a broad, digitated palmation. But the relative length of the beam (and hence the spread of the antlers as a whole) has undergone drastic reduction (Fig. 1). In studying this or other changes in antler form, two factors need to be taken into account. The first is body size differences between successive populations, to which antler size and complexity are allometrically linked (Huxley, 1931; Gould, 1974). The second is individual ontogenetic variation: through the lifespan of a male deer, the antlers, which are shed and regrown each year, increase in size and complexity with individual age, until they reach a plateau in the mature adult, before degenerating somewhat in old age. Species must thus be compared either at a particular individual age, or preferably, along the entire ontogenetic trajectory (Lister, 1987). In Fig. 2, antler beam length is plotted against the circumference at the base of the beam, for samples of the three nominal *Alces* species. Basal circumference is a measurement known from studies of living deer to increase year by year through the life of the animal (e.g. Hattemer & Dreschler, 1976), as well as being correlated to the intrinsic body size of the individual. It thus forms a suitable abscissa against which to plot other antler dimensions, clarifying variation due to age and size (Lister, 1990). In Fig. 2, the younger and smaller individuals plot to the left of the graph, older and larger ones to the right.

Figure 2 clearly shows a reduction in relative beam length from *gallicus* to *latifrons* to *alces*. The three scatters each span a wide range of basal circumferences, but are displaced from each other in the length dimension, indicating that the species

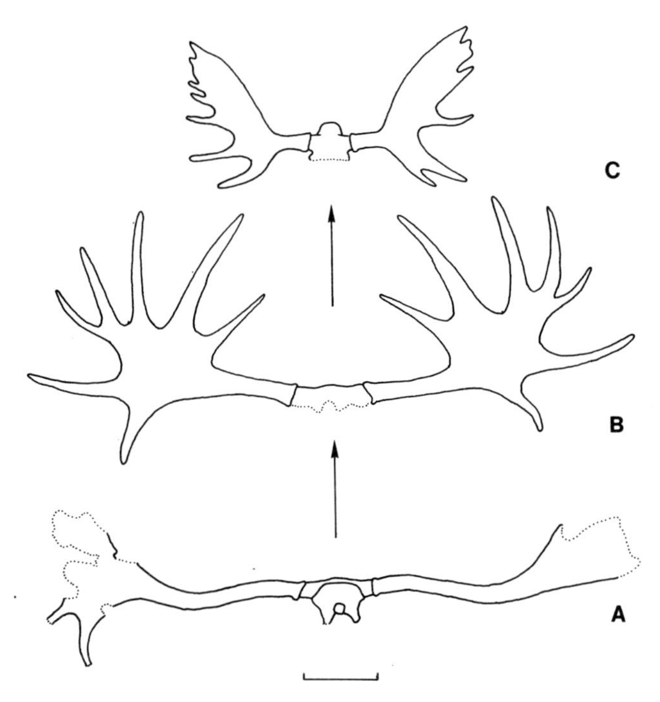

Figure 1 Reduction in antler beam length of Eurasian Quaternary moose. A: *Alces gallicus,* Lower Pleistocene, Senèze, France, in posterior view (from Azzaroli, 1952). B: *A. latifrons,* early Middle Pleistocene, Goldshöfer Sands, Germany, in anterior view (from a slide courtesy of Professor W. von Koenigswald). C: *A. alces,* modern, Sweden, in anterodorsal view (original, from a specimen in the University Museum of Zoology, Cambridge). Scale bar 25 cm.

differences are valid across a range of ages and sizes, and are not due to bias in the ontogenetic composition of the samples. At the 'young' end of the graph, the three scatters overlap, their differentiation becoming more evident with increasing ontogenetic age of the animals. In Fig. 3, these data have been summarized, by removing juvenile individuals, and plotting the ratio of length to circumference for the adult antlers of each sample. The wide range of each sample reflects its broad age composition (length / circumference ratio varies allometrically with age) but the shift in mean values clearly indicates the evolutionary trend between the three sampling points. The chronological gaps between these samples are rather large, and well-dated remains of moose antlers pertaining to the intervening periods are unfortunately very scarce. Two Upper Pleistocene specimens are available from the period 0.2–0.12 million years BP, and are plotted in Figs 2 and 3. Although this sample is very small, the beam proportions are intermediate between the *latifrons* and *alces* ranges, so that we may provisionally speak of four samples in chronological order, showing a unidirectional trend.

The pattern of evolution revealed by this study indicates that the antlers of modern moose arose through intermediate stages over about 2 million years – a long period relative to the lifetime of Quaternary mammalian species. In this sense, the transition was 'gradualistic'. On the other hand, spacing between sampling points is large: *c.* 1 and 0.5 million years respectively for the first two steps – so we are not justified in drawing a 'gradual' line between them; the transitions might have been relatively

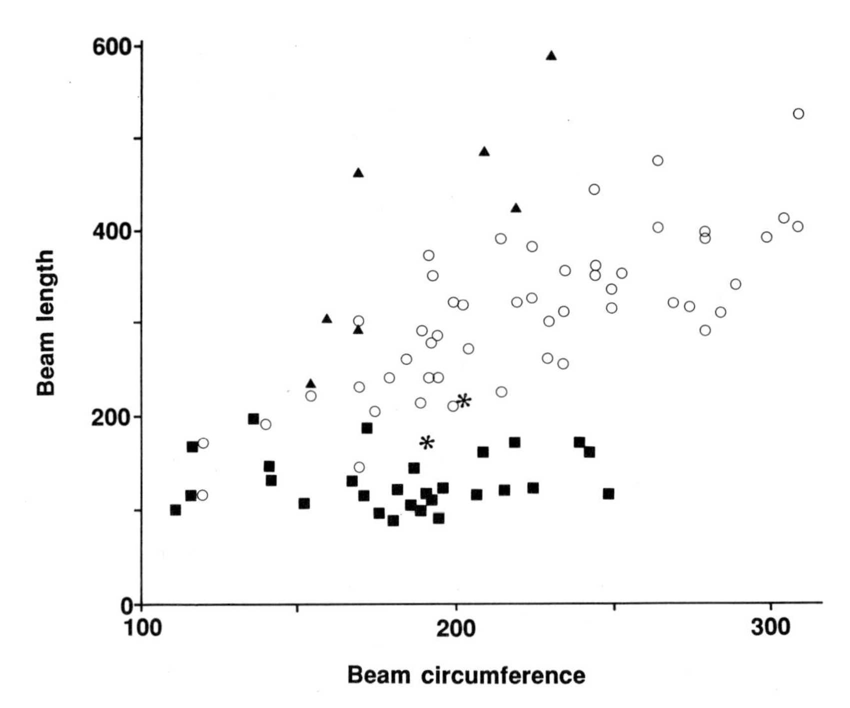

Figure 2 Beam length of Eurasian Quaternary moose antlers, plotted against beam circumference as an index of individual age and size. Triangles: Lower Pleistocene *Alces gallicus,* combined sample from Senèze, France and East Runton and Sidestrand, U.K. Circles: early Middle Pleistocene *A. latifrons,* combined sample from Mosbach III and Süssenborn, Germany. Asterisks: late Middle or early Upper Pleistocene *Alces sp.,* Ehringsdorf and Taubach, Germany. Squares: Modern *A. alces*, Scandinavia. Beam length measured from the burr to the beginning of the palmation (defined as the point at which beam circumference has increased by 50%). Beam circumference measured directly above the burr. See Table 1 for details and dating of sites. All data original, except some modern *A. alces* from Erdbrink (1954).

rapid in the gaps between samples (Fig. 4). As for cladogenetic speciation events, we can only say that there is no evidence for them: no evidence of temporal overlap between ancestral and descendant types in Eurasia at any point in the sequence. Some Siberian Middle Pleistocene populations have been described as local variants, for example *A. latifrons postremus* Vangengeim & Flerov, a small-bodied, late form of *A. latifrons*; and an unnamed form contemporaneous with Eurasian *A. latifrons* but with rather slender beams (Sher, 1987b). There is no reason to regard these as other than local populations (perhaps subspecies) of the main, widely distributed lineage, analogous to variation observed within *Alces alces* at the present day (Heptner, Nasimovich & Bannikov, 1988). The only candidate for temporal overlap between ancestral and descendent forms is the persistence of North American *Alces latifrons* until well after the (probably Eurasian) origin of *A. alces*. This hardly provides grounds for regarding the origin of *A. alces* as a cladogenetic event, since we have no reason to believe that the Eurasian origin of this species had anything to do with the split between *latifrons* populations on the two continents. Cladogenetic evolution cannot, with such a sparse record, be ruled out, especially for the *latifrons–alces* transition, where significant cranial changes occurred in addition to antler shortening (Azzaroli, 1981; Sher, 1987a), but the evidence is lacking.

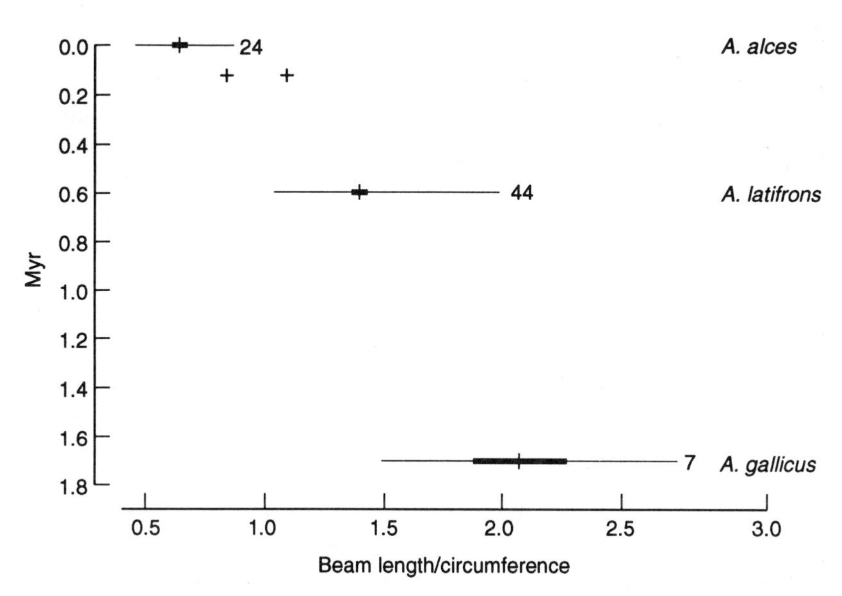

Figure 3 Change of proportion in antler beams of Eurasian Quaternary moose, plotted from the data in Fig. 2. For each sample, the mean, range, standard error of the mean and sample size are shown. Juvenile antlers (beam circumference <150 mm for *A. gallicus* and *A. alces*, <175 mm for *A. latifrons*) have been excluded.

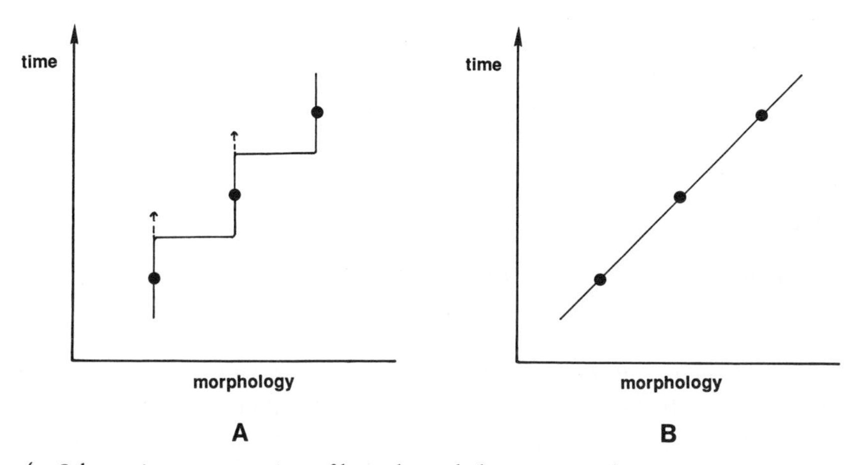

Figure 4 Schematic representation of hypothetical alternative evolutionary patterns based on three dated morphological points. A: Extreme punctuated pattern (also showing hypothetical cladogenetic speciation events). B: Extreme gradualistic pattern (constant-rate evolution). Intermediate patterns between these extremes are possible.

Possible adaptive reasons for the reduction in relative antler length of moose will be briefly discussed. In the first step, from *gallicus* to *latifrons*, the explanation may lie in the great increase in body size, which occurred for unknown reasons. The type skeleton of *A. gallicus* from Senèze, France is the largest known individual of the species, and stands 140 cm at the shoulder (Azzaroli, 1952; personal measurements); skeletal dimensions of *A. latifrons* are on average 50% larger (Mauser, 1990; Lister, in press b), suggesting a shoulder height of *c.* 210 cm. Total antler span of the Senèze *gallicus* specimen (Fig. 1) was *c.* 2.2 m with palmations restored; if *latifrons* antlers had simply been scaled up isometrically along with body size, antler spans of some 3.3 m would have resulted, or 4 m or more if the positive allometry of antler size to body size normal among deer (Gould, 1974) is allowed. It is likely that the biomechanical implications and/or energy requirements of such huge antlers would have resulted in selection against them. Thus, the total antler spread in *A. latifrons* remained similar to that of *A. gallicus* at *c.* 2 m, despite the great increase in body size, producing the relative antler reduction illustrated in Figs 2 and 3.

In the second step, from *A. latifrons* to *A. alces*, body size was somewhat reduced (Mauser, 1990), but not nearly enough to account for the two-thirds reduction in antler beam length which occurred, from 40–50 cm to typically 15 cm in adult individuals (Fig. 2). This transition may have been linked to a change in habitat. *A. latifrons*, to judge from associated flora and fauna, lived in a relatively open, steppe-like habitat (Sher, 1974:198). At some point in the transition to *A. alces*, the lineage moved into the coniferous forest habitat which the moose occupies today. An antler span of 2 m or more would have been manageable in an open habitat, but in forest could impede movement. The antler span of modern moose is smaller than 'expected' for a deer of its size (Gould, 1974:195); even so, male animals sometimes have trouble moving rapidly through forest because of snagging their antlers (R.D. Guthrie, personal communication).

These adaptive explanations cannot be regarded as proven, but have the status of hypotheses. They are not simply 'Just-So' stories of which any number could be posited, as they are based on reasoned deduction from known aspects of morphology and environment, aided especially by reference to related modern forms. The particular interest of the suggested explanations for moose antler reduction, if they are correct, is that different selective forces may have been responsible for two phases of a unidirectional morphocline.

Mammoths

The mammoth lineage is believed to have originated in Africa, spreading north into Europe at around 2.5 million years BP, and subsequently expanding across Eurasia into North America (Maglio, 1973; Lister, in press c). Abundant fossils of the Eurasian mammoth lineage, spanning the time from *c.* 2.5 million years ago to the extinction of the lineage at *c.* 10 000 years BP, have been classified into a series of three chronospecies: Pliocene to Lower Pleistocene *Mammuthus meridionalis* (Nesti), Middle Pleistocene *M. trogontherii* (Pohlig) and Upper Pleistocene *M. primigenius* (Blumenbach). Derived morphological features of these forms (such as the high skull and twisted tusks), their chronological replacement, and the lack of alternative ancestors, imply that they represent an approximate evolutionary line of descent. Striking evolutionary changes occur through time in this lineage, including heightening and shortening of the cranium, doubling of molar crown height, and doubling the number of enamel bands (lamellae) in the molars (Maglio, 1973; Lister, in press c). The increased 'tooth life' resulting from these changes is generally regarded as an adaptation to the coarse vegetation of the ice age 'steppe–tundra' biome, in keeping with palaeoenvironmental evidence that *Mammuthus* shifted from warmer, partly wooded habitats to cold, open regions through the Pleistocene (Lister, in press c).

Molar evolution in this lineage has long been posited as an example of gradualistic change, based essentially on a comparison of 'typical' samples of the three nominal chronospecies (Fig. 5), with assumed gradualism inbetween (e.g. Adam, 1961; Aguirre, 1969; see Lister, in press, c for discussion). The accumulation of well-dated samples in recent years allows this model to be tested. In Fig. 6, two key variables are plotted for ten chronologically ordered samples: hypsodonty index, a measure of molar crown height; and lamellar frequency, an index of the number of enamel bands in the molar.

Figure 6 indicates that for the first *c.* 2 million years of Eurasian mammoth evolution (samples 1–5), a gradual pattern is apparently supported. Although there is still up to 0.5 million years between sampling points, the morphological overlap between successive levels, and relatively small mean shift at each step, leave little room for unseen jumps in morphology between levels.

However, this directional pattern is not maintained after *c.* 0.5 million years BP. Hypsodonty index reaches its maximum value by *c.* 0.5 million years BP, lamellar frequency by *c.* 0.4 million years BP (Fig. 6). Thereafter, there is fluctuation in mean values, but no net trend – an apparent pattern of stasis. This period of time has been examined in more detail by Lister (in press c), using data additional to those of Fig. 6. The more 'primitive' samples (such as sample 7 in Fig. 6) are concentrated in the period 0.4–0.2 million years ago, and tend to be associated with more temperate,

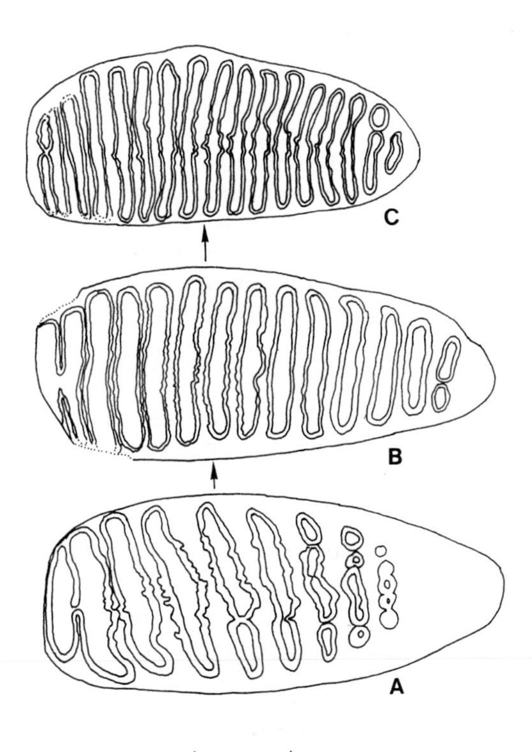

Figure 5 Right upper third molars of Eurasian Quaternary *Mammuthus,* illustrating changes in lamellar frequency and enamel thickness. A: *M. meridionalis,* Lower Pleistocene, Siniaya Balka, Russia (Palaeontological Institute, Moscow, no. 1249/234). B: *M. trogontherii,* early Middle Pleistocene, Mosbach III, Germany (Natural History Museum, Mainz). C: *M. primigenius,* Upper Pleistocene, Gydan Peninsula, Russia (Palaeontological Institute, Moscow). There was also a marked increase in crown height between A and B. Scale bar 5 cm.

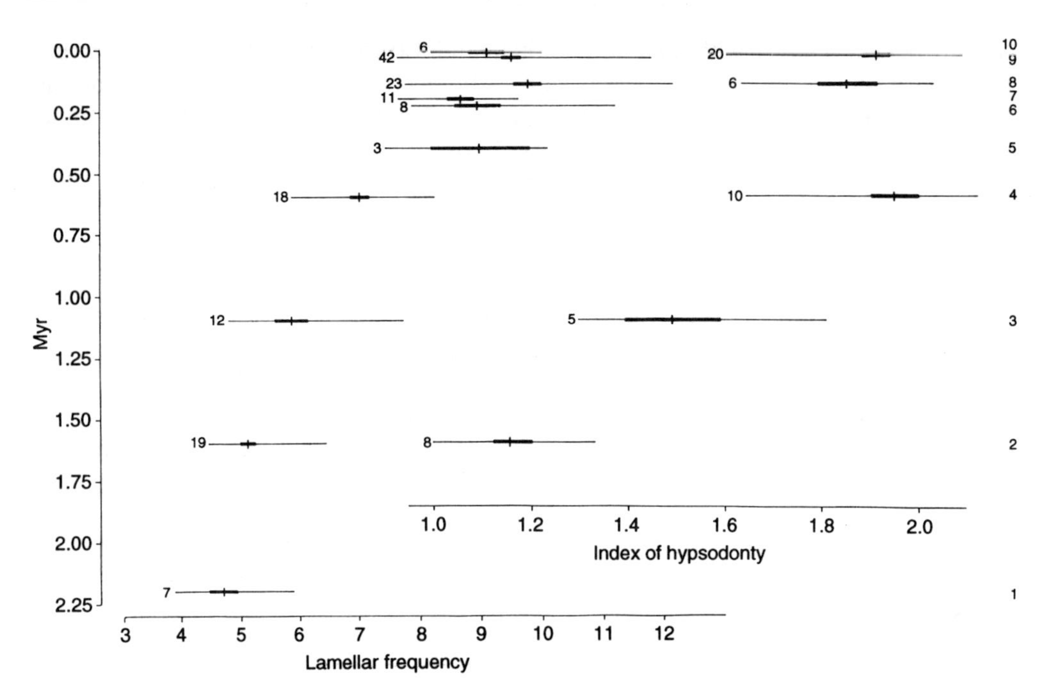

Figure 6 Changes in lamellar frequency and hypsodonty index of third upper molars in European Quaternary mammoth samples. Lamellar frequency indicates the number of lamellae (enamel plates) in a typical 10 cm length of tooth. Hypsodonty index was measurable on fewer samples, and is the ratio between the maximum height and maximum width of the molar crown. All data original; measurements taken according to the method of Maglio (1973). For each sample, the mean, range, standard error of the mean and sample size are shown. Samples 1–3 are conventionally named *M. meridionalis;* sample 4, *M. trogontherii;* samples 5–10, *M. primigenius.* To the right of the graph, samples are numbered as follows: 1: Khapry/Liventsovka, Russia; 2: Upper Valdarno, Italy; 3: Siniaya Balka, Russia; 4: Mosbach III, Germany; 5: Homersfield, U.K.; 6: Tourville, France; 7: Ilford, U.K.; 8: Balderton, U.K.; 9: Předmosti, Czechoslovakia; 10: Kostenki I, Russia. See Table 1 for details and dating of samples.

partly wooded palaeoenvironments. They alternate with more 'advanced' samples which are associated with colder, more open habitats, and which come to predominate after 0.2 million years BP.

The pattern of change over the first *c.* 2 million years of Fig. 6 has the appearance of anagenetic evolution, and it would be unparsimonious to posit four speciation events between samples one to five in the sequence (cf. Fortey, 1985). Nor is there any direct evidence for cladogenetic speciation in the dental material, which would be indicated by a bimodal morphological distribution within a site sample, reflecting co-existence of ancestral and descendent forms. On the other hand, Azzaroli (1977) has suggested a speciation event for the *meridionalis–trogontherii* transition (between samples 3 and 4), on the basis of some crania of late *meridionalis* from Italy (*c.* 1.0 million years BP) said to show specialized features ruling out ancestry to *trogontherii.* This remains a possibility, but it seems equally plausible that late *meridionalis* could have been differentiated into local populations of which the Italian form was one, while one or more others, perhaps less specialized, could have been ancestral to *trogontherii.* Significant here is sample 3 in Fig. 6, from the Siniaya Balka locality in Russia, also *c.* 1.0 million years BP (Dubrovo, 1977), with hypsodonty index intermediate between typical *meridionalis* (sample 2) and *trogontherii* (sample 4). Although crania are lacking from this sample, the high tooth crowns imply a deepening of the maxillary region, one of the significant changes of cranial architecture in the *meridionalis–trogontherii* transition. Also relevant to this question are indications of a morphological inversion around *c.* 0.7–0.6 million years BP, the conventional *meridionalis/trogontherii* boundary. Some European molars classifiable as late *meridionalis* appear in higher strata than others referable to early *trogontherii*: see Lister (in press c) for details. This again would be consistent with a cladogenetic model, but could equally represent fluctuation within a lineage, perhaps due to population movements. The degree of morphological difference is certainly consistent with intraspecific variation.

The upper part of the sequence provides no direct evidence of speciation. Kotsakis, Palombo & Petronio (1978) discuss the possibility that the alternation between more 'primitive' and more 'advanced' samples in Europe after 400 000 years BP might represent two species shifting their ranges in accordance with climatic change, but it is at least as likely that these were ecotypal populations within a single species. The latter interpretation is perhaps favoured by existing data, which reveal considerable variability among the samples, rather than a simple two-way split. Together with the greater preponderance of 'primitive' samples in the earlier part of this phase, this suggests a picture akin to the 'shifting balance' model of evolution, with different populations at somewhat different morphological levels, but progressive modernization of the species as a whole over time. The categorization of this period as 'stasis', based on the early appearance of some populations as advanced as later ones, may therefore to some extent disguise a situation of dynamic interplay between populations, leading in this case to the eventual loss of the more 'primitive' ones (Lister, in press c).

Water voles

The living water vole *Arvicola terrestris* has a broad Palaearctic distribution extending from western Europe to eastern Siberia, and as far south as Iran. The evolution of this and related species over several million years has been the subject of intensive research, based mostly on molar morphology (Fejfar & Heinrich, 1990). The most recent part of the *A. terrestris* lineage, over the past 0.3 million years or so, provides a relatively high-resolution picture of evolutionary change, which has been investigated by Heinrich (1987), van Kolfschoten (1990) and others. The molars of these voles comprise a series of enamel loops, each loop being formed of a concave 'leading' edge and a convex 'trailing' edge (Fig. 7). The leading edges are those which first come into play during mastication.

The earliest *Arvicola*, from the European early Middle Pleistocene *c.* 0.5 million years ago, inherited from its immediate ancestor an unusual thickening of the trailing edge (Fig. 7A). Through the Middle and Upper Pleistocene in Europe, there was a reduction in the thickness of the trailing edge until it became thinner than the leading edge (Fig. 7B). The histological basis for this change, and possible adaptive reasons for it, have been discussed by von Koenigswald (1980).

Heinrich (1987) quantified this transition by calculating an 'index of enamel differentiation': essentially the ratio between trailing and leading edge thickness, summed and averaged over all enamel loops of the tooth. He measured molars from

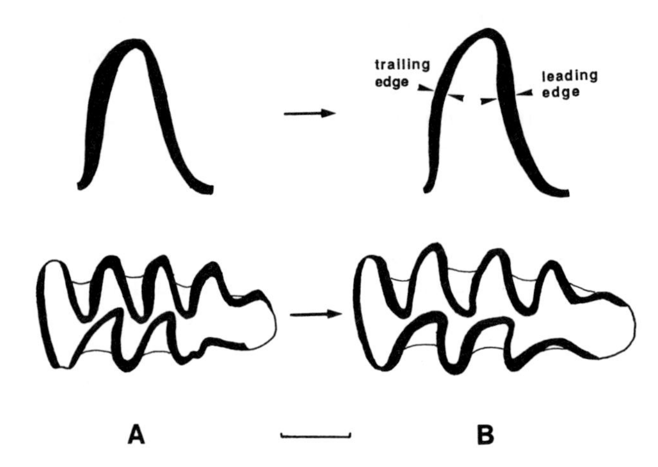

Figure 7 First lower molars of water voles, *Arvicola terrestris*, in occlusal view. A: Early form, showing thicker enamel on the convex (trailing) edges of the enamel loops. B: Later form, showing thicker enamel on the leading edges. Above: enlargement of a single enamel loop, indicating the positions of measurement. After Stuart (1982) and van Kolfschoten (1990). Scale bar 1 mm.

a series of dated deposits in central and western Europe, and found a unidirectional trend of reduction in the index over about ten samples between *c.* 0.3 million years BP and the present day. Subsequently, van Kolfschoten (1990) added further samples to the dataset, making a total of 22 sampling points over *c.* 300 000 years (Fig. 8), an average density of one sample per 14 000 years. The combined data illustrate an overall trend of reduction in the index, but include several fluctuations, including an important reversal between the late Saalian cold stage and Eemian interglacial, evidenced by three late Saalian samples not available to Heinrich (1987).

This reversal might have been due to evolutionary tracking within a single population, or alternatively to population replacement. As shown by Röttger (1987) and discussed by van Kolfschoten (1990), present-day *A. terrestris* displays a marked cline in enamel differentiation, with lower indices in north-west and central Europe, higher indices to the south and east. The fossil samples plotted in Fig. 8 are all from the central and north-western region, so it is possible that the 'advanced' late Saalian population was replaced in the Eemian by more 'primitive' animals migrating from the south or east, perhaps linked to the warming of climate in the Eemian interglacial.

The detection of a morphological reversal in this sequence is a reflection of its relatively high stratigraphic resolution. Such reversals quite likely occurred within longer sequences such as that of the moose and the early mammoths, described above, but are not perceived because of coarser sampling. As in those examples, there is no evidence of speciation within the water vole sequence, although the possible occurrence of undetected sibling species cannot be discounted when dealing with small mammals (Lister, 1992).

Red deer

The isolation of mammalian populations on small islands provides valuable, if idiosyncratic, evolutionary case histories. One of the commonest results of such isolation is extreme size reduction, seen in Pleistocene populations of deer, elephants,

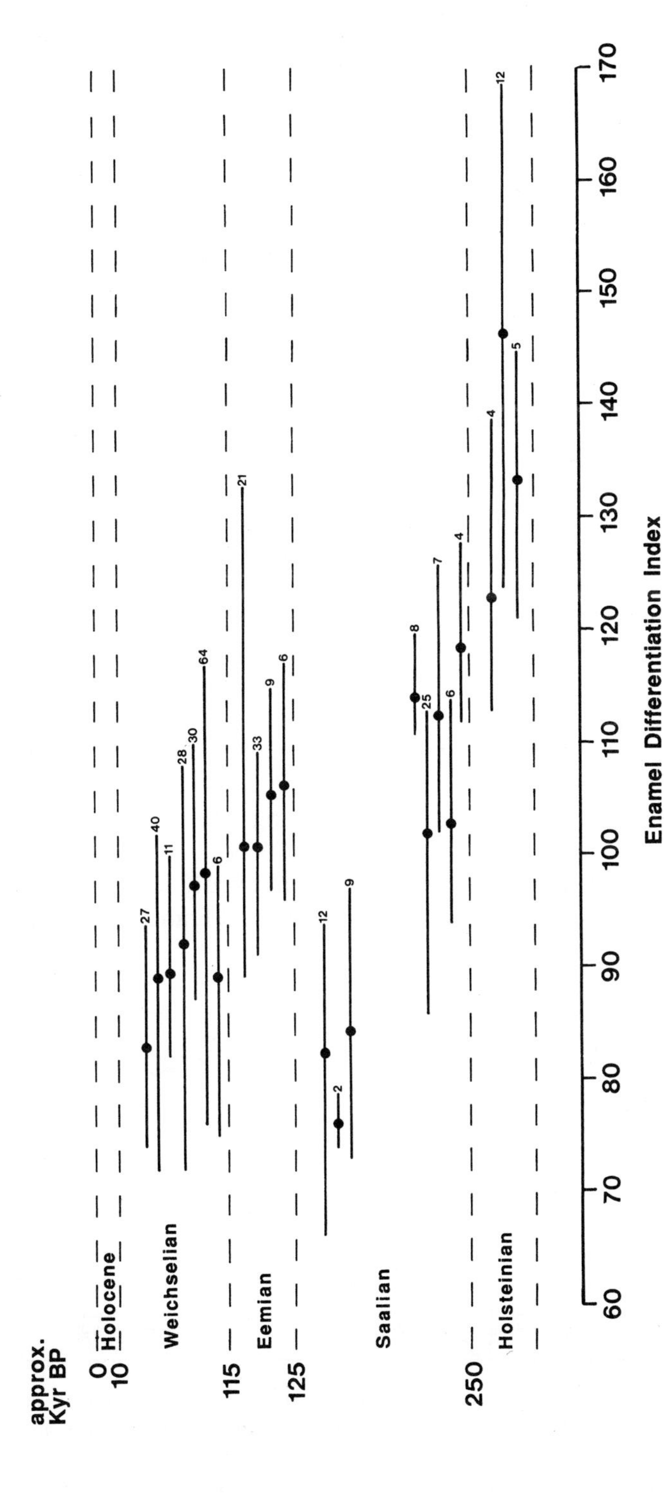

Figure 8 Changes in enamel thickness of water vole molars through the European late Middle and Upper Pleistocene (cf. Fig. 7). The enamel differentiation index is calculated as the ratio between the enamel thicknesses on the trailing and leading edges, computed for each of the seven loops and then averaged between them. The samples are from The Netherlands, Germany and central Europe; each is plotted by its mean, range, and sample size. Data from Heinrich (1987) and van Kolfschoten (1990), where sample details can be found.

hippopotami and other taxa, isolated on islands in the Mediterranean, South-East Asia, and elsewhere (Sondaar, 1977). The adaptive significance of these changes has been widely discussed (e.g. Heaney, 1978), but the potential for studying evolutionary rates barely exploited because the dating of the fossils is usually poor or absent.

One relatively well-documented example concerns the dwarfing of red deer (*Cervus elaphus* L.) on the island of Jersey in the English Channel (Zeuner, 1946; Lister, 1989). Remains of at least five dwarfed individuals have been found within the coastal cave of Belle Hougue (Fig. 9), in a fossil beach deposit above the reach of present-day tides, and dated to the last (Eemian) interglacial (Keen, Harmon & Andrews, 1981) which spanned the time range *c.* 126 000–115 000 years BP (Shackleton & Pisias, 1985).

The Belle Hougue remains can be identified as *Cervus elaphus* on the basis of antler, dental and bone characters, but they are much smaller than contemporaneous remains of this species from mainland Britain and France, showing a size reduction to about 56% of the limb-bone diameters of the parent population (Fig. 10). This measure is strongly correlated to body weight among deer, following an approximately geometric (cube-law) relationship (Scott, 1987) which suggests a weight reduction of 0.56^3, or around one-sixth, in the Jersey population. Mainland red deer bones of the last interglacial are comparable in size to the largest present-day European populations, indicating male body weights of *c.* 200 kg and shoulder heights of *c.* 125 cm, so the Jersey animals, by contrast, averaged *c.* 36 kg and 70 cm. Although body size in red deer is known to be quite ecophenotypically labile, this process is not known to take average male weights below about 100 kg in Europe today (Suttie & Hamilton, 1983), so the Jersey dwarfing can be assumed to have had a significant genetic component. In addition, as is often the case among dwarfed mammals, their teeth were relatively less reduced than their postcrania, and there is some suggestion of relatively shortened distal limb elements (Lister, 1989). Antler branching was considerably reduced (Zeuner, 1946).

For much of the Pleistocene, global sea-level was markedly lower than today, and Jersey was part of a broad land mass connecting Britain and France (Fig. 9A). Deposits

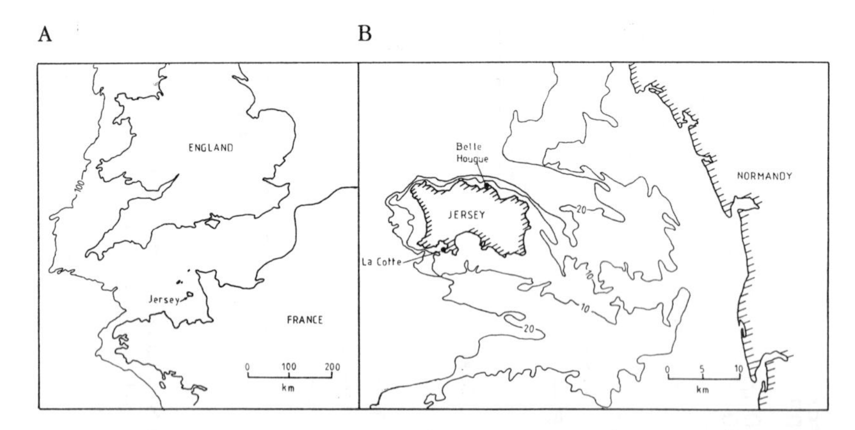

Figure 9 A: location map of Jersey, showing the 100 m submarine contour (the approximate position of the shoreline during Pleistocene glacial periods). B: Submarine contours around Jersey (20- and 10-m), from Admiralty Charts (1989). Depths are reduced to Chart Datum, which is approximately the level of the lowest astronomical tide. A lowering of sea-level by 10 m would therefore produce a narrow isthmus at low tide.

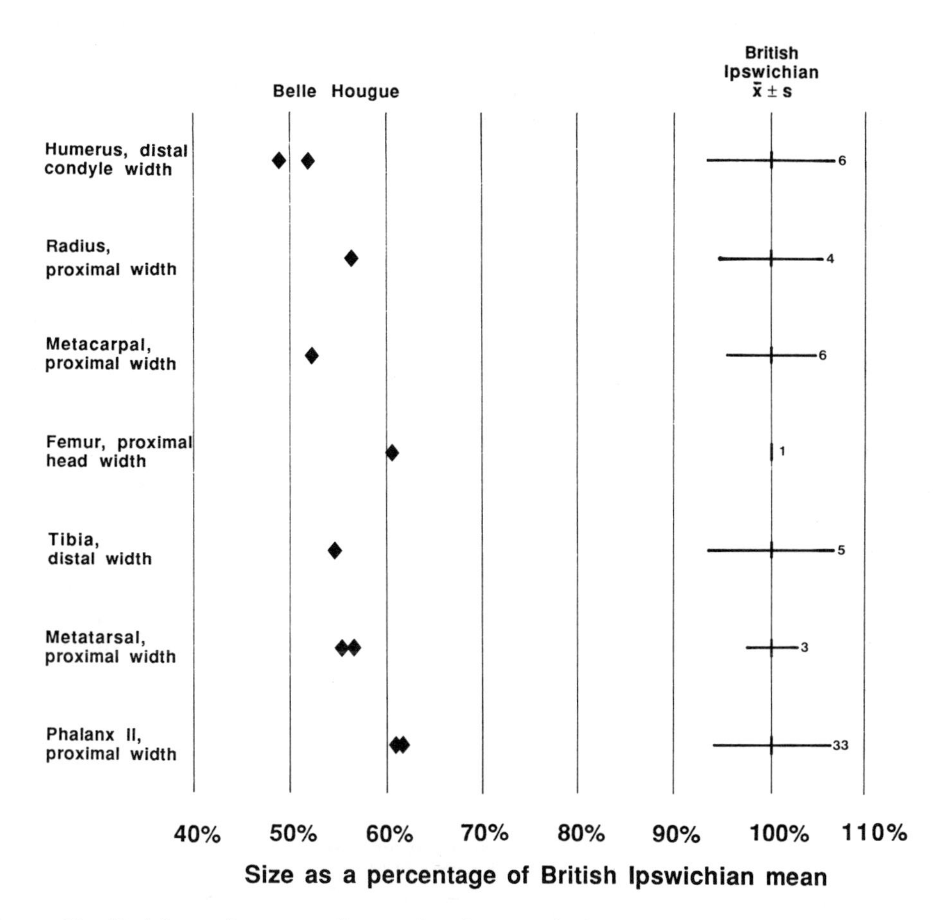

Figure 10 Limb bone diameters of Jersey dwarf *Cervus elaphus,* in comparison with a contemporary Ipswichian (Eemian) sample from mainland Britain. Each mainland mean is plotted as 100%, and shown plus or minus one standard deviation, with sample size. The ten Jersey bones average 55.9% the size of their mainland counterparts. For details of mainland sites and raw data, see Lister (1989).

at the site of La Cotte de St Brelade on Jersey, dating to the later part of the Saalian cold stage, include red deer remains of 'normal', large size (Scott, 1986; Lister, 1989). As sea-level rose in the succeeding Eemian interglacial, Jersey was isolated and the deer became dwarfed, while normal-size populations continued on mainland Britain and France.

The time taken for the dwarfing cannot be determined by absolute dating of the fossil deer, because we do not have a sequence of remains through the Eemian, and because in any case the process was too fast to be resolved given the error range of existing dating methods. Instead, assuming that the dwarfing process could not begin until Jersey became isolated, we can calculate, using estimates of sea-level changes, the time available between the isolation of the island and the occurrence of the dwarf form.

The pattern of global sea-level change has been established by the study of oceanic oxygen isotope ratios (Shackleton, 1987), tropical coral reefs (Chappell & Shackleton, 1986) and marine transgressions and regressions in low-lying terrestrial areas (Zagwijn,

1983). As the Saalian ice-caps melted, the sea rose rapidly, and would have reached the point of isolating Jersey (about 10 m below present sea-levels) about 1000 years after the beginnning of the Eemian interglacial (Lister, 1989; Fig. 11). Maximum sea-level was reached at around 3000 years into the interglacial, and lasted for *c.* 4000 years. It is likely that the dwarf deer fossils date to some time within this maximum stand, since the preservation of the beach deposit and its molluscan fauna imply deposition during the peak of the interglacial (Keen *et al.*, 1981). This means that the dwarfing of the deer must have occurred some time between the isolation of Jersey at 1000 years into the Eemian, and the fall from maximum sea-level at 7000 years: a span of *c.* 6000 years. The dwarfing process therefore took 6000 years or less (Lister, 1989).

We do not know how long within the 6000 years available the deer took to become dwarfed, but even the maximum duration is very short geologically: about 1% of the *c.* 600 000 years that *Cervus elaphus* has existed in mainland Europe (Lister, 1986). And since the Jersey population was an allopatric isolate, the process has some of the features of the punctuated equilibrium model (Gould & Eldredge, 1977). On the other hand, we have no evidence whether the dwarf deer had speciated in a reproductive sense. Later deposits at La Cotte de St Brelade on Jersey (Marrett, 1911), dating to the Devensian cold stage after the reincorporation of Jersey into the mainland, include red deer fossils of 'normal' size but no sign of the dwarf form, which had presumably died out either by introgression (if not a distinct species), or by competition from mainland forms.

The 6000 years available for the dwarfing process represents up to perhaps 2000

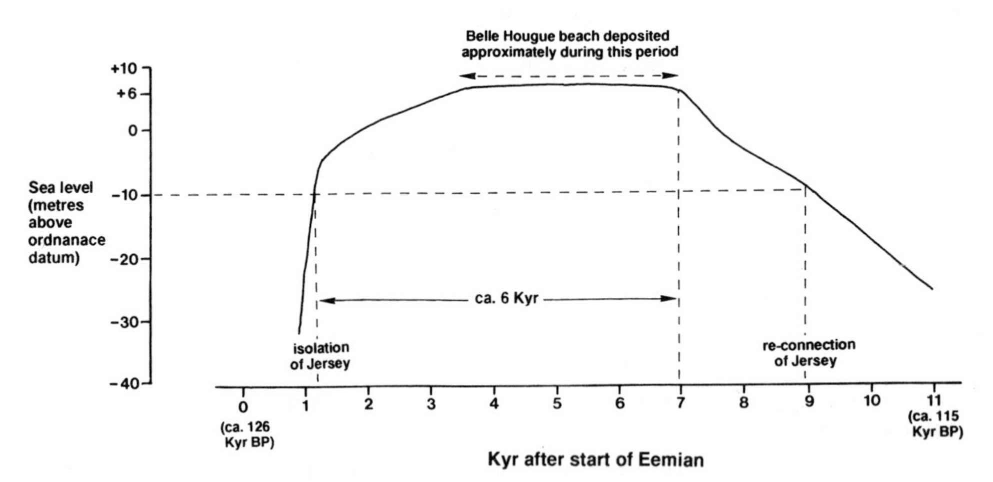

Figure 11 Simplified estimate of sea-level change through the Last (Eemian or Ipswichian) Interglacial (Lister, 1989). The sea-level curve is based on pollen-dated marine transgressions and regressions in The Netherlands (Zagwijn, 1983), calibrated onto an absolute time-scale provided by varved Eemian lake sediments (Müller, 1974), and scaled to an Eemian global maximum of + 6 m (Chappell & Shackleton, 1986; Shackleton, 1987). A similar timescale and sea-level curve is independently provided by deep-sea oxygen isotope data (Shackleton, 1987). The sea-level required for the isolation/re-connection of Jersey is based on the present displacement between the raised beach (at +6 m) and the sea floor (−10m: cf. Fig. 9). The deer are assumed to have become dwarfed some time during the 6000 years between the isolation of Jersey, and the end of the maximum sea-level stand during which the beach containing their remains was deposited.

generations of the deer, plenty of time for 'gradual' change on a biological timescale. This does not prevent the rate of change from qualifying as a 'punctuation' event in geological time.

DISCUSSION

Rates of change

When considering the length of time over which an evolutionary transition has been observed to occur, it is essential to view it in relation to typical species durations in the group concerned. A transition which took a few tens or even hundreds of thousands of years may be regarded as relatively rapid in an invertebrate lineage which shows millions of years of stasis before and/or after the transition. In Quaternary mammals, however, species durations, whether defined morphologically or in terms of the time from immigration to extinction, are typically of the order of 500 000 years; few species endure more than 1 million years (Kurtén, 1968). From this perspective, the transitions in the mammoth and water vole lineages (and also that of the moose, although we could wish for fuller sampling), correspond to a gradualistic rather than punctuated tempo. Over *c.* 2 million years, the mammoth and moose lineages each underwent a morphological transition of magnitude clearly equivalent to two or three typical species differences, by a largely directional process with overlapping sequential sample ranges and thus far no clear evidence of cladogenetic speciation events.

The water vole transition may appear morphologically less significant, no doubt partly due to our inability to enter the voles' world and empathize with changes in enamel bands measured in tenths of a millimetre. Nevertheless, differences of equivalent magnitude to the entire 300 000 year transition are found as geographical variation in *Arvicola terrestris* today, and may therefore objectively be regarded as only subspecific in nature (van Kolfschoten, 1990). The net directionality of this microevolutionary transition is nonetheless worthy of note.

Although the punctuated equilibrium model allows that morphological adaptations arise within populations at the level of individual selection (Stanley, 1982; Eldredge & Gould, 1988), it suggests that this process takes up only a small proportion of the total duration of the species, which demonstrates stasis for the rest of the time. Stanley (1981:15), for example, saw little change of form or adaptation in the Quaternary mammoth lineage. The data on mammoth, moose and water voles do not support this contention, at least for these taxa. Various authors (e.g. Gould & Eldredge, 1977; Gould, 1990) have argued for 'punctuated' change because 'gradualistic' evolution over even such (geologically relatively short) periods of time as 0.3–2 million years would require excessively low selection pressures and are hard to envisage. Although this would be true if the evolution were perfectly constant-rate, the observed 'gradualistic' period may well have included, at fine timescales, either long stretches of little change, or perhaps more realistically, constant fluctuation of mean phenotype with only occasional periods of selected change in the direction of the net trend. Further, even if the net trend is mathematically within the capabilities of genetic drift (Lande, 1976), it seems implausible that drift alone could be responsible for such major anatomical changes as occurred in the moose antlers and mammoth teeth, or for the net directionality observed in the vole teeth.

The dwarfing of the red deer, by contrast, represents a much more rapid transition.

Gingerich (1983) has pointed out that high rates of evolution are generally measured only on short timescales; however, this does not negate the possible importance of such brief, fast episodes in evolution. The Jersey deer represented a local adaptation which did not yield further evolutionary potential. This case history illustrates the potential of the allopatric isolate for yielding evolutionary novelty, but we do not know whether the pattern observed on an offshore island can be generalized to the continental masses, where most of life's diversity has arisen. Although 'islands' of habitat occur in mainland ecosystems, it is unfortunately much more difficult to demonstrate palaeontologically allopatric speciation there, than in the ideal situation of an offshore island.

Evolutionary trends

There are various biases in the fossil record, or our perception of it, which may cause us falsely to 'see' a predominance of either punctuated or gradualistic evolution where none exists. Biases in favour of 'seeing' punctuations include: stratigraphic unconformities or non-sequences; the tendency to see variation as clustered into a series of static species (Fortey, 1985); and mistaking rapid immigration with rapid origin. Biases in favour of 'seeing' gradualism include: the greater ease with which an evolutionary lineage can be recognized the more gradational its links; and the selection for study of sequences showing clear directional trends. Since three of the case histories presented here appear to show gradualistic trends, some discussion of the possible biases which may have influenced these findings will be given.

The mammoth and moose lineages were selected for detailed research because, among other reasons, each appears to form a simple lineage, and shows significant directional evolution; and I would be surprised if the same were not true of the water vole study by Heinrich (1987) and van Kolfschoten (1990). Moreover, the characters highlighted here (moose antlers, mammoth teeth and vole molars) are those which show clear morphological trends. Any evolving taxon will display any number of characters which are invarying or show no net direction of change. In a study of ten dental characters across the three main levels of the moose lineage, Lister (1981) found that seven showed no net change and only three a directional trend, no more than would be expected by chance (cf. Cheetham, 1987:295), although this does not rule out their having been the product of selection.

Since we are primarily interested in how evolution produces significant modification, it would seem reasonable to concentrate on those cases and characters where this can actually be observed. On the other hand, we cannot ignore the many Quaternary mammalian taxa which 'appear' in the record, without clear ancestry or a sequence of chronospecies behind them. Examples include the woolly rhinoceros *Coelodonta antiquitatis* (Blum.) and the reindeer *Rangifer tarandus* (L.). Whether this reflects punctuated equilibrium, or simply our ignorance (evolution – at an unknown rate – in an unsampled area) is not known. This highlights the asymmetry by which gradualistic transitions are more likely than punctuated ones to be recorded. Some workers might be biased to describe the appearance of species such as *Coelodonta* as punctuations, but the evidence is lacking. Lister (1987) speculated that the origin of distinct antler plans (important species-defining characters) in many species of deer might have been rapid, largely stochastic events, but here too, the evidence was circumstantial.

A related 'trap' has been described by Gould (1990): the selection of a terminal member of a clade, followed by the picking-out of a sequence of fossils showing an evolutionary 'trend' toward that form. If selected from a bush of species showing morphological excursions in various directions, but no overall trend, this procedure gives a misleading impression of an anagenetic trend. Trends are rather, for Gould (1990), the product of directional sorting among the array of variously modified species.

This model deserves to be tested on supposed cases of lineage anagenesis. In the case of the moose and mammoth sequences, we would expect to find, through the period of the evolutionary trend, an array of moose species of differing antler lengths, and of mammoth species of different molar heights and lamellar frequencies. The species more 'advanced' in the direction of the trend would have the greater tendency to survive and themselves speciate than those less 'advanced', and/or there would be a bias in the direction of speciation toward the 'advanced' morphology. At present, the Eurasian Quaternary record of mammoth and moose does not demonstrate such a pattern. Admittedly, Gould (1990) chose as an example of his model the horse lineage, spanning some 60 million years of Cainozoic time. The lineages discussed here are no more than 2 million years long, so it could be argued that they are too short to show a pattern of species sorting: they rather illustrate the pattern of origin of species, which might themselves then be subject to sorting. Nonetheless, these case histories do indicate that significant anatomical modification can originate in an apparently unbranching, anagenetic mode. Variation at each horizon appears in the form of intraspecific demes, rather than species arrays, corresponding perhaps to a 'shifting balance' mode of evolution (Wright, 1977).

REFERENCES

ADAM, K.-D., 1961. Die Bedeutung der pleistozänen Säugetier-Faunen Mitteleuropas für die Geschichte des Eiszeitalters. *Stuttgarter Beiträge zur Naturkunde, 78:* 1–34.

ADMIRALTY CHARTS, 1989, No. 2669. *The Channel Islands and Adjacent Coast of France.* Taunton: Hydrographer of the Navy.

AGUIRRE, E.E., 1969. Revisión sistemática de los Elephantidae por su morfología y morfometría dentaria. Parts I–III. *Estudios Geologicos, 24:* 109–167; *25:* 123–177, 317–367.

AZZAROLI, A.A., 1952. L'alce di Senèze. *Palaeontographia Italica, 47:* 133–141.

AZZAROLI, A.A., 1977. Evolutionary patterns of Villafranchian elephants in central Italy. *Atti della Accademia nazionale dei Lincei. Memorie. Classe di Scienze Fisiche, Matematiche e Naturali, Sez. 2a 14 (Ser. 8):* 149–168.

AZZAROLI, A.A., 1981. On the Quaternary and recent cervid genera *Alces, Cervalces, Libralces. Bollettino della Società Paleontologica Italiana, 20:* 147–154.

AZZAROLI, A. DE GIULI, C., FICCARELLI, G. & TORRE, D., 1988. Late Pliocene to early mid-Pleistocene mammals in Eurasia: faunal succession and dispersal events. *Palaeogeography, Palaeoclimatology, Palaeoecology, 66:* 77–100.

BAIGUSHEVA, V.S., 1971. The fossil theriofauna of the Liventsov quarry, north-eastern Azov area. *Trudy Zoologicheskogo Instituta, A N SSSR, 49:* 35–42.

BOUT, P., 1970. Absolute ages of some volcanic formations in the Auvergne and Velay areas and chronology of the European Pleistocene. *Palaeogeography, Palaeoclimatology, Palaeoecology, 8:* 95–106.

BOWEN, D.Q., HUGHES, S., SYKES, G.A. & MILLER, G.H., 1989. Land–sea correlations in the Pleistocene based on isoleucine epimerisation in non-marine molluscs. *Nature, 340:* 49–51.

BRANDON, A. & SUMBLER, M.G., 1991. The Balderton Sand and Gravel: pre-Ipswichian cold stage fluvial deposits near Lincoln, England. *Journal of Quaternary Science, 6:* 117–138.

BRÜNING, H., 1978. Zur Untergliederung der Mosbacher Terrassenabfolge und zum klimatischen Stellenwert der Mosbacher Tierwelt im Rahmen des Cromer-Komplexes. *Mainzer Naturwissenschaftliches Archiv, 16:* 143–190.

CALLOW, P. & CORNFORD, J.M. (Eds), 1986. *La Cotte de St Brelade 1961–1978: Excavations by C.B.M. McBurney.* Norwich: Geo Books.

CARPENTIER, G. & LAUTRIDOU, J.-P., 1986. The low terrace of the Seine: the alluvium, periglacial deposits, interglacial fluviomarine deposits, slope deposits and palaeosols, fauna. In J.-P. Lautridou, D.H. Keen & J.L. Monnier (Eds), *The Loess and Other Pleistocene Periglacial Deposits of Northwest Europe, Including Their Relationships with Marine Formations and Features:* 64–68. Caen: CNRS.

CHAPPELL, J. & SHACKLETON, N.J., 1986. Oxygen isotopes and sea level. *Nature, 324:* 137–140.

CHEETHAM, A.H., 1987. Tempo of evolution in a Neogene bryozoan: are trends in single morphologic characters misleading? *Paleobiology, 13:* 286–296.

COXON, P., 1982. The terraces of the River Waveney. In P. Allen (Ed), *Field Guide to the Gipping and Waveney Valleys, Suffolk:* 80–94. Cambridge: Quaternary Research Association.

DUBROVO, I.A., 1964. Elephants of the genus *Archidiskodon* in the USSR. *Paleontologicheskij Zhurnal, 3:* 82–94.

DUBROVO, I.A., 1977. A history of elephants of the *Archidiskodon–Mammuthus* phylogenetic line on the territory of the USSR. *Journal of the Palaeontological Society of India, 20:* 33–40.

DUBROVO, I.A., 1989. Systematic position of the Khapry elephants. *Paleontologicheskij Zhurnal, 1989/1:* 78–87.

ELDREDGE, N. & GOULD, S.J., 1988. Punctuated equilibrium prevails. *Nature, 332:* 211–212.

ERDBRINK, D.P., 1954. On one of the oldest known remains of the common elk, *Alces alces* L., found recently in the Netherlands. *Geologie en Mijnbouw, N.S.16:* 301–309.

FEJFAR, O. & HEINRICH, W.-D., 1990. *International Symposium on the Evolution, Phylogeny and Biostratigraphy of Arvicolids.* Prague: Pfeil-Verlag.

FLEROV, K.K., 1952. Musk deer and deer. In *Fauna of the USSR: Mammals,* 1/2. Jerusalem: Israel Program for Scientific Translation.

FORTEY, R.A., 1985. Gradualism and punctuated equilibria as competing and complementary theories. *Special Papers in Palaeontology, 33:* 17–28.

GHISELIN, M.T., 1974. A radical solution to the species problem. *Systematic Zoology, 25:* 536–544.

GIBBARD, P.L., WEST, R.G., ZAGWIJN, W.H., BALSON, P.S., BURGER, A.W., FUNNELL, B.M., JEFFERY, D.H., DE JONG, J., VAN KOLFSCHOTEN, T., LISTER, A.M., MEIJER, T., NORTON, P.E.P., PREECE, R.C., ROSE, J., STUART, A.J., WHITEMAN, C.A., & ZALASIEWICZ, J.A., 1990. Early and early Middle Pleistocene correlations in the southern North Sea basin. *Quaternary Science Reviews, 10:* 23–52.

GINGERICH, P.D., 1980. Evolutionary patterns in early Cenozoic mammals. *Annual Review of Earth and Planetary Science, 8:* 407–424.

GINGERICH, P.D., 1983. Rates of evolution: effect of time and temporal scaling. *Science, 222:* 159–161.

GOULD, S.J., 1974. The origin and function of 'bizarre' structures: antler size and skull size in the 'Irish Elk' *Megaloceros giganteus. Evolution, 28:* 191–220.

GOULD, S.J., 1990. Speciation and sorting as the source of evolutionary trends, or 'Things are seldom what they seem'. In K.J. McNamara (Ed.), *Evolutionary Trends:* 3–27. London: Belhaven.

GOULD, S.J. & ELDREDGE, N., 1977. Punctuated equilibria: the tempo and mode of evolution reconsidered. *Paleobiology, 3:* 115–151.

HATTEMER, H.H. & DRESCHLER, H., 1976. Rosenstock- und Geweihmerkmale von Harz-Hirschen und ihr Zusammenhang mit dem Alter. *Zeitschrift für Jagdwissenschaft, 22:* 36–50.

HEANEY, L.R., 1978. Island area and body size of insular mammals: evidence from the tri-colored squirrel (*Callosciurus prevosti*) of Southeast Asia. *Evolution, 32:* 29–44.

HEINRICH, W.-D., 1987. Neue Ergebnisse zur Evolution und Biostratigraphie von *Arvicola* (Rodentia, Mammalia) im Quartär Europas. *Zeitschrift für Geologische Wissenschaften, 15:* 389–406.

HEPTNER, V.G., NASIMOVICH, A.A. & BANNIKOV, A.G., 1988. *Mammals of the Soviet Union, volume 1, Artiodactyla and Perissodactyla.* Washington, D.C.: Smithsonian Institution.

HUXLEY, J.S., 1931. The relative size of antlers in deer. *Proceedings of the Zoological Society of London, 1931:* 819–864.

IGEL, W., 1984. Paläogeographie durch Sediment-Morphoskopie. Grundlagen, Methodik und Anwendung. Unpublished Ph.D. dissertation, University of Mainz, 430 pp.

KAHLKE, H.-D., 1969 (Ed.). *Das Pleistozän von Süssenborn. Paläontologische Abhandlungen A (Paläozoologie),* 3. Berlin: Deutsche Gesellschaft für Geologische Wissenschaften.

KAHLKE, H.-D., 1975. Die Cerviden-Reste aus den Travertinen von Weimar-Ehringsdorf. *Abhandlungen des Zentralen Geologischen Instituts, 23:* 201–249.

KAHLKE, H.-D., 1976. Die Cervidenreste aus den Travertinen von Taubach. *Quartärpaläontologie, 3:* 113–122.

KEEN, D.H., HARMON, R.S. & ANDREWS, J.T., 1981. U series and amino acid dates from Jersey. *Nature, 289:* 162–164.

KLEIN, R.G., 1969. *Man and Culture in the Late Pleistocene.* San Francisco: Chandler.

KOTSAKIS, T., PALOMBO, M.R. & PETRONIO, C., 1978. *Mammuthus chosaricus* e *Cervus elaphus* del Pleistocene Superiore di Via Flaminia (Roma). *Geologica Romana, 17:* 411–445.

KURTÉN, B., 1968. *Pleistocene Mammals of Europe.* London: Weidenfeld & Nicholson.

LANDE, R., 1976. Natural selection and random genetic drift in phenotypic evolution. *Evolution 30:* 314–334.

LISTER, A.M., 1981. 'Evolutionary studies on Pleistocene deer'. Unpublished Ph.D. thesis, University of Cambridge, U.K.

LISTER, A.M., 1986. New results on deer from Swanscombe, and the stratigraphical significance of deer in the Middle and Upper Pleistocene of Europe. *Journal of Archaeological Science, 13:* 319–338.

LISTER, A.M., 1987. Diversity and evolution of antler form in Quaternary deer. In C.M. Wemmer (Ed.), *Biology and Management of the Cervidae:* 81–98. Washington, D.C.: Smithsonian Institution.

LISTER, A.M., 1989. Rapid dwarfing of red deer on Jersey in the Last Interglacial. *Nature, 342:* 539–542.

LISTER, A.M., 1990. Critical reappraisal of the Middle Pleistocene deer species '*Cervus*' *elaphoides* Kahlke. *Quaternaire, 1:* 175–192.

LISTER, A.M., 1992. Mammalian fossils and Quaternary biostratigraphy. *Quaternary Science Reviews, 11:* 329–344.

LISTER, A.M., in press a. The stratigraphical distribution of deer species in the Cromer Forest-bed Formation. *Journal of Quaternary Science.*

LISTER, A.M., in press b. Evolution of mammoths and moose: the Holarctic perspective. In R.A. Martin & A.D. Barnosky (Eds), *Morphological Change in Quaternary Mammals of North America.* New York: Cambridge University Press.

LISTER, A.M., in press c. Evolutionary history of mammoths in Eurasia. *Palaeontology.*

LISTER, A.M. & BRANDON, A., 1991. A pre-Ipswichian cold stage mammalian fauna from the Balderton Sand and Gravel, Lincolnshire, England. *Journal of Quaternary Science, 6:* 139–157.

MAGLIO, V.J., 1973. Origin and evolution of the Elephantidae. *Transactions of the American Philosophical Society, 62:* 1–149.

MALMGREN, B.A., BERGGREN, W.A. & LOHMAN, G.P., 1983. Evidence for punctuated gradualism in the Late Neogene *Globorotalia tumida* lineage of planktonic foraminifera. *Paleobiology, 9:* 377–389.

MARRETT, R.R., 1911. *Pleistocene Man in Jersey.* Oxford: Society of Antiquaries.

MAUSER, M., 1990. *Alces latifrons* (Johnson, 1874) (Mammalia, Cervidae) von der altpleistozänen Säugetier-Fundstelle Würzburg-Schalksberg (Unterfranken, BRD). *Quartärpaläontologie, 8:* 205–214.

MÜLLER, H., 1974. Pollenanalytische Untersuchungen und Jahresschichtungzählungen an der eem-zeitlichen Kieselgur von Bispingen/Luhe. *Geologisches Jahrbuch, A21:* 149–169.

MUSIL, R., 1968. Die Mammutmolaren von Predmostí (CSSR). *Paläontologische Abhandlungen A (Paläozoologie), 3:* 1–192.

RÖTTGER, U., 1987. Schmelzbandbreiten an Molaren von Schermäusen (*Arvicola* Lacépède, 1799). *Bonner zoologische Beiträge, 38:* 95–105.

SCOTT, K., 1986. The large mammal fauna. In P. Callow & J.M. Cornford (Eds), *La Cotte de St Brelade 1961–1978: Excavations of C.B.M. McBurney:* 159–183. Norwich: Geo Books.

SCOTT, K.M., 1987. Allometry and habitat-related adaptations in the postcranial skeleton of Cervidae. In C.M. Wemmer (Ed.), *Biology and Management of the Cervidae:* 65–80. Washington, D.C.: Smithsonian Institution.

SHACKLETON, N.J., 1987. Oxygen isotopes, ice volume and sea level. *Quaternary Science Reviews, 6:* 183–190.

SHACKLETON, N.J. & OPDYKE, N.D., 1973. Oxygen isoptope and palaeomagnetic stratigraphy of equatorial Pacific core V28-238: oxygen isotope temperatures and ice volumes on a 10^5 and 10^6 year scale. *Quaternary Research, 3:* 39–55.

SHACKLETON, N.J. & PISIAS, N.G., 1985. Atmospheric carbon dioxide, orbital forcing, and climate. *Geophysical Monograph, 32:* 303–317.

SHER, A.V., 1974. Pleistocene mammals and stratigraphy of the far north-east USSR and North America. *International Geological Review, 16:* 1–206.

SHER, A.V., 1987a. History and evolution of moose in the USSR. *Swedish Wildlife Research Supplement, 1:* 71–97.

SHER, A.V., 1987b. Olyorian land mammal age of northeastern Siberia. *Palaeontographia Italica, 74:* 97–112.

SMART, P.L. & FRANCES, P.D., 1991. *Quaternary Dating Methods – a User's Guide.* Quaternary Research Association Technical Guide no. 4. Cambridge: Quaternary Research Association.

SONDAAR, P.Y., 1977. Insularity and its effect on mammalian evolution. In M.K. Hecht, P.C. Goody & B.M. Hecht (Eds), *Major Patterns in Vertebrate Evolution:* 671–707. New York: Plenum.

STANLEY, S.M., 1981. *The New Evolutionary Timetable.* New York: Basic Books.

STANLEY, S.M., 1982. Macroevolution and the fossil record. *Evolution, 36:* 460–473.

STEINMÜLLER, A., 1972. Die Schichtenfolgen von Süssenborn und Voigtstedt und die Gliederung des Mittelpleistozäns. *Geologie, 21:* 129–142.

STUART, A.J., 1976. The history of the mammal fauna during the Ipswichian/Last Interglacial in England. *Philosophical Transactions of the Royal Society of London, B, 276:* 221–250.

STUART, A.J., 1982. *Pleistocene Vertebrates in the British Isles.* London: Longman.

SUTCLIFFE, A.J., 1976. The British glacial-interglacial sequence. *Quaternary Newsletter, 18:* 1–7.

SUTTIE, J.M. & HAMILTON, W.J., 1983. The effect of winter nutrition on growth of young Scottish red deer (*Cervus elaphus*). *Journal of Zoology (London), 201:* 153–159.

THOUVENY, N. & BONIFAY, E., 1984. New chronological data on European Plio-Pleistocene faunas and hominid occupation sites. *Nature, 308:* 355–358.

VANGENGEIM, E.A. & PEVSNER, M.A., 1991. The Villafranchian of the USSR: bio- and magneto-stratigraphy. In *Palaeogeography and Biostratigraphy of the Pliocene and Anthropogene:* 124–145. Moscow: Geological Institute, USSR Academy of Sciences.

VANGENGEIM, E.A., VEKUA, M.L., ZHEGALLO, V.I., PEVSNER, M.A., TAKTAKISHVILI, I.G. & TESAKOV, A.C., 1991. Position of the Taman faunal complex on the stratigraphic and magnetochronological scales. *Byulletin Komissii po Izucheniyu Chetvertichnogo Perioda, 60:* 41–52.

VAN KOLFSCHOTEN, T., 1990. The evolution of the mammal fauna in The Netherlands and the middle Rhine area (West Germany) during the Late Middle Pleistocene. *Mededelingen Rijks Geologische Dienst, 43:* 1–69.

VON KOENIGSWALD, W., 1980. Schmelzstruktur und Morphologie in den Molaren der Arvicolidae (Rodentia). *Abhandlungen der Senckenbergischen Naturforschenden Gesellschaft, 539:* 1–129.

VRBA, E.S. & ELDREDGE, N., 1984. Individuals, heirarchies and processes: towards a more complete evolutionary theory. *Paleobiology, 10:* 146–171.

WEST, R.G., 1977. *Pleistocene Geology and Biology, with Especial Reference to the British Isles.* London: Longman.

WRIGHT, S., 1977. *Evolution and the Genetics of Populations, volume 3, Experimental Results and Evolutionary Deductions.* Chicago: University of Chicago Press.

ZAGWIJN, W.H., 1983. Sea-level changes in The Netherlands during the Eemian. *Geologie en Mijnbouw, 62:* 437–450.

ZEUNER, F.E., 1946. *Cervus elaphus jerseyensis* and other fauna in the 25-ft beach of Belle Hougue Cave, Jersey, C.I. *Bulletin de la Société Jerseiaise, 14:* 238–254.

Part II

From Pattern to Process

CHAPTER

7

Postglacial distribution and species substructure: lessons from pollen, insects and hybrid zones

GODFREY M. HEWITT

Orbits and ice ages	**98**
Postglacial hybrid zones	**99**
Width and postglacial gene flow	**101**
Initial contact and mixing	**103**
Postglacial reinforcement and parapatric speciation	**105**
Movement and position	**106**
Postglacial advance	**108**
Rate of spread	**109**
Adaptation and divergence	**110**
Southern edge	**111**
Genetics of advance	**113**
Genome reassortment	**116**
Repeated range shifts	**117**
Acknowledgements	**119**
References	**119**

Keywords: Hybrid zones – parapatry – gene flow – range change – ice ages – invasion genetics – founder effects – refugia – adaptation – speciation.

Evolutionary Patterns and Processes
ISBN 0-12-440895-8

Abstract

The sun and the earth describe orbital changes
which drive climate cycles and modify ranges.
The shape of the land forms a number of places
which allow the survival of different races.
When enclaves advance with the ice in retreat
some form hybrid zones where the two ranges meet.
Such regions are common and not very wide
so the mixing of genes affects neither side.
They divide up the range in a patchwork of pieces
with echoes and glimpses on the nature of species.
A brief rendez-vous and the ice comes again.

When the glaciers melt so that ranges expand
some plants will spread quickly where there's suitable land.
Those insects which eat them will follow this lead
some flying, some walking to establish their breed.
Those that try later meet a resident band,
they must somehow be better to make their own stand.
But the mixture will change as more types arrive
and warming conditions allow new species to thrive.
Some will move on to fresh places ahead,
those that remain must adapt, or are dead.
And then the tide turns and the ice comes again.

Each refuge could foster a deviant form,
new neighbours, chance changes and drift from the norm.
When the warm breakout comes, those few in the van
disperse from the edge and breed where they can.
Pioneer pockets grow to large populations,
a very good place to strike new variations.
Some may not work well with their parental kind
so stopping the spread of those from behind.
Continental theatres provide plenty of chances
to establish new morphs in both retreats and advances.
New species may form when the ice comes again.

ORBITS AND ICE AGES

As the earth circles the sun its orbit shows three regular variations: axial precession, axial tilt and orbital eccentricity. These occur with cycles of about 23 000 years, 41 000 years and 100 000 years, respectively, and cause variation in insolation and hence in climate. Recent evidence shows that these variations are ultimately the pacemaker of the ice ages (Hays, Imbrie & Shackleton, 1976). This model is known as the Milankovitch theory, and was promulgated in the 1930s by Milutin Milankovitch, a Yugoslav astronomer, although it was first put forward by a Scot, James Croll, who published such an idea in the 1860s (Gribbin, 1989). These cycles and their effects can be traced by a number of measures, for example, O_2 isotope ratios in marine sediments of foraminifera, CO_2 from ice and carbon isotopes from foraminifera, while pollen and beetle exoskeletons from soil sediments provide particularly detailed information over the last ice age to the present (Coope, 1977; Huntley & Birks, 1983; Martinson *et al.*, 1987). Whilst the most dramatic effects of these orbital peculiarities are the several Pleistocene ice ages, with their extensive glaciations occurring approximately every 100 000 years, the combinations of the driving cycles have produced series of lesser changes in climate which have repeatedly altered the habitats and ranges of living organisms (e.g. Bartlein & Prentice, 1989).

At the height of the last ice age some 18 000–20 000 BP, the northern parts of America and Europe were greatly affected by glaciation (Denton & Hughes, 1981). The massive Laurentide ice sheet covered North America down to a latitude of 40°N, covering the Great Lakes, while the Scandinavian ice sheet came down to 52°N, covering most of Britain and northern Europe as far as Warsaw. In addition, Europe had ice sheets on the high mountains of Cantabria, the Pyrenees, the Alps, Transylvania and the Caucasus around 45°N. In northern Asia the ice cover in the last glaciation was not so great as it was in previous advances. In general, this would have moved vegetation zones and associated faunas further south, compressing them towards the tropics. It was thought that the tropics were relatively stable through the Quaternary, but a number of recent studies demonstrate that in fact considerable changes occurred. For example, sites in Guatemala considered to contain primaeval rain forest, apparently had xeric vegetation at the end of the last glaciation and were not Pleistocene refugia for mesic vegetation: these forests have developed since 11 000 BP (Leyden, 1984). Furthermore, the insolation differences caused by the precession cycle (~23 000 years) seem to be important in determining the monsoons, and so the tropical climate and vegetation would not change in concert with that of temperate regions (Bartlein & Prentice, 1989). The climates of Australia and South America are also affected by these changes, but were not subject to glaciation to any great extent.

As the amplitude of the insolation cycle increased, with more in the summer and less in the winter, the ice began to melt around 18 000 BP. The speed of this melt is still somewhat uncertain, but by 9000 BP, when summer insolation was maximal, there was still some ice over northern Scandinavia, and the Laurentide ice sheet was north of Hudson Bay and the Great Lakes. It had retreated to the Arctic by 6000 BP. The concomitant changes in climatic conditions were not simple, comprising complex combinations of temperature, wind and precipitation that we are just beginning to understand (Webb, 1986). Indeed the field of climate and vegetation history over the last 20 000 years is attracting a great deal of interest and making considerable progress (e.g. COHMAP members, 1988; Huntley & Webb, 1988). The overall effect of this general warming was to allow species living in the south to expand northward into previously inhospitable territory, across what had been steppe and tundra, and across areas uncovered by the retreating ice.

POSTGLACIAL HYBRID ZONES

The present distributions of species are determined by suitable climatic and biotic conditions, and also by their history, which determines whether or not they reach a particular place. It is quite common to find the range of a species or species complex divided into subspecies and races, forming a patchwork of distinct forms. Often where such forms meet they mate and produce hybrids, generating a hybrid zone (Barton

& Hewitt, 1985). These hybrid zones are frequently very narrow relative to the size of the species range; they can be a few hundred metres wide and can run for hundreds of kilometres. Furthermore, they have been found to exist for a wide variety of organisms with different lifestyles, including mammals, birds, reptiles, fish, insects, flowering plants and trees. The hybridizing taxa may differ for any number of characters, and to differing degrees. There is selection against the hybrids in most cases. Whilst they are usually first described on obvious morphological and behavioural characteristics, on further examination they often show physiological, chromosomal, allozymic and other molecular differences as well. Now a considerable number of hybrid zones have been discovered through chromosomal differences where there was no prior morphological or behavioural indication, e.g. *Vandiemenella viatica* (Fig. 1) (White *et al.*, 1967; White, 1978). This is significant because large parts of the genome may differ without obvious morphological effect, which suggests that many more 'cryptic' hybrid zones may exist, and species genomes may be much more subdivided geographically than we currently appreciate (Hewitt, 1988).

Many hybrid zones within these geographically subdivided species occur in places which would have been inhospitable to the organism during the last ice age, and consequently they must have formed since then (Hewitt, 1985, 1989). For a number

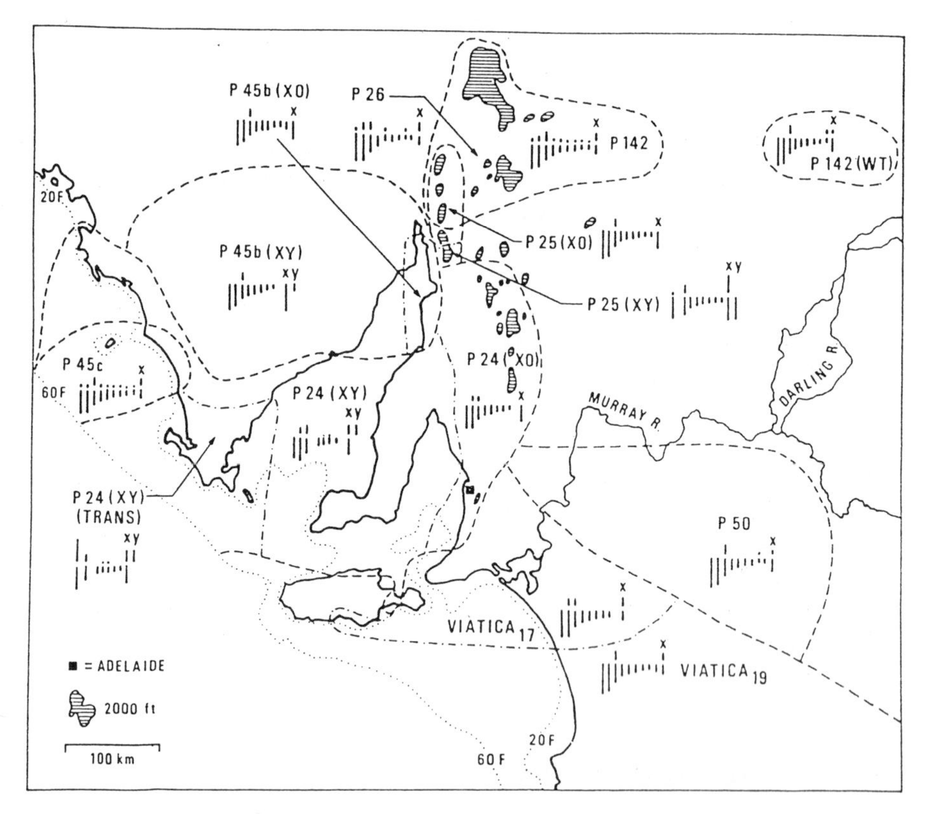

Figure 1 The distribution of the chromosomal races of *Vandiemenella viatica* in South Australia. The coastline during the ice ages is indicated by 60-fathom and 20-fathom lines. The main mountains are shown hatched. The karyotypes are shown as haploid except that both X and Y are diagrammed in XY races. The lighter 'dot–dash' boundaries represent parapatry dependent on only one major factor. (Data by courtesy of Michael White, Graham Webb and Margaret Mrongovius, see Hewitt, 1979 for further details.)

of reasons it is most probable that this occurred through secondary contact between two diverged genomes, rather than selection for these differences within an essentially continuous population (Barton & Hewitt, 1985; Hewitt, 1989). The meeting in the hybrid zone of so many genetic differences, many of which affect hybrid fitness, is difficult to explain in any other way. Using postglacial secondary contact as a starting point, we may examine a number of questions concerning species structure and the role of the ice ages in producing this.

WIDTH AND POSTGLACIAL GENE FLOW

First, we can deduce the approximate date when the two forms met, and measure how much flow of alleles there has been across the zone from one form to the other. This gives us a gauge of the rate of postglacial mixing of genomes. For example, the European meadow grasshopper *Chorthippus parallelus* comprises a complex of subspecies, and two of these, *C. p. parallelus* (north-west Europe) and *C. p. erythropus* (Iberia) form a hybrid zone along the Pyrenees (Fig. 2) (Hewitt *et al.*, 1988). This grasshopper can live up to 2100 m in these high mountains, so the two subspecies meet only in the cols lower than this. The glaciers and permanent snow would have remained relatively late in these high parts and conditions probably did not allow them to hybridize until around 8000–10 000 BP (Jalut, 1977; Hewitt, 1990). We are particularly fortunate in this study since pollen counts have been made recently on cores taken within 3 km of the centre of one of these high hybrid zones at Col de la Quillane (Reille, 1990). These indicate that following the cold Younger Dryas period the Preboreal vegetation did not establish until around 9000 BP. From then till now the hybridization will have allowed genes to diffuse from one subspecies into the other, and this has produced clines of introgressing alleles of varying widths (Table 1). Dispersal in this grasshopper is about 30 m year^{-1} (Virdee & Hewitt, 1990), and if the two diffusing alleles from different races are selectively equivalent (neutral), this should produce in 9000 years a cline some 5 km wide, using Endler's (1977) diffusion equation for neutral clines. A number of character differences show clines of this order, but there are some that are narrower, and others that are broader. The narrow clines can be explained if there is selection against the heterozygotes for the loci controlling these morphological and behavioural characters; while the broader clines suggest that the grasshopper's dispersal can be faster on some occasions than we have measured. In any event the broadest cline is still only 40 km wide and most are about 8 km, so that after 9000 generations there has been little exchange and introgression into the main bodies of the genomes of these two subspecies which range for thousands of kilometres. Many other hybrid zones are similarly relatively narrow compared with their species range, and in such cases the two genomes may remain distinct for some time, maybe until the next ice age!

As mentioned, two genomes forming a hybrid zone usually differ for a range of characters and involve many genes (Barton & Hewitt, 1985). Each set of alleles will have evolved some coadaptation; there will be epistasis between loci and also linkage. Along with heterozygote disadvantage, these combine to produce a significant barrier to gene flow (Barton & Hewitt, 1989). This has been clearly demonstrated and elegantly analysed in the hybrid zone between the toads *Bombina bombina* and *B. variegata* in Poland (Szymura & Barton, 1991). This shows that the strength of the

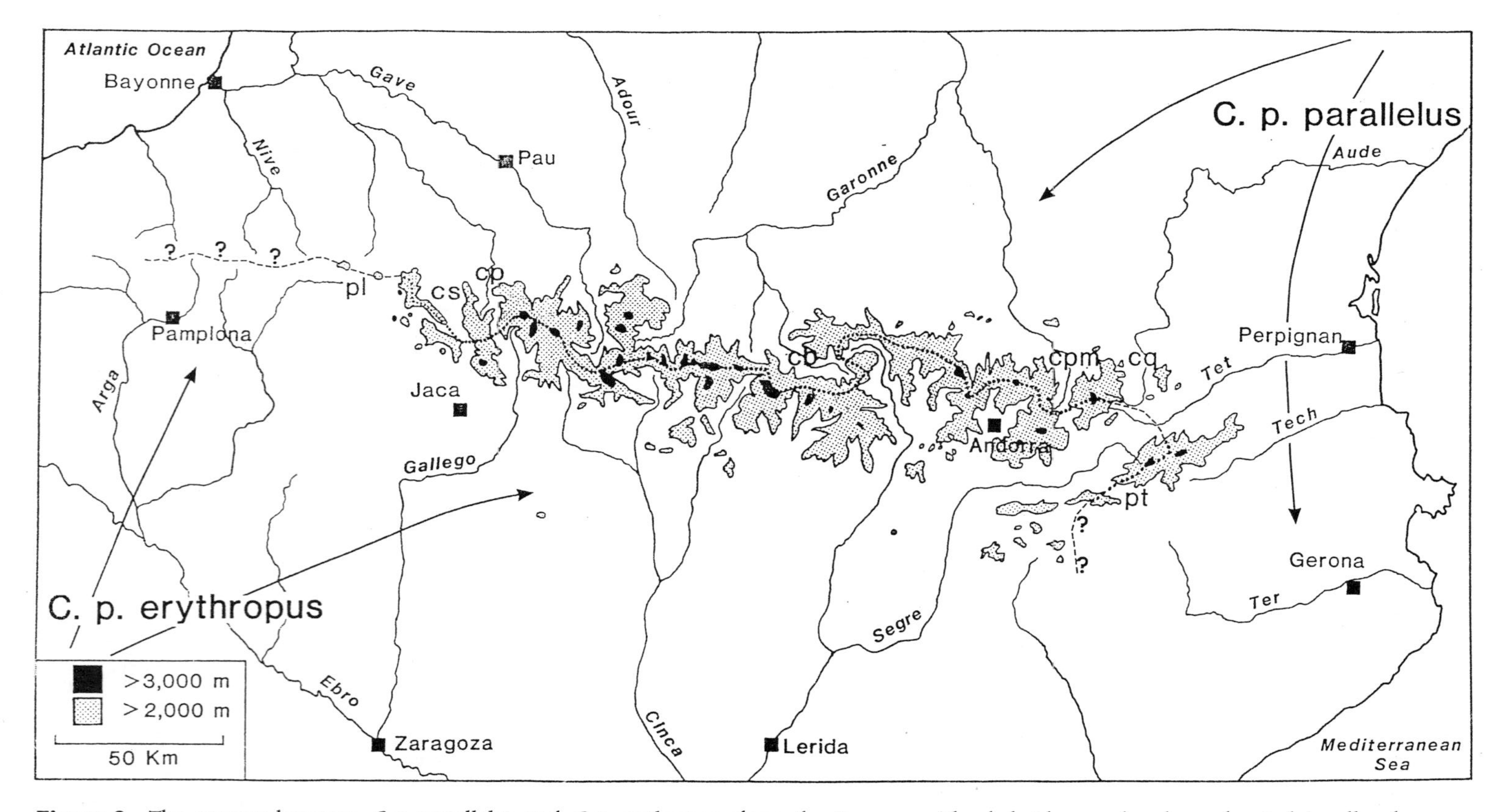

Figure 2 The contact between *C. p. parallelus* and *C. p. erythropus* along the Pyrenees. The hybrid zone has been located in all cols below 200 m. pl, Puerto de Larrau; cs, Col de Somport; cp, Col de Portalet; cb, Col de Beret; cpm, Col de Puymorens; cq, Col de la Quillane; pt, Puerto de Tosas. Its exact course at the extreme west and east ends is under investigation.

Table 1 The widths of clines (km) for some character differences between *C. p. parallelus* and *C. p. erythropus* across the hybrid zone at Col de la Quillane and Col de Portalet in the Pyrenees (see Butlin, 1989; Butlin, Ritchie & Hewitt, unpublished).

Character	Quillane	Portalet
Pronotum ratio	0.90 ± 3.20	0.74 ± 1.34
X-nucleolar organizer	narrow	~ 0.6
Wing ratio	2.6 ± 1.14	11.2 ± 4.98
Song echeme interval	2.8 ± 1.30	wider?
Courtship index	4.2 ± 4.56	~ 28
Peg number	8.4 ± 1.14	18.0 ± 2.84
Femur ratio	8.4 ± 3.40	21.2 ± 6.30
Esterase-2	~ 15	wider?
Peg rows length	16.0 ± 3.72	11.8 ± 3.20
Song syllable length	39.0 ± 6.58	wider?
1st canonical variate for 11 male morphological measures	6.92 ± 0.846	19.42 ± 2.76

barrier in this toad would delay neutral alleles for some 2700 generations, which is similar to the time since the zone was formed after the last glaciation.

INITIAL CONTACT AND MIXING

Thus the rate of genetic mixing in most hybrid zones should be relatively slow even for neutral alleles, and the widths of most clines fit within a dispersal model with strong to very weak selection against heterozygotes (Barton & Hewitt, 1985). However, even supposing simple neutral diffusion there are some that seem rather wide for the organism's dispersal, for example in some *Orthoptera, Lepidoptera* and *Rodentia.* The case of *C. parallelus* in the Pyrenees is an example where some clines are wider than expected; thus to produce a cline 20 km wide in 9000 years requires an average dispersal of some 120 m year^{-1}. This dispersal figure is the standard deviation of parent–offspring distances for the whole population, so even though a long-distance migrant could achieve this, it is inconceivable that, under normal conditions, large numbers would. An attractive explanation for this greater introgression comes from considering what may have happened when the cols were first colonized by *Chorthippus* after the ice retreated.

The advance up the mountainside of *C. p. parallelus* from France and *C. p. erythropus* from Spain was probably in a series of advances and retreats dictated by the climatic oscillations. The colonization would not have been by a solid phalanx of grasshoppers, but by pioneers, long-distance migrants that established small colonies in suitable places. These then expanded as the climate improved and more habitat became available. On secondary contact such an advance would have produced an interdigitated mixing of the two forms. Pioneer patches of both pure genomes would be interspersed over some distance and then mix and hybridize as they expanded. Selection against heterozygotes at certain loci would produce narrow clines for those genes, while neutral alleles would show a broad cline which would continue to spread, albeit more slowly by diffusion. We are currently engaged in modelling such processes by computer simulation in order to quantify the importance of the various parameters

(Fig. 3) (Nichols & Hewitt, unpublished). Clearly the particular combination of dispersal, habitat spacing and climatic change will produce different forms and rates of mixing; and these will vary between organisms with different lifestyles, be they grasshoppers, crows, mice or pines. However, wide clines can be produced by such processes and they may well have been important in determining the width and structure of other hybrid zones.

Whilst it seems unlikely on present evidence that *C. p. parallelus* met *C. p. erythropus* in the high cols of the Pyrenees before about 9000 BP, they probably could have met earlier in the lower hills at both ends of the range. Once again using Endler's (1977) simple diffusion model of secondary contact, to produce a cline of 20 km with dispersal of 30 m year^{-1} would take 155 000 years. Since at the end of the last ice age (<18 000 BP) there was ice and tundra all over this region, this is clearly not a sufficient explanation, and so we still require something like pioneer dispersal. Interestingly, the course of the zone is more tortuous at the ends of the Pyrenean range. In the east it bends back from Col de la Quillane across the upper valley of the river Tet, and runs south-west along the Sierra del Cadi into Spain. To the west in the Basque country it is also more complex and appears wider, and we are currently collecting more data from this area. Whilst this grasshopper was not in

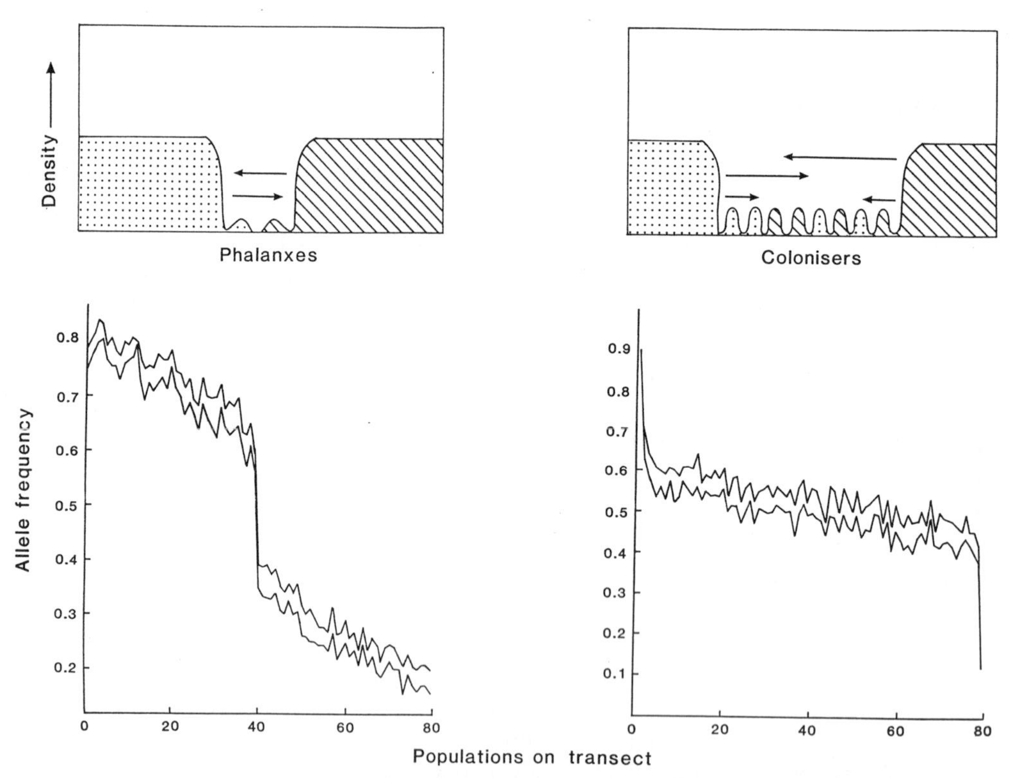

Figure 3 Two types of secondary contact and their consequences for zone width. In the first type, contact is made by two advancing phalanxes containing different alleles, while in the second, colonizing migrants set up localized interspersed population of each allele. After 1000 generations the cline produced by advancing waves is stepped and much narrower than that produced by colonizers. The clines drawn are each the maximum and minimum of 10 replicate simulation (Nichols & Hewitt, unpublished).

these regions during the last ice age it may possibly have reached the Pyrenees from the south during the initial warming around 11 500 BP. However, the intervention of the cold Younger Dryas period 10 700–9500 BP with associated tundra vegetation would have pushed it back again. It could have advanced into the lower foothills soon after this, somewhat earlier than in the high mountains. It was probably then that the pioneer mixing occurred. The contours of valleys and ridges are more complicated at the ends of the Pyrenees than in the straight central spine, and may well have promoted greater interdigitation of the two subspecies.

POSTGLACIAL REINFORCEMENT AND PARAPATRIC SPECIATION

Since hybrid zones separate taxa that have diverged to various extents, producing forms, races, subspecies and species, they allow us to examine some of the genes and processes involved in generating various levels of reproductive isolation. They are particularly relevant to reinforcement, a major theory of speciation (Wallace, 1889; Mayr, 1963; Dobzhansky, 1970). Secondary contact of differentiated genomes where the hybrids are less fit present what would seem ideal conditions for selection to exaggerate differences to prevent hybridization, and thus produce full species. There has been some disquiet and debate recently about the likelihood of such a process occurring (Paterson, 1978, 1985; Templeton, 1981; Barton & Hewitt, 1985; Butlin, 1987, 1989; Spencer, McArdle & Lambert, 1986; Coyne & Orr, 1989; Sanderson, 1989). The theoretical conditions under which reinforcement could occur are quite limited, and within a hybrid zone the process would be swamped by the influx from the parental populations. Furthermore, a careful examination of over 150 hybrid zones for evidence of reinforcement provides little support; in only three cases is it possible and these are far from conclusive (Butlin, 1987, 1989). The Pyrenean hybrid zone in *C. parallelus* has been used for a number of experiments searching for divergence in the mating system across it, which would indicate that reinforcement had occurred. It has the classical attributes of postglacial contact, strong F1 hybrid sterility, gene flow, distinctive mating behaviour and significant assortative mating between the two subspecies. However, despite a number of extensive experiments measuring morphometrics, fecundity, assortative mating and female preference, no good evidence for reinforcement has emerged (Butlin, 1989; Ritchie, Butlin & Hewitt, 1989, 1990; Butlin & Ritchie, 1991).

It appears, therefore, that contrary to what was expected, hybrid zones are not likely places in which speciation by reinforcement could take place. It is still possible, however, that secondary contact has occurred on occasion between races that have diverged in just the right way for reinforcement to proceed (e.g. very few genes producing significant differences in the mating system and linked to alleles showing heterozygote unfitness). In such cases, the process of speciation may proceed very rapidly and the two new mating systems can become sympatric. If this occurs we might expect to find pairs of sibling species with partly overlapping ranges. The likelihood of this would depend on the lifestyle and ecology of the two forms. They probably would have similar ecological requirements, and sympatry should not be possible in a fairly continuous distribution due to interaction and competition (Bull, 1991). But sympatry might occur in a distinctly patchy habitat (Shorrocks *et al.*, 1984; Shorrocks & Swingland, 1990). It is difficult to say just how far such a process could penetrate, and how large the sibling overlap would be. This would depend on the dispersal powers of the organism, population density and the patchiness of its habitat. For many organisms it seems unlikely that postglacial overlap of sister species would be great, but dispersing ephemerals like *Drosophila* may be different.

It could also be argued that when two ecologically different forms meet they could overlap where both niches occur and thus provide a broad theatre in which reinforcement could occur. Of course, they must not be so ecologically distinct that they do not mate, if this were so they would already be species. There are a number of hybrid zones where the taxa show strong ecological differences, e.g. *Bombina bombina/variegata* (Szymura & Barton, 1991), *Gryllus pennsylvanicus/firmus* (Harrison & Rand, 1989), and *Triturus cristatus/marmoratus* (Arntzen & Wallis, 1991), and these regions of overlap are a mosaic of the two forms. These are postglacial zones with various local widths from 6 km to 20 km, although the total width of interspersion is up to 250 km for *Triturus* in north-west France. So far there is no evidence of reinforcement in such cases. It is worth recalling here the general point that in situations where few hybrids are produced there will be little reinforcing selection; it is a process of diminishing returns.

Remington (1968) listed many examples of pairs of species from adjacent regions of North America that were hybridizing in a local zone. He argued that a number of these hybridizations were the result of recent changes in climate and the European argricultural practices allowing the species to make contact. Some of these would seem to be candidates for detailed analysis of their mating system and hybridization like that performed on *C. parallelus*. It would seem particularly relevant to examine those where little hybridization is occurring for differences in their ecology and behaviour, and to test allopatric and sympatric isolation.

In the light of such considerations, it could be argued that in order for taxa making secondary contact to become fully sympatric, they must have evolved behavioural, lifestyle and ecological differences that effectively prevent mating and hybridization.

MOVEMENT AND POSITION

As stated before, many hybrid zones show evidence of hybrid unfitness, and this is probably so in the great majority of cases (Barton & Hewitt, 1985, 1989). The width of the zone is maintained by a balance between dispersal and selection against hybrids, and so they are aptly called tension zones (Key, 1981). Bazykin (1969) modelled a simple tension zone between two alleles with heterozygote disadvantage, and in a homogeneous environment this is free to move across the surface. Where one allele is superior as a homozygote then this will drive the zone forward. However, the situation in the field is complicated by variations in the organism's environment, so that genotypes may be better adapted to different habitats, and some places will support greater or lesser numbers of individuals. Consequently a tension zone will tend to be held at an ecotone where the two genotypes are equally fit, and it will also be held in a density trough where the local habitat supports few individuals or reduces their dispersal (Hewitt, 1975; Barton, 1979; Barton & Hewitt, 1985, 1989). Now for most organisms the environment varies greatly from place to place, producing many density differences and barriers. These will trap the hybrid zone very effectively and only if one genotype is very much fitter will it advance. Where the hybridizing genotypes are similar in fitness very slight differences in density will hold the zone.

Distinct environmental features like mountain ridges, hot valleys, river courses, arid flats and dense forests would act as major barriers, and a number of hybrid zones are located at inhospitable regions (Hewitt, 1989). Furthermore, detailed mapping of density in the grasshopper *Podisma pedestris* in the Alpes Maritimes shows how very local variations in distribution can account for the exact course of the zone between the obvious barriers and troughs along which it runs (Nichols & Hewitt, 1986; Barton & Hewitt, 1989).

The consequence of this is that once formed, hybrid zones should not move far until there is a major environmental change which radically alters the organism's distribution. An ice age or interstadial could do this, and so can man! Lesser climatic fluctuations could also modify conditions sufficiently to eliminate and recreate hybrid zones in those places where the particular species is near to its ecological limits. Two grasshopper cases provide evidence that zones formed after the last ice age have not moved far (Hewitt, 1988). *Vandiemenella viatica* forms hybrid zones on Kangaroo Island which continue on the South Australian mainland just as though the sea had not risen and cut them off as the ice melted. *Podisma pedestris* in the Alpes Maritimes has a zone which similarly runs across islands of suitable habitat provided by isolated mountain blocks. It is difficult to explain such distributions other than by postulating that they have survived in the same region since the end of the last ice age when both species moved to approximately their present positions.

In his compendial article on the distribution of animals and trees in North America, Remington (1968) describes "suture-zones of hybrid interaction between recently joined biotas". These are where a large number of hybrid zones from different species are found in the same region; he identified six major suture zones and seven minor ones. Several of these are associated with major physical and ecological barriers, for example, the Sierra Nevada and the Blue Mountains, the Rocky Mountains, the Texas Plateau, the Florida Borders and the Appalachian Mountains. These are seen as interfaces of regional biotas, which met and hybridized after the last ice age. For the Pacific–Rocky Mountains Suture Zone he cites as a cause the warming in the last 100 years, for the Rocky Mountains–Eastern Suture Zone the recent lack of burning of the Great Plains, and for the Northern Florida Suture Zone a change in the forest due to cooling since 2000 BP. Whilst recent events, including man, may well have been important in changing the opportunity for hybridization, such broad distributions were most likely set up earlier, just after the last ice age, so there has probably been earlier hybridization in many cases, for example across the Sierra Nevada Range in the Hypsithermal (~5000 BP). Other possible suture zones exist in central Mexico, the Arctic tree line and northern New South Wales. Clusters of contact zones can also be found in the tropics of the Americas and Africa (Prance, 1982).

Remington (1968) also suggests the Urals, northern Manchuria, the Yenisey region of Siberia, and central Europe as possible suture zones. There are indeed a number of hybrid zones running down central Europe, those in the toad *Bombina*, the crow *Corvus*, the mouse *Mus* and the grass snake *Natrix* being well-documented (Barton & Hewitt, 1985). Remington feels that "suture-zones in the Palearctic and Africa may be blurred due to the antiquity of extensive human cultivation and chronic deforestation". It is true that each one follows an individual track, but those from North America are no more co-incident in places. Clearly, the particular course followed by a hybrid zone will be the product of several things, viz, where the species came from after the last ice age, the organism's particular attributes and adaptations which prescribe its 'niche', the availability of physical and biotic conditions to provide that niche, and the variability of those conditions in space and time. Whilst the last major range changes in North America and Europe occurred following the retreat of the ice, some plant species certainly changed their distributions during the Hypsithermal 7000–5000 BP (Huntley & Webb, 1988). Some are apparently doing so today driven by climate or affected by agriculture. As such advances and retreats occur, the biotic conditions in the community will change, and so too may the chance of hybridization for the other cohabiting species.

POSTGLACIAL ADVANCE

A considerable body of information has now been collected and collated on pollen, lake levels and marine plankton for the Holocene (COHMAP members, 1988), so we now have a much more detailed knowledge of climate changes and particularly of distributions of plant and tree pollen remains in North America and Europe (see articles by B. Huntley, R.S.T. Thompson and T. Webb in Huntley & Webb, 1988). As conditions began to warm and the ice sheets retreated (~ 16 000 BP) large areas of glaciated soil, tundra and previously inhospitable land were colonized by plants and animals that had been living further south during the cold spell. By 12 000–13 000 BP only a tongue of ice covered most of Scandinavia and the Baltic, while the North American sheet still extended down to the Great Lakes, 60°N and 46°N respectively. In Europe, wormwoods, grasses and sedges are recorded first in the advance, followed by pioneering birch and pine trees. In the eastern U.S.A., scattered spruce with forbs and creeping cypress were found south of the ice. Conditions across these vast areas would have differed in temperature, moisture, wind and soil, and different species in various combinations occupied the regions.

Coope (1990) reports studies of beetle remains showing that species which today have a Mediterranean distribution had reached Great Britain by 13 000 BP. Clearly conditions this far north (53°N) were warm enough for those insects which could survive on this open vegetation, and had the dispersal capability to get here. It is quite possible that the climate at that time was warmer than now (Atkinson, Briffa & Coope, 1987), and mean annual temperature must have risen rapidly by perhaps some 16°C from the end of the glaciation.

This rapid advance was sharply reversed around 10 500 BP for up to 1000 years in the Younger Dryas period. The readvance of the Marine Polar Front in the Atlantic from north of the British Isles down to the coast of Iberia brought much colder conditions and a re-extension of the ice in places. Tundra spread down through France and the birch trees of northern Europe died out. The pine and oak in southern Europe retreated, and the Mediterranean dung and tree beetles vanished from Britain. The vegetation responded quickly to the changes in climate, and the beetles were even quicker. Indeed, such insects are very sensitive indicators of sudden and relatively brief climatic changes (Coope, 1977), indicating that their advance and retreat is rapid when the conditions allow.

When the Marine Polar Front shifted north again around 10 000 BP, the climate warmed and the vegetation advanced rapidly over Europe, so that by 6000 BP the vegetation pattern broadly resembled today's. Several smaller vegetational changes occurred after that and it was probably around 3000 BP that most places had their current vegetation. In North America by 10 000 BP the vegetation south of the ice was modern in its main components, and by 6000 BP when the Laurentide ice sheet largely disappeared, those species now in these northern parts had moved in. There are obviously major differences between North America and Europe, the largest being the Laurentide ice sheet, so that vegetational and faunal advances in North America were farther south initially. The western Mediterranean is at the same latitude as the Appalachian Mountains, and Florida is level with the Sahara. The Rocky Mountains and Great Plains run north–south down North America and the conditions and vegetation history in these and on either side were quite different. In Europe the Mediterranean Sea and mountains to the north of it (the Pyrenees, the Alps, the Transylvanian Alps, the Caucasus) run east–west, with a large broad plain to the north. As North America narrows into Mexico, the southern deserts were less extensive in the ice age, whilst Old World deserts like the Sahara were still formidable barriers to many species. The species involved in the vegetation advances differed between regions. Each species has different properties and requirements, consequently in each region a different sequence of colonization and replacement occurred.

Because of topography and regional differences in climate, some areas would also have opened up earlier to colonization. The pollen maps of Europe (Huntley & Birks, 1983) show that the postglacial vegetation advanced much more quickly up the eastern side of Europe between the Caspian Sea and the White Sea (35°E) than in central and western Europe. Thus alder and spruce reached 60°N by 13 000 BP in the east. It is worth noting that at this time the north-western Atlantic coasts of Scotland and Ireland had pine, the south-west of Ireland had oak and elm, and Brittany had oak, elm and alder. This Atlantic fringe with nothing inland suggests that they had been transported north from Iberia by ocean currents or animals. Similarly, in North America the persistence of the Laurentide ice sheet and consequent climatic conditions meant that the advance north of the Great Lakes was slower than to the west. In the far west the mountains, depressions and a wetter climate produced a more intricate pattern of species movement and distribution, and provide excellent evidence for the role of mountainous terrain as local refugia with species moving up and down them (Thompson, 1988).

RATE OF SPREAD

The rate of spread after the Younger Dryas cold spell has been well-documented from pollen cores, which have been collated to produce isopoll maps for a series of times determined by radiocarbon dating. In Europe, trees expanded their ranges northward at quite astounding speeds on occasion (Huntley & Birks, 1983). Most estimates fall between 50 and 500 m year^{-1}, but pine and hazel reach 1500 m year^{-1} and alder 2000 m year^{-1}. The North American beech spread some 1400 km from Tennessee (16 000 BP) to its northern limit in Canada (7000 BP) at an average of 150 m year^{-1}. Its rate of expansion varied greatly from place to place; it was much faster at times and probably slowed down considerably as it approached its northern limits (Bennett, 1985). As mentioned before, the response of certain beetles in tracking climatic change is probably even faster so long as the vegetation is adequate. Those species which are scavengers or feed on grasses and herbs may spread more quickly than trees, but they would still be limited by the climate and its rate of change.

We may deduce from the presence of insects like *Chorthippus parallelus* in Britain but not Ireland that it reached England after the ice age, just before the sea-level rose to cut the English Channel and Irish Sea. This also apparently stopped the invasion by a number of later arriving species that can presently be found across the Pas de Calais. Rates of spread similar to those recorded for trees (300–500 m year^{-1}) are necessary to explain such range expansion from southern European refugia (Hewitt, 1990).

Invading species that have been recorded in historical times provide some startling rates of spread and useful details of the causes. For example, the cheat grass *Bromus tectorum* spread over 2000 km to dominate grassland from British Columbia to Nevada in 30 years following the introduction of cattle and new agricultural practices (Mack, 1981). The collared dove began extending its range into Europe from Turkey around AD 1900 and by the 1980s it had covered most of the Continent from Ireland to western Siberia (Hengeveld, 1988). Hengeveld argues that this invasion was triggered by climatic change which just tipped the balance of the bird's reproduction. Many other examples of extensive invasion are known, and whilst most are affected or caused by man's activities they demonstrate that individual species can expand at a rapid rate when the climate and ecology becomes suitable for them. In most cases this will allow them to track the changing conditions driven by the climatic cycles.

ADAPTATION AND DIVERGENCE

As it expands, a species will colonize new areas where it can survive and reproduce adequately. With further climatic amelioration, a particular location may well become even better for its reproduction, but it will also become accessible to new species which will change the biotic environment. The first species may be advantaged or disadvantaged by this, and it will tend to adapt to these changes. During a time of relative environmental stability these adaptations may proceed apace, and there is considerable evidence on character modification under selection in the laboratory and in nature (Endler, 1986; Falconer, 1989). Morphological, physiological and behavioural characters can be changed, and important components of life history show variation in the wild and respond to selection in the laboratory (Denno & Dingle, 1981; Dingle, 1990).

Of particular interest in considering range expansion of insects are a number of cases where there are geographic differences showing genetic variance that can be modified by laboratory selection. Two particularly nice examples are described in the pitcher plant mosquito *Wyeomyia smithii,* which ranges between Florida and northern Ontario (Istock, 1981), and in the cornborer moth *Ostrinia nubilalis,* which was introduced from Europe to U.S.A. in about AD 1910 and has spread over most of the eastern half of the U.S.A. (Showers, 1981). Both of these show north–south geographic differentiation in diapause which responds to selection. We can therefore expect populations once they can become established to adapt to their new local habitat, and then be able to track slowly changing conditions. During a slow advance the populations established in the recently acquired range will receive reinforcements of their own species from behind, and these may have genotypes more suited to the ameliorating conditions and changing mixture of species. Of course, some environmental changes will go beyond their limits of individual survival and replacement reproduction. It is also possible that some adaptations are not possible without major restructuring of the genotype that would require the dismembering of previously adaptive gene complexes to allow selection to build a genotype suited to the new conditions; some adaptive peaks may be separated by chasms. Furthermore, during a time of rapid climatic and biotic change they may not be able to adapt quickly enough to the many changes required of them.

Recent work on the genetics of stress provides useful insight into the nature of species boundaries and also suggests that evolution may occur rapidly under changing climatic conditions. Towards the edge of a species range environmental stress and perturbation may demand so large an allocation of metabolic energy that the amount remaining for growth and reproduction becomes critical and limiting. In non-marginal populations of *Drosophila melanogaster* there is a rapid response to selection for desiccation resistance involving lowered metabolic rate, but this is associated with lower behavioural activity and fecundity. Similar genetic associations between lower metabolic rate and stress resistance have been found in other animals (Hoffman & Parsons, 1989). Thus there would seem to be significant genetic variance to adapt towards marginal conditions at the edge of the range. During climatic changes as a species advances, the pioneer populations will be subjected to selection in various stressful environments which may accelerate their divergence, while at the trailing edge increases in the severity of the conditions will select for different genotypes until the critical extinction point is reached.

It is difficult to judge what the outcome of these processes would be; it will clearly depend on the particular genetic attributes of the species, which other species are also present, the physical conditions and the rates at which they change. It seems most unlikely that optimal fitness in that environment will be achieved. If populations evolve in different ways this may support higher genetic variation within the species. If in the end a particular species cannot tolerate or adapt to these challenges then it dies out in that area. This will determine the rear edge of the species range as it advances.

It has been emphasized that in different geographic regions there will be different combinations of species, so expanding species may well meet a different biota, for example, competitors, food plants, host plants and predators, as well as physical conditions. It will tend to adapt to these differences and produce different modifications. Consequently where parts of an advancing species follow different paths they could diverge into distinct forms. For example, a species with ice age survivors in southern Europe would meet very different conditions as it spread north-west and north-east. Likewise a species from southern U.S.A. may take very different expansion routes.

SOUTHERN EDGE

Having advanced northward, the rear edge of a species to the south will also have moved northward. For example, the southern extent of *C. parallelus* in western Europe, in Spain, Italy and Greece, is probably where the species survived the ice age. Obviously the conditions here now are very different from the northern edge in Scotland or Sweden, and one would expect different adaptations to develop for many aspects of its biology. It occurs in local moist places high up mountains like the Sierra Nevada in Spain or the southern Appenines of Italy, while in northern Scotland it is found only occasionally in low sheltered warm meadows. So during the postglacial expansion some locations in the south will have seen pioneer, established and restricted phases of the range change of this species, with great differences in the form of natural selection caused by climatic and biotic variation.

This raises again the question of the direct genetic origin of the present population in any particular place. Is it comprised of descendants of the first pioneers from the northern edge of the refugium that moved in and then adapted to changing conditions, or is it descended from more southerly ice age populations that moved north later with the warmer climate? Details of the molecular phylogeography (Avise *et al.*, 1987) may give useful clues in organisms where there are enough remnants of the range of the old species. We may hope that fossil DNA studies can also help, where a species leaves behind usable remains (Pääbo, 1989; Golenberg *et al.*, 1990). In this context the pack rat middens of the south-west U.S.A. seem worthy candidates, given the considerable background in botanical palaeoecology of these parts (e.g. Thompson, 1988; Spaulding, 1990).

For many European species ice age refugia have necessarily been postulated in Mediterranean locations, which were south of the ice, tundra and steppe. Pollen records provide evidence of the distribution of grasses, shrubs and trees in these regions at the end of the full glacial 20 000-13 000 BP (Huntley & Webb, 1988). Likewise, North American refugia were south of the ice and several locations for plant species have been discovered (Webb, 1986; Thompson, 1988). In western Europe today the northern distribution of broadleaf deciduous forest roughly corresponds with that of *C. parallelus* and a number of other grasshoppers, although in the east different local conditions allow it into the boreal forest. This deciduous vegetation was restricted to the south-west of Iberia, the tip of Italy and the southern Balkans in the ice age, and *C. parallelus* most likely survived there. The Mediteranean Sea forms a large east-west barrier to southerly migration, and whilst the species is found in the mountains of Turkey it is apparently not in Africa or Arabia. When species like *C. parallelus* expanded from these limited peninsular refuges they could go up the mountains and across the plains of Europe. The east-west transverse mountains of the Balkans, Alps and Pyrenees may have been some environmental barrier to early migration, but pollen data indicate that any effect was relatively short-lived as species expanded round these blocks. None the less, very different subspecies of *C. parallelus* are described for Spain and Greece than for northern Europe, and the forms from Italy and the Balkans show some differentiation. There is, of course, a hybrid zone along the Pyrenees and it will be interesting to examine detailed distributions in the Alps and Balkans.

The wingless alpine grasshopper *Podisma pedestris* is only found high up in southern European mountains between 1400 and 2600 m in the right conditions. At high altitude it is very local where the insolation and patchy vegetation allow, while at lower altitude it occurs in clearings amongst trees where it is locally cooler. Its main range is in Siberia, where it can be a pest, and orthopterists argue for a Pleistocene invasion of Europe during the ice ages. As it warmed, *Podisma* would go up the mountains of southern Europe, and as it cooled could come down again to cover large areas south of the ice during the glacial maxima.

The topographic diversity of southern Europe means that species have more chance of finding suitable habitats during these changes than to the north (Fig. 4). North

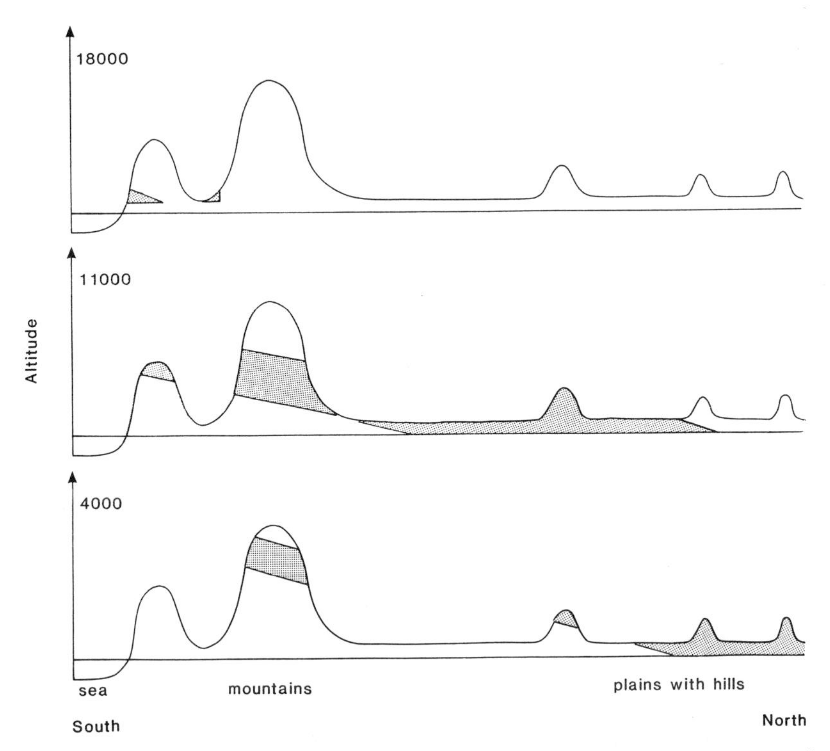

Figure 4 A diagrammatic representation of Europe with the Mediterranean Sea and mountains in the south and plains to the north, showing the distribution of a species (e.g. *Chorthippus parallelus*) from its ice age distribution to the Hypsithermal. It can move up and down the mountains and across the plains.

America is different; there the east is very different from the west. The eastern plains and Florida are much further south than Europe with greater insolation, while in the west are the north-south mountains with a long run south to the deserts, Mexico and the tropics. The survival of a species in a region depends on the topographical diversity, the rate of climatic and habitat change, the species dispersal ability and its adaptability. Its range changes will also be determined by these, and will be very different for distinct combinations of geographic features such as large plains, mountains and seas.

GENETICS OF ADVANCE

Given such rapid and major changes in climate, and hence habitat, large areas of northern Europe and North America were opened up for recolonization after the ice age. This process probably proceeded through long-distance dispersal from populations at the leading edge. These pioneers would find pockets of unpopulated suitable habitat in which they could breed rapidly and establish a colony. In time these colonies would meet and exchange individuals. A number of studies indicate that such spreading consists of long-distance dispersal and an establishing growth phase (Bennett, 1986; Hengeveld, 1989). When an expansion is over large distances then the process of establishment from long-distance dispersal will be repeated many times, with the populations at the leading edge having a greater probability of providing the migrants. Thus, those individuals at the leading edge will go through a series of bottlenecks followed by some growth each time. Since the climate will be oscillating within this general amelioration, further bottlenecking will occur after small colonies have been founded.

The theoretical effect of such repeated bottlenecking will be a reduction in genetic variability due to chance sampling. There will be a loss of alleles and a tendency to homozygosity (Nei, Maruyama & Chakraborty, 1975). This loss of alleles will be ameliorated by the rapid population growth after the founding of the colony, so that rare selected alleles may recover in frequency, but nevertheless loss should occur. There is some evidence for this process from the Nearctic wild sheep *Ovis dalli* (Sage & Wolff, 1986), where a study of 27 allozyme loci showed that beasts from the north were nearly monomorphic and less variable than other wild or domestic sheep. Furthermore they were only one-third as variable as the average mammalian species (H_{Nei} = 0.015 vs. 0.041). A particularly fine example has been reported in the lodge pole pine, which is found all down western North America (Cwynar & MacDonald, 1987). The subspecies *Pinus contorta latifolia* has spread from its glacial refugia south of the ice sheet some 2200 km to the central Yukon, where its most northern populations have become established in the last 100 years. Northern populations show less allelic diversity than southern ones, and there is a correlation over 15 sites between number of alleles per locus and the time since founding, as determined by radiocarbon dating of fossil pollen records. Also during this expansion the dispersal ability of the seed has been increased by reduction of the wing loading; a nice example of adaptation. A number of other plants and animals that must have expanded into previously glaciated areas also show reduced variability in their northern populations (Critchfield, 1984; Sage & Wolff, 1986), and there is a real need for more such pertinent studies.

One may also look at recent invasions for evidence of the genetic consequences of range expansion (Barrett & Richardson, 1986; Barrett & Husband, 1990). A number of studies bear up the expectation of loss of genetic variation with the invasion, caused by founder effects and bottlenecking. The evidence for homozygosity and progressive population differentiation is clearest amongst weedy species such as the North American barnyard grass *Echinochloa microstachya,* and the introduction of the European rabbit *Oryctolagus cuniculus* to Australia provides another example. However, the picture is not so clear in most cases and more experiments and data are required to determine the importance of the different factors in the colonizing process. This is particularly true of quantitative characters such as size, growth, behaviour and reproduction, which are obviously very important. Recent theory and experiment (Goodnight, 1988; Bryant & Maffert, 1988; Carson & Wisotzkey, 1989) indicate an increase in genetic variance following a severe bottleneck, probably due to the reassortment of epistatic interactions. This means that, despite the loss of allelic variation in allozymes, there may be phenotypic variation allowing response to selection produced by changing conditions during the species' advance.

The long-distance colonizing from the leading edge of an expansion may have other consequences besides a reduction in genetic diversity, differentiation among populations and reorganization of epistatic interactions. In a large population new mutants have little chance of survival, and virtually none at all if they reduce fitness in the heterozygote, even if their homozygote is normal and fit. However, in the small founding populations with repeated bottlenecking there is much greater opportunity for such negatively heterotic mutants to become fixed by drift and establish a colony. Such isolated leading edge colonies would grow rapidly to fill the habitat until they met other expanding populations. The heterozygote disadvantage between the new mutant population and the parental genotype would form a hybrid zone of some type, thus tending to keep the two genomes parapatric. Consequently, the parental genotype which spawned the mutant colony will not be able to advance through the now extensive new genotype. Since the new genotype is at the leading edge it will be more likely to colonize farther ahead through long-distance dispersants (Fig. 5). It may expand to the ecological limits of the species range, or until another negatively heterotic mutation becomes established ahead of it.

Such hybrid zones may involve one gene with only a small reduction in fitness, and would be a very weak barrier to gene flow. On the other hand, genes with major effects may be established, or chromosomal rearrangements involving blocks of genes, and these would produce stronger hybrid zones. Consequently within races separated by very strong hybrid zones, we might expect to find genome areas and patches of varying distinctness where the species range expansion has promoted their production. Obtaining decisive data on this will require some effort, but a number of cases of weak hybrid zones exist and several species have multiple geographic genotypes separated by distinct clines in gene frequency. For example, some of the chromosomal hybrid zones in shrews, mice and grasshoppers divide up parts of the species range on a fine scale, and would have been formed after the last ice age (White, 1978; Corti, Capanna & Estabrook, 1986; Searle, 1988).

Because of the difference between the rapid population growth of the leading colonists and the replacement dynamics of the established standing population behind, it will be much more difficult for a migrant from the main body to become established, whether or not it produces hybrid unfitness. It is most likely to disperse into an already colonized habitat and its rate of reproduction will be logistically low, rather than the exponentially high rate of the colonizers at the front (Fig. 6). It will need to be distinctly advantageous to become established. Even advantageous alleles and genomes advancing from the main body behind will also move at a much slower rate through a cover of established populations than genomes at the leading edge, which may spread by long-distance dispersal and grow rapidly in the absence of competition. In *C. parallelus* the dispersal each generation is some 30 m year^{-1},

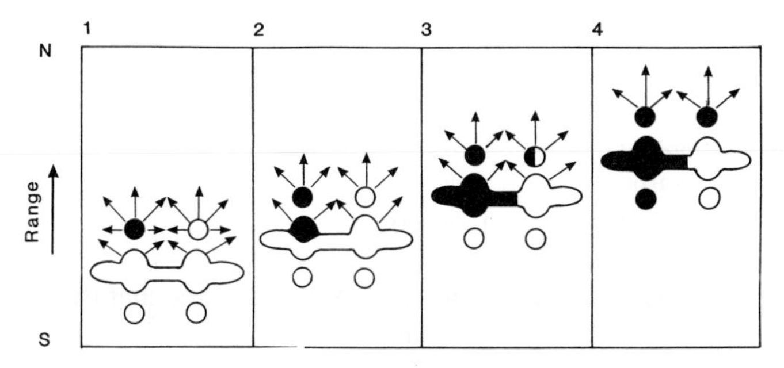

Figure 5 A model of the range shift from south to north through times (1–4). A mutant conferring heterozygous disadvantage establishes in a small founder population at the leading edge (1). Through successive founder events it advances, restricting the spread of the parental genotype type from behind.

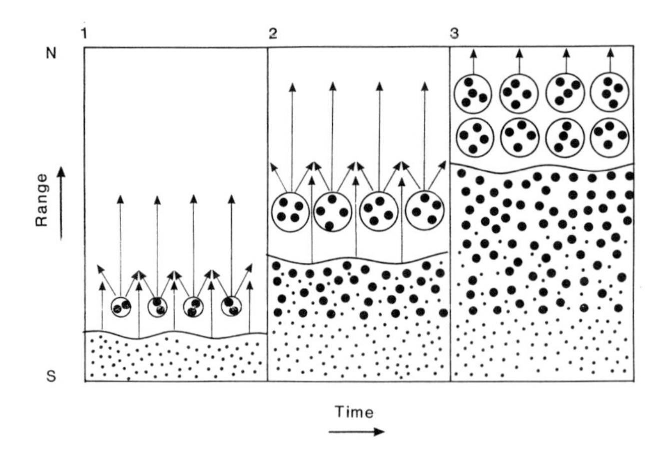

Figure 6 The leading edge of a range change: Alleles and genotypes in the small colonies at the front (large spots) reproduce much faster than those behind in the main body (small spots) because they are colonizing unoccupied habitat. Those behind reproduce more slowly and invade by diffusion.

while in the postglacial expansion it seems to have spread at some 300 m year^{-1} on average (Hewitt, 1990). Such differences in rates would apply to many species, and trees are a good example (Bennett, 1986). It is therefore not clear how fast an advantageous gene could spread in reality; the established main body may be viscous to gene flow relative to the fluid expansion of the leading edge genomes. These latter may thus come to cover large areas of the range.

Furthermore, at a hybrid zone the two genomes involved may well have different epistatic gene interactions, and in zones between well-differentiated genomes there is likely to be considerable co-adaptation and even ecological adaptation. Consequently, whilst an allele may be advantageous in one genome, when it meets another genome it may not be so, due to its different interactions with other genes. We have seen that species are subdivided, maybe extensively so, and the diffusion of new alleles will depend on the extent of that subdivision and its concomitant genetic adaptations.

GENOME REASSORTMENT

In the small populations which occur around or between refugia, or at the edge of a species range, there is a greater chance for hybrid genomes to survive. If two genomes meet and hybridize to produce occasional small isolated colonies, for example by long-distance migrants from the edge making contact or when two genomes climb the same refugial mountain during a warm stage, then segregants and recombinants may survive and by chance establish a colony. This may spread over a considerable area in a period of expansion. Since different genomes and hybrid zones should not move through each other, this mechanism was proposed to explain the reassortment of chromosomal rearrangements distinguishing taxa of the Australian Eumastacid grasshopper *Vandiemenella viatica* (Hewitt, 1979).

Similarly, whilst the hybrid zone between *Mus domesticus* and *Mus musculus* runs across Jutland, north of the zone in Sweden *M. musculus* have mtDNA like *M. domesticus* (Gyllensten & Wilson, 1987). The unhitching of mitochondrial and nuclear genomes probably occurred by the chance survival of a reassorted genome in a bottlenecked population, as the hybrid zone formed progressively northward up the middle of Europe after the glaciation. Gyllensten & Wilson (1987) suggest that this occurred on East Holstein close to where the hybrid zone meets the Baltic Sea, and that a small propagule founded the house mouse population in Sweden. Conversely, in the Australian grasshopper *Caledia captiva* Marchant, Arnold & Wilkinson (1988) argue that the hybrid zone between Torresian and Moreton karyotypes has moved south since the last ice age leaving a trail of Moreton markers some distance behind in the present range of the Torresian subspecies. These include mtDNA ~250 km, allozymes ~350 km and rDNA ~450 km. A series of bottlenecks and expansions involving reassortment could also explain this fascinating situation. In both *Mus* and *Caledia* other sections of their long hybrid zones show coincidence of these subspecific characters, and hybrid zones in general show a broad coincidence of character clines (Barton & Hewitt, 1985). However, few zones have been studied over large areas for many characters, and very few include DNA markers, so we may well find more examples like these.

It is possible to study the detailed dynamics of isolated hybrid zone colonies, as for example with the grasshopper *Podisma pedestris* in its southern edge mountain-top islands in the Alpes Maritimes. Near Seyne-les-Alpes at the lower edge of the species range, the hybridizing genomes are much more patchy, with colonies intermingled over a considerably greater distance than in the denser regions of the zone (Nichols & Hewitt, 1986; Nichols, Humpage & Hewitt, 1990). It will be just this sort of population that will lead the expansion when the climate cools again.

REPEATED RANGE SHIFTS

During the last 2 million years of ice ages there have been a number of warm spells like today that have moved species ranges repeatedly south and north across continents, and up and down mountains for thousands of metres. The rapid movement each of these entailed probably means that the range is largely occupied by descendants of the small populations at the edge of the refugia which have gone through repeated founder-flush events. There probably would have been little mixing of genomes from the front and the back, and with major shifts a species would have become extinct over the entire refugial range unless there had been a convenient high mountain to climb (Fig. 7). Population bottlenecking would have altered the genetic architecture, negatively heterotic mutants probably would have established and hybrid zones been set up, and the changing mixture of species during the advance would have generated selection for different adaptations through a host of interactions. Responses to this selection may be very rapid and considerable, as witnessed by present-day invaders like the cornborer moth and the dispersal adaptation of the lodge pole pine during its postglacial expansion. When conditions reversed the southern edge populations would have led the way, repeating the process, but under different climatic conditions and species mix. This time the northern range would have become extinct, unless it could find a low warm enclave south of the ice.

Each time this happens we can expect the genome to be modified and evolve; how much of this differentiation survives the subsequent extinctions will depend on

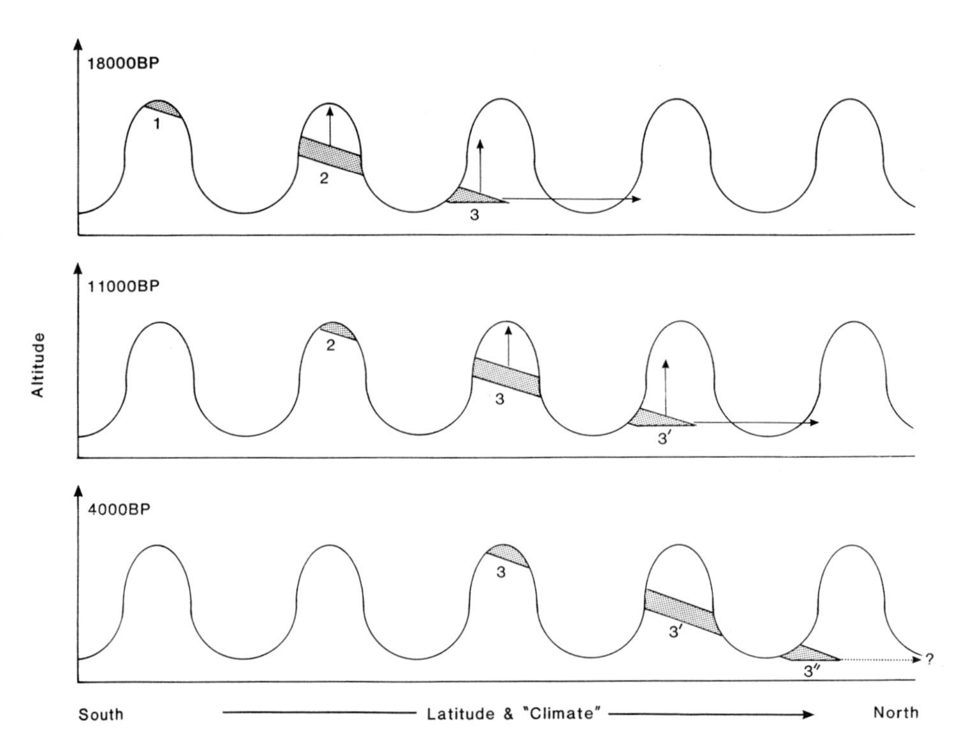

Figure 7 A diagram of species range change with warming after the last ice age across variable topography and habitat. Numbers 1, 2 or 3 identify genomes in different regions. After a northward advance only genome 3 remains, which may have produced variant 3′ and 3″ during its expansion. The process could be reversed during a cooling period.

the particular circumstances of topography and climatic change. Where a species has several refugia, different forms may survive in them; this would explain the number of geographic subspecies in *Chorthippus* and other groups across southern Europe. Furthermore, the conditions in each shift will be different, since the orbital cycles are not synchronous and produce different climatic effects. Consequently, the species mix each timc will be different, as is clearly seen from pollen data (e.g. West, 1980), and so selection will favour other adaptive changes. Each advance and retreat may follow a different geographic route and involve different refugia, as the beetle data clearly show (Coope, 1977, 1990). All of this should accumulate genetic differences between lineages occupying different parts of the species range, and a number of these derived genomes may survive in different refugia. When two of these diverged genomes expand and meet they will form a hybrid zone, which will be quite strong if a number of fitness characters are involved. In many species there will be little gene flow through such zones, so that the two parts remain distinct, and repeated cycles may accumulate this divergence until subspecies, and ultimately species, are formed.

It is worth highlighting here the possible effects of these events on the mate recognition system (MRS) (Paterson, 1985). Such range changes with small pioneer populations expanding into different environments will promote selection for changes in morphology, physiology and behaviour that may well modify parts and processes involved in mating and reproduction. Subsequent changes in the environment may add to this, and in a period of relative stasis selection may stabilize a modified MRS. If the MRS in two lineages diverges sufficiently they will become new species.

Speciation can be rapid and frequent under certain circumstances, as Hawaiian *Drosophila* amply demonstrate (Carson & Yoon, 1982). Similar evidence is found in other groups from these islands, which have been seminal in the development of the founder principle of speciation (Giddings, Kaneshiro & Anderson, 1989). Farther afield, Stanley (1979), in his book on macroevolution, describes a number of examples of divergence and speciation that occurred during the Pleistocene, particularly in fish in inland waters. There is striking evidence from plants for rapid, even explosive evolution during and after the last ice age (e.g. *Clarkia,* Lewis, 1966; *Larrea,* Wells & Hunziker, 1976; and *Atriplex,* Stutz, 1978). Many other species complexes, including the *Chorthippus* groups, are thought to have diverged during the Pleistocene.

A number of recent studies using mtDNA have provided data on the degree of divergence between populations, subspecies and species, some involving hybridization (e.g. Echelle *et al.*, 1989; Wallis & Arntzen, 1989; Riddle & Honeycutt, 1990; Vyas *et al.*, 1990; Bernatchez & Dodson, 1991; Brown & Chapman, 1991). If the rate of mtDNA divergence is taken to be in the order of 2% per million years, these results support considerable subspeciation and speciation in the Pleistocene, with large areas recolonized after the ice by the same or very similar mitochondrial genotypes. As measures of DNA sequence divergence become available over the range of a species complex, they may be put into a phylogeographic framework (Avise *et al.*, 1987; Moritz, Dowling & Brown, 1987). This should help us to clarify how the genetic and geographic processes have operated in the past to produce the species structures we now observe.

ACKNOWLEDGEMENTS

It is a pleasure to thank my several colleagues who have worked with me on hybrid zones and contributed greatly to our understanding of them. Roger Butlin kindly suggested some improvements to the manuscript. I am also grateful to the N.E.R.C. and S.E.R.C. for financial support and to Shirley Wilkins who drew the figures.

REFERENCES

ARNTZEN, J.W. & WALLIS, G.P., 1991. Resticted gene flow in a moving hybrid zone of the newts *Tritutus cristatus* and *T. marmoratus* in western France. *Evolution, 45:* 805–826.

ATKINSON, T.C., BRIFFA, K.R. & COOPE, G.R., 1987. Seasonal temperatures in Britain during the past 22 000 years, reconstructed using beetle remains. *Nature, 325:*587–592.

AVISE, J.C., ARNOLD, J., BALL, R.M., BERMINGHAM, E., LAMB, T., NEIGAL, J.E., REEB, C.A. & SAUNDERS, N.C., 1987. Intraspecifc phylogeography: the mitochondrial DNA bridge between population genetics and systematics. *Annual Review of Ecology and Systematics, 18:* 489–522.

BARRETT, S.C.H. & HUSBAND, B.C., 1990. The genetics of plant migration and colonisation. In A.H.D. Brown, M.T. Clegg, A.L. Kahler & B.S. Weir (Eds), *Plant Population Genetics, Breeding, and Genetic Resources:* 254–277. Sunderland, Massachusetts: Sinauer Associates.

BARRETT, S.C.H. & RICHARDSON, B.J., 1986. Genetic attributes of invading species. In R.H. Groves & J.J. Burden (Eds). *Ecology of Biological Invasions:* 21–33. Cambridge: Cambridge University Press.

BARTLEIN, P.J. & PRENTICE, I.C., 1989. Orbital variations, climate and paleoecology. *Trends in Ecology and Evolution, 4:* 195–199.

BARTON, N.H., 1979. The dynamics of hybrid zones. *Heredity, 43:* 341–359.

BARTON, N.A. & HEWITT, G.M., 1985. Analysis of hybrid zones. *Annual Review of Ecology and Systematics, 16:* 113–148.

BARTON, N.H. & HEWITT, G.M., 1989. Adaptation, speciation and hybrid zones. *Nature, 341:* 497–503.

BAZYKIN, A.D., 1969. Hypothetical mechanism of speciation. *Evolution, 23:* 685–687.

BENNETT, K.D., 1985. The spread of *Fagus grandifolia* across eastern North America during the last 18,000 years. *Journal of Biogeography, 12:* 147–164.

BENNETT, K.D., 1986. The rate of spread and population increase of forest trees during the postglacial. *Philosophical Transactions of the Royal Society, London, B, 314:* 523–531.

BERNATCHEZ, L. & DODSON, J.J., 1991. Phylogeographic structure in mitochondrial DNA of the Lake Whitefish (*Coregonus clupeaformis*) and its relation to Pleistocene glaciations. *Evolution 45:* 1016–1035.

BROWN, B.L. & CHAPMAN, R.W., 1991. Geneflow and mitochondrial DNA variation in the Killifish, *Fundulus heteroclitus. Evolution, 45:* 1147–1161.

BRYANT, E.H. & MEFFERT, L.M., 1988. Effect of an experimental bottleneck upon quantitative genetic variation in the housefly. *Genetics, 114:* 1191–1211.

BULL, C.M., 1991. Ecology of parapatric distributions. *Annual Review of Ecology and Systematics, 22:* 19–36.

BUTLIN, R.K., 1987. Speciation by reinforcement. *Trends in Ecology and Evolution, 2:* 8–13.

BUTLIN, R.K., 1989. Reinforcement of premating isolation. In D. Otte & J.A. Endler (Eds), *Speciation and its Consequences:* 158–179. Sunderland, Massachusetts : Sinauer Associates.

BUTLIN, R.K. & RITCHIE, M.G., 1991. Variation in female mate preference across a grasshopper hybrid zone. *Journal of Evolutionary Biology, 4:* 227–240.

CARSON, H.L. & WISOTZKEY, R.G., 1989. Increase in genetic variance following a population bottleneck. *The American Naturalist, 134:* 668–673.

CARSON, H.L. & YOON, J.S., 1982. Genetics and evolution of Hawaian *Drosophila.* In M. Ashburner, H.L. Carson & J.N. Thompson, Jr. (Eds), *The Genetics and Biology of Drosophila, volume 3b:* 297–344. London : Academic Press.

COHMAP MEMBERS, 1988. Climatic changes of the last 18,000 years : observations and model simulations. *Science, 241:* 1043–1052.

COOPE, G.R., 1977. Fossil coleopteran assemblages as sensitive indicators of climatic change during the Devension (last) cold stage. *Philosophical Transactions of the Royal Society, London, B, 280:* 313–340.

COOPE, G.R., 1990. The invasion of Northern Europe during the Pleistocene by Mediterranean species of Coleoptera. In F. di Castri, A.J. Hansen & M. De Bussche (Eds), *Biological Invasions in Europe and the Mediterranean Basin:* 203–215. Dordrecht: Kluwer.

CORTI, M., CAPANNA, E. & ESTABROOK, G.F., 1986. Microevolutionary sequences in house mouse chromosomal speciation. *Systematic Zoology, 35:* 163–175.

COYNE, J.A. & ORR, H.A., 1989. Patterns of speciation in *Drosophila. Evolution, 43:* 362–381.

CRITCHFIELD, W.B., 1984. Impact of the Pleistocene on the genetic structure of North American conifers. In R.M. Lanner (Ed.), *Proceedings 8th North American Biophysics Workshop:* 70–118. Utah State University.

CWYNAR, L.C. & MACDONALD, G.M., 1987. Geographical variation of lodgepole pine in relation to population history. *The American Naturalist, 129:* 463–469.

DENNO, R.F. & DINGLE, H., 1981. *Insect Life History Patterns: Habitat and Geographic Variation.* New York : Springer-Verlag.

DENTON, G.H. & HUGHES, T.J., 1981. *The Last Great Ice Sheets.* New York : Wiley & Sons.

DINGLE, H., 1990. The evolution of life histories. In K. Wöhrmann & S.K. Jain (Eds), *Population Biology:* 267–289. Berlin: Springer-Verlag.

DOBZHANSKY, T., 1970. *Genetics of the Evolutionary Process.* New York: Columbia University Press.

ECHELLE, A.A., DOWLING, T.E., MORITZ, C.C. & BROWN, W.M., 1989. Mitochondrial - DNA diversity and the origin of the *Menidia clarkhubbsi* complex of unisexual fishes (Atherinidae). *Evolution, 43:* 984–993.

ENDLER, J.A., 1977. *Geographic Variation, Speciation and Clines.* Princeton: Princeton University Press.

ENDLER, J.A., 1986. *Natural Selection in the Wild.* Princeton : Princeton University Press.

FALCONER, D.S., 1989. *Introduction to Quantitative Genetics.* Harlow: Longman.

GIDDINGS, L.V., KANESHIRO, K.Y. & ANDERSON, W.W., 1989. *Genetics, Speciation and the Founder Principle.* Oxford: Oxford University Press.

GOLENBERG, E.M., GIANNASI, D.E., CLEGG, M.T., SMILEY, C.J., DURBIN, M., HENDERSON, D. & ZURAWSKI, G., 1990. Chloroplast DNA sequence from a miocene *Magnolia* species. *Nature, 344:* 656–658.

GOODNIGHT, C.J., 1988. Epistasis and the effect of founder events on the additive genetic variance. *Evolution, 42:* 441–454.

GRIBBIN, J., 1989. The end of the ice ages? *New Scientist,* 17 June 1989: 48–52.

GYLLENSTEN, U. & WILSON, A.C., 1987. Interspecific mitochondrial DNA transfer and the colonisation of Scandinavia by mice. *Genetic Research, 49:* 25–29.

HARRISON, R.G. & RAND, D.M., 1989. Mosaic hybrid zones and the nature of species boundaries. In D. Otte and J.A. Endler (Eds), *Speciation and its Consequences:* 111–133. Sunderland, Massachusetts : Sinauer Associates.

HAYS, J.D., IMBRIE, J. & SHACKLETON, N.J., 1976. Variations in the Earth's orbit : pacemaker of the ice ages. *Science, 194:* 1121–1132.

HENGEVELD, R. 1988. Mechanisms of biological invasions. *Journal of Biogeography, 15:* 819–828.

HENGEVELD, R. 1989. *Dynamics of Biological Invasions.* London: Chapman & Hall.

HEWITT, G.M., 1975. A sex chromosome hybrid zone in the grasshopper *Podisma pedestris* (Orthoptera : Acrididae). *Heredity, 35:* 375–385.

HEWITT, G.M., 1979. *Animal Cytogenetics III, Orphoptera.* Stuttgart, Gebruder Borntraeger.

HEWITT, G.M., 1985. The structure and maintenance of hybrid zones – with some lessons to be learned from alpine grasshoppers. In J. Gosalvez, C. Lopez-Fernandez and C. Garcia de la Vega (Eds), *Orthoptera:* 15–54. Madrid: Fundacion Ramon Areces.

HEWITT, G.M., 1988. Hybrid zones – natural laboratories for evolutionary studies. *Trends in Ecology and Evolution, 3:* 158–167.

HEWITT, G.M., 1989. The subdivision of species by hybrid zones. In D. Otte & J. Endler (Eds), *Speciation and its Consequences:* 85–110. Sunderland, Massachusetts: Sinauer Associates.

HEWITT, G.M., 1990. Divergence and speciation as viewed from an insect hybrid zone. *Canadian Journal of Zoology, 68:* 1701–1715.

HEWITT, G.M., GOSALVEZ, J., LOPEZ-FERNANDEZ, C., RITCHIE, M.G., NICHOLS, R.A. & BUTLIN, R.K., 1988. Differences in the nucleolus organisers, sex chromosomes, and Haldane's Rule in hybrid zone. In P.E. Brandham (Ed.), *Kew Chromosome Conference III:* 109–119. Norwich: H.M.S.O.

HOFFMAN, A.A. & PARSONS, P.A., 1989. An integrated approach to environmental stress tolerance and life-history variation: desiccation tolerance in *Drosophila. Biological Journal of the Linnean Society, 37:* 117–136.

HUNTLEY, B. & BIRKS, H.J.B., 1983. *An Atlas of Past and Present Pollen Maps for Europe.* Cambridge: Cambridge University Press.

HUNTLEY, B. & WEBB, T., 1988. *Vegetation History.* Dordrecht: Kluwer.

ISTOCK, C.A., 1981. Natural selection and life history variation: theory plus lessons from a mosquito. In R.F. Denno & H. Dingle (Eds), *Insect Life History Patterns:* 113–127. New York : Springer-Verlag.

JALUT, G., 1977. Végétation et climat des Pyrénées méditeranéennes depuis quinze mille ans. *Archives d'Écologie Préhistorique No. 2,* Toulouse (2 vols).

KEY, K.H.L., 1981. Species, parapatry, and the morabine grasshoppers. *Systematic Zoology, 30:* 425–458.

LEWIS, H., 1966. Speciation in flowering plants. *Science, 152:* 167–172.

LEYDEN, B.W., 1984. Guatemalan forest synthesis after Pleistocene aridity. *Proceedings of the National Academy of Sciences, U.S.A., 81:* 4856–4859.

MACK, R.N., 1981. Invasion of *Bromus tectorum* L. into western North America : an ecological chronicle. *Agro-Ecosystems, 7:* 145–165.

MARCHANT, A.D., ARNOLD, M.L. & WILKINSON, P., 1988. Gene flow across a chromosomal tension zone. 1. Relicts of ancient hybridization. *Heredity, 61:* 321–328.

MARTINSON, D.G., PISIAS, N.G., HAYS, J.D., IMBRIE, J., MOORE, T.C. & SHACKLETON, N.J., 1987. Age dating and the obital theory of the ice ages: development of a high resolution 0–300,000 year chronostratigraphy. *Quaternary Research, 27:* 1–29.

MAYR, E., 1963. *Animal Species and Evolution.* Cambridge, Massachusetts: Harvard University Press.

MORITZ, C., DOWLING, T.E. & BROWN, W.M., 1987. Evolution of animal mitochondrial DNA : relevance for population biology and systematics. *Annual Review of Ecology and Systematics, 18:* 269–292.

NEI, M., MARUYAMA, T. & CHAKRABORTY, R., 1975. The bottleneck effect and genetic variability in populations. *Evolution, 29:* 1–10.

NICHOLS, R.A. & HEWITT, G.M., 1986. Population structure and the shape of a chromosomal cline between two races of *Podisma pedestris* (Orthoptera : Acrididae). *Biological Journal of the Linnean Society, 29:* 301–316.

NICHOLS, R.A., HUMPAGE, E.A. & HEWITT, G.M., 1990. Gene flow and the distribution of karyotypes in the alpine grasshopper *Podisma pedestris* (L.) (Orthoptera : Acididae). *Boletin de Sanidad Vegetal (Fuera de Serie), 20:* 373–379.

PÄÄBO, S., 1989. Ancient DNA : Extraction, characterization, molecular cloning and enzymatic purification. *Proceedings of the National Academy of Sciences, U.S.A., 86:* 1939–1943.

PATERSON, H.E.H., 1978. More evidence against speciation by reinforcement. *South African Journal of Science, 74:* 369–371.

PATERSON, H.E.H., 1985. The recognition concept of species. In S.Vrba (Ed.), *Species and Speciation, Transvaal Museum Monograph No. 4:* 21–29. Pretoria : Transvaal Museum.

PRANCE, G.T., 1982. *Biological Diversification in the Tropics.* New York: Columbia University Press.

REILLE, M., 1990. La Tourbiére de la Borde (Pyrénées orientales, France): Un site clé pour l'etude du Tardiqlaciare Sud-Européan. *Comptes rendus Académie des sciences, Paris, Série II, 310:* 823–829.

REMINGTON, C.L., 1968. Suture-zones of hybrid interaction between recently joined biotas *Evolutionary Biology, 2:* 321–428.

RIDDLES, B.R. & HONEYCUTT, R.L., 1990. Historical biogeography in North American arid regions: an approach using mitochondrial DNA phylogeny in grasshopper mice (Genus *Onychomys*). *Evolution, 44:* 1–15.

RITCHIE, M.G., BUTLIN, R.K. & HEWITT, G.M., 1989. Assortive mating across a hybrid zone in *Chorthippus parallelus* (Orthoptera : Acrididae). *Journal of Evolutionary Biology, 2:* 339–352.

RITCHIE, M.G., BUTLIN, R.K. & HEWITT, G.M., 1990. Mating behaviour across a hybrid zone in *Chorthippus parallelus* (Zetterstedt). *Boletin de Sanidad Vegetal (Fuera de Serie) 20:* 269–275.

SAGE, R.D. & WOLFF, J.O., 1986. Pleistocene glaciations, fluctuating ranges, and low genetic variability in a large mammal (*Ovis dalli*). *Evolution, 40:* 1092–1095.

SANDERSON, N., 1989. Can gene flow prevent reinforcement? *Evolution, 43:* 1223–1235.

SEARLE, J.B., 1988. Karyotypic variation and evolution in the common shrew, *Sorex araneus.* In P.E. Brandham (Ed.), *Kew Chromosome Conference III:* 97–107. Norwich: H.M.S.O.

SHORROCKS, B. & SWINGLAND, I., 1990. *Living in a Patchy Environment.* Oxford: Oxford University Press.

SHORROCKS, B., ROSEWELL, J., EDWARDS, K. & ATKINSON, W.D., 1984. Interspecific competition is not a major organising force in many insect communities. *Nature, 310:* 310–312.

SHOWERS, W.B., 1981. Geographic variation of the diapause response in the European Corn Borer. In R.F. Denno & H. Dingle (Eds), *Insect Life History Patterns:* 97–111. New York: Springer-Verlag.

SPAULDING, W.G., 1990. Vegetation dynamics during the last glaciation, south eastern Great Basin, U.S.A. *Quaternary Research, 33:* 188–203.

SPENCER, H.G., McARDLE, B.H. & LAMBERT, D.M., 1986. A theoretical investigation of speciation by reinforcement. *American Naturalist, 128:* 241–261.

STANLEY, S.M., 1979. *Macroevolution, Pattern and Process.* San Francisco: W. H. Freeman.

STUTZ, H.C., 1978. Explosive evolution of perennial *Atriplex* in Western America. *Great Britain Naturalist Memoirs, 2:* 161–168.

SZYMURA, J.M. & BARTON, N.H., 1991. The genetic structure of the hybrid zone between the fire-bellied toads *Bombina bombina* and *B. variegata* : comparisons between transects and between loci. *Evolution, 45:* 237–261.

TEMPLETON, A.R., 1981. Mechanisms of speciation: a population genetics approach. *Annual Review of Ecology and Systematics, 12:* 23–48.

THOMPSON, R.S., 1988. Western North America. In B. Huntley & T. Webb (Eds), *Vegetation History:* 415–458. Dordrecht: Kluwer.

VIRDEE, S.R. & HEWITT, G.M., 1990. Ecological components of a hybrid zone in the grasshopper *Chorthippus parallelus* (Zetterstedt) (Orthoptea: Acrididae). *Boletin de Sanidad Vegetal (Fuera de serie), 20:* 299–309.

VYAS, D.K., MORITZ, C., PECCININI-SEALE, D., WRIGHT, J.W. & BROWN, W.M. 1990. The evolutionary history of parthenogenetic *Cnemidophorus lemniscatus* (Sauria: Teiidae). II. Maternal origin and age inferred from mitochondrial DNA analysis. *Evolution, 44:* 922–932.

WALLACE, A.R., 1889. *Darwinism.* London: Macmillan.

WALLIS, G.P. & ARNTZEN, J.W., 1989. Mitochondrial-DNA variation in the crested newt superspecies: limited cytoplasmic geneflow among species. *Evolution, 43:* 88–104.

WEBB, T., 1986. Is vegetation in equilibrium with climate? How to interpret late-quaternary pollen data. *Vegetatio, 67:* 75–91.

WELLS, P.V. & HUNZIKER, J.H., 1976. Origin of the creosote bush (*Larrea*) deserts of Southwestern North America. *Annals of the Missouri Botanic Garden, 63:* 843–861.

WEST, R.G., 1980. Pleistocene forest history in East Anglia. *New Phytologist, 85:* 571–622.

WHITE, M.J.D., 1978. *Modes of Speciation.* San Francisco: W.H. Freeman.

WHITE, M.J.D., BLACKITH, R.E., BLACKITH, R.M. & CHENEY, J., 1967. Cytogenetics of the *viatica* group of morabine grasshoppers. I. The "coastal" species. *Australian Journal of Zoology, 15:* 263–302.

CHAPTER

8

Holocene mollusc and ostracod sequences: their potential for examining short-timescale evolution

J.G. EVANS & H.I. GRIFFITHS

Introduction **126**

Conventional use of molluscs and ostracods in Holocene palaeoecology **126**

Conditions necessary for Holocene microevolutionary studies **128**

Time: scale and resolution 128

Contexts: the importance of population data 129

Suitable groups 129

Some examples of infraspecific change **130**

The proposal **131**

Acknowledgements **133**

References **134**

Keywords: Microevolution – Quaternary period – terrestrial – freshwater – Mollusca – Ostracoda – lake sediments – timescales – ancient DNA.

Abstract

Evolution is usually studied at two timescales. Modern studies benefit from detailed temporal resolution but being of only a few decades or, in exceptional cases, centuries, lack a long-term overview. Geological studies on the other hand, although they have the long-term overview, lack short-term temporal detail. Quaternary deposits offer a suitable intermediate scale, in which there are continuous sequences covering centuries or millennia. Mollusca and Ostracoda are probably the best potential taxa because they occur in sufficient abundance to allow former populations to be identified. Lake sediments are probably the best contexts because they are often deposited gently, sometimes in recognizable annual layers, and sometimes remain

Evolutionary Patterns and Processes
ISBN 0-12-440895-8

undisturbed. Also they are often organic, with the shells of Mollusca and Ostracoda preserving a protein component which allows the potential for molecular studies.

INTRODUCTION

In this paper we set out a proposal for using land and freshwater molluscs and ostracods from Holocene deposits in microevolutionary studies. The two groups have unique benefits in this area of study which are not afforded by any other, particularly in the continental (as opposed to the oceanic) arena.

By microevolutionary, we mean infraspecific directional change in shell morphology with time. We do not specify whether this is due to direct environmental effects on the phenotype (ecophenotypic) or whether this is a genomic component brought about by selection. Nor are we particularly concerned as to whether such changes are adaptive. In the first instance we are solely concerned with tracking change.

CONVENTIONAL USE OF MOLLUSCS AND OSTRACODS IN HOLOCENE PALAEOECOLOGY

Land and freshwater Mollusca and freshwater Ostracoda from archaeological and related deposits have generally been used to interpret former environments (Evans, 1972; Bradbury *et al.*, 1990). This is usually the main goal.

Characterization of sub-fossil assemblages is almost always done on a morphologically based concept of species. The species composition of the assemblage and numbers of individuals within each species form the basis for analysis and data presentation, whether it be by presence/absence, by some estimate of quasi-absolute abundance, or by abundance relative to total numbers. More sophisticated numerical methods such as clustering, ordination or diversity indices also use species-numbers data, even though in some cases (e.g. diversity indices) particular species identities are irrelevant (Magurran, 1988).

Species-numbers data from a single sample are referred to as an 'assemblage', although for a restricted taxonomic group the term 'taxocene' is more appropriate. This term emphasizes the asymmetry of much palaeontological data in that, because they come from particular types of preservational contexts and because human specialists tend to concentrate on particular taxonomic groups, the data cannot be considered as a reflection of former complete communities.

Environmental interpretation is generally based on present-day analogy through the ecologies and distributions of the species themselves (Evans, 1991). Although in a few species there is evidence for changes in ecology or at least habitat preferences, for example, the Holocene colonization of interstitial habitats by the ostracod *Nannocandona faba* Ekman (Marmonier & Danielopol, 1988), usually we are on fairly safe ground in supposing that morphological identity between modern and sub-fossil individuals implies species identicality, or at least some degree of genomic continuity. In one example of an apparent clash, the disparate palaeo- and neo-ecological biotype associations of the ostracod *Scottia browniana* (Jones), further research led to the discovery that a genuine species pair was involved, one of which became extinct during the last glacial (Kempf, 1971). However, as the techniques of modern systematics increase in resolution and sensitivity, there may be an increasing separation between neontological studies and the morphological concept of species delimitation used in Quaternary palaeobiology. These differences reflect the material available for study. Holocene material is essentially a record of hard-part anatomies, whilst the chaetotaxic and genitalia-based taxonomies of the zoomorphological taxonomist remain occult (Fig. 1). Nevertheless, confusion can be avoided by the description of both hard- and soft-part anatomies during revisory work.

An important feature of palaeoecology, whether it involves molluscs, ostracods, pollen or Coleoptera, is that assemblages are generally interpreted not individually but in the context of a temporal sequence. Individual data sets are related to progression through time. As a consequence, it is often the case that species that are temporally patchy are of the greatest interest and value. Introductions, dyings outs, prominent peaks, troughs and steps are of far more use in Quaternary palaeoecology than continuous presence.

Admittedly there has been some diversification of both aims and methods. Early work in pollen and molluscan analysis was particularly concerned with the climatic basis of Quaternary zonation, especially prior to radiocarbon dating. Our own work, in contrast, is concerned with changes at the local, site or sampling column scale, a consequence of the fact that most localities and contexts of archaeological interest are at habitat boundaries, for example, fields, trackways, settlements or points on a hydrosere. Other recent studies have concentrated on a broader, biogeographical scale. Sequences of faunal change, usually of introductions, but occasionally of regional or (in rare cases) country-wide dyings outs, have allowed a return to the sorts of zonation schemes proposed in the early years of this century but on a more precise basis (Absolon, 1973; Kerney, 1977).

But whatever the approach, three unifying themes underlie this work: (1) the application of an essentially pragmatic and strongly morphologically based

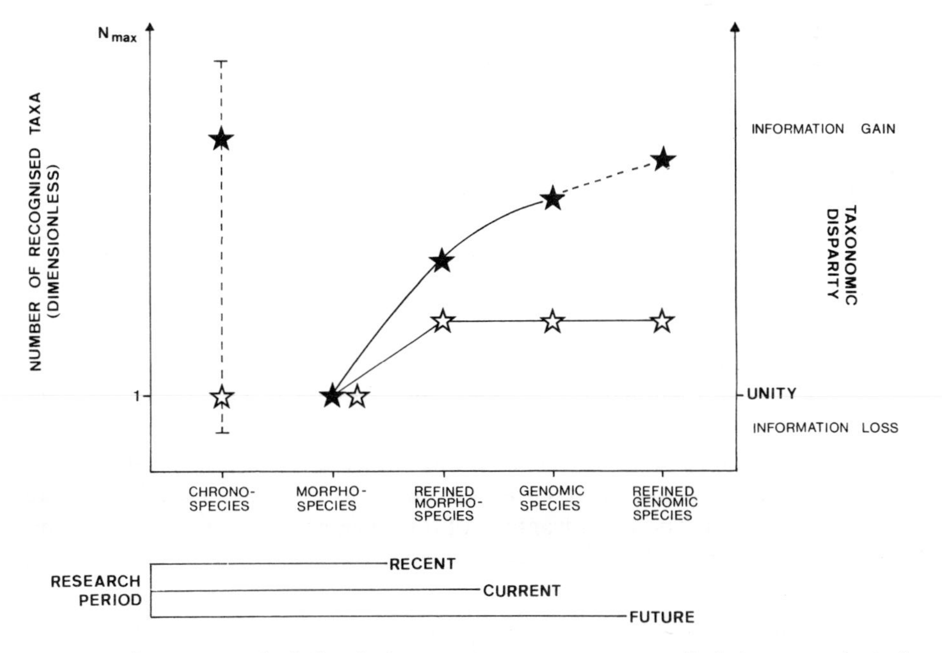

Figure 1 Development and drift of the species concept as applied in neontological and palaeoecological research. Open stars, palaeontology; closed stars, neontology.

species concept, (2) the use of data to assist in environmental interpretation and palaeobiogeographical studies, and (3) the use of stenotopic 'indicator' species, that is, taxa that are stenotopic in local ecology, broader temporal/regional distributions, or both.

By contrast, across the board of the different species (i.e. looking at all molluscs or all ostracods in an assemblage irrespective of species), there have been no studies on Quaternary material of features such as size, form, biomass, colour, hairyness or sculpturing even though the environmental and adaptive significances of many of these characters are known. Different growth stages of the same species are often different in many of these characters and this reflects different adaptations during development, yet there have been practically no attempts to use these in environmental interpretation, the work of Gordon & Ellis (1985) on *Trichia hispida* (L.) being an exception. Equally there have been no detailed studies on intraspecific variation of individual species through sequences (see p. 130).

These points have been made in order to demonstrate how research can become polarized along certain avenues whilst others of potential value for different fields remain unexplored.

CONDITIONS NECESSARY FOR HOLOCENE MICROEVOLUTIONARY STUDIES

Time: scale and resolution

There are two main timescales by which evolution is studied. In the life sciences, short-timescale change or stability can be observed over a few decades (Cain, 1983), assuming that research continuity is maintained, although documentary evidence of domestication can push this back a few centuries (Clutton-Brock, 1987). Observations of different forms of *Lymnaea peregra* (Müller) in lakes in south-west Ireland going back to 1834 provide one of the longest records for Mollusca (Boycott, Oldham & Waterston, 1932; Evans, 1982). Cain (1983) describes an indirect approach, operating over 50–1000 years in which morph frequencies of *Cepaea nemoralis* (L.) are related to historically documented land-use changes. At the other extreme is the timescale of the solid geological record of several or tens of millions of years, although this is being narrowed down in certain favourable (although generally oceanic) cases (see Chapter 2). Thus in modern biological studies temporal resolution is fine but the resulting record is short; in traditional palaeontology a long overview is provided, but temporal resolution is coarse.

The Quaternary offers a different scenario again. Data from individual horizons have, of course, been known for a long time, and have allowed us to track, for example, the morphological changes of animal and plant domestication during the Holocene. But although on a shorter timescale than the solid geological record, these data are still disjunct in that each set is separated by several hundreds or thousands of years from the next. The big advantage of the Quaternary is that it offers continuity and this is important in that it allows a detailed record of the nature and rate of change. Useful thicknesses of deposits span hundreds to tens of thousands of years, and are sometimes laid down in recognizable annual increments as in the varved or laminated sediments of some European and African lakes (Cohen, 1984; Schmidt & Simola, 1991) or the loess deposits of central Europe (Ložek, 1972).

Contexts: the importance of population data

In studying microevolutionary change from geological sequences we need data that are as close as possible to the structure and density of former populations. Reasons for this can be summarized as follows:

1. To assess the range of variation within individual species, both to obtain the norm of the characters under consideration and to allow accurate delimitation of species boundaries.
2. To estimate the age composition of a population so that ontogenetic and allometric changes can be assessed. Knowledge of reproductive strategies is also important in assessing environmental stress (Calow, 1981, 1983). For example, Holopainen & Jónasson (1989) have shown how *Pisidium nitidum* Jenyns in Lake Esrom, Denmark, produces two broods in summer at 2 m depth where conditions are warm, but only one at 11 m where they are cooler.
3. To estimate population density so that the relationship between it and body size can be assessed, cf. *Cepaea hortensis* (Müller) (Perry & Arthur, 1991).
4. To obtain data which are as close as possible to a panmictic population in terms of genetic composition.

Assemblages derived from geological deposits have been subjected to a variety of post-mortem and depositional processes, sorting and preservational bias (Danielopol, Casale & Olteanu, 1986; Weigelt, 1989; Briggs, Gilbertson & Harris, 1990; Allison, 1991). Such taphonomic processes arguably continue until data presentation, with only certain aspects of the data being selected for interpretation, as we have already shown. The types of deposits that are most suitable, therefore, are those which are laid down so gently that the biota are not disrupted, only sealed, and remain undisturbed. The main types are lacustrine sediments, fine riverine overbank flood deposits, loess and peat.

Suitable groups

In addition we need to consider the organisms themselves: what groups are most suitable? Only a few offer the possibility for sampling in sufficient abundance to obtain the necessary population data while in a sufficiently small lateral area to obtain environmental, populational and genetic uniformity. This immediately disqualifies a number of groups. Mammals (and vertebrates generally) are usually impossible to study from localized contexts because the density of remains is too low and the record too coarse. Large samples cover too large a spatial and/or temporal domain, including individuals mostly not in their places of life, and derived from a variety of local biotypes. Concentrations of vertebrate remains almost invariably derive from natural traps (e.g. Sutcliffe, 1985) or predator-mediated assemblages (e.g. Boylan, 1981; Andrews, 1990). The record of macroscopic plant remains – seeds, leaves, tree stumps – although sporadically autochthonous does not encompass long continuous sequences, with the exception of autogenic plant remains in peat bogs. Insects (particularly Coleoptera) have an extensive record throughout the Quaternary, are present in the numbers and contexts required (Coope, 1987), but show little or no evidence of microevolutionary change (Coope, 1987; Buckland & Coope, 1991). A possible exception is the case of chironomid head capsules from lake deposits.

However, these (as well as diatom and cladoceran remains) undergo spatial averaging during deposition (Frey, 1988). Pollen analysis, the most widespread and successful of the biological tools used in Quaternary studies, is similarly unhelpful in evolutionary studies. Pollen is often impossible to identify to species level, and even then its context is always derived, reflecting the fall-out from a wide geographical catchment area (Lowe & Walker, 1984).

The only two groups that really qualify are the Mollusca and the Ostracoda. In both terrestrial and freshwater contexts they are ideal for the study of microevolutionary change. Their taxonomy is reasonably well-known (at least in the western Palaearctic) as are their distributions. In Quaternary contexts shells of both groups are well-preserved and relatively easy to identify to species. They also occur in sufficient numbers for sample sizes to allow the study of intraspecific and interspecific variation in shell morphology. Different life stages are also recognizable. Population structure (age-abundance class data) not only allows the assessment of taphonomic grain (closeness to life-assemblage) but also the reproductive strategy. In the case of ostracods, many hundreds of shells may be recovered from a 5-cm-thick section of a 7.5-cm diameter lake core, which include both the sexually dimorphic shells of adults and a range of the (usually) eight sub-adult moult stages (De Deckker, 1988). In addition, as many benthic species are semi-infaunal, sub-fossil assemblages correlate well with the living equivalents from which they derive (Löffler, 1969). Overall, both groups show a wide range of intraspecific variation and ecophenotypic lability (e.g. Goodfriend, 1986). For lakes, apart from those few whose origin goes back millions of years (Boss, 1978), one of the main genetic properties of pulmonate Mollusca is that selection has produced genotypes which can show phenotypic flexibility (Russell-Hunter, 1978). There is thus potential for identifying shell morphometric changes alongside changes in reproductive strategy, and against a background of independently derived environmental change.

At this stage it is only fair to point out that no one has yet isolated anything approaching a single population from Quaternary sediments. This is partly because of time-averaging and the unlikelihood of ever isolating a single annual event because of the thickness of the deposit in which such an event is preserved (maximum *c.* 2 mm, and this is exceptional). But it also reflects a lack of research effort, which has instead concentrated upon assemblages of species and environmental interpretation.

SOME EXAMPLES OF INFRASPECIFIC CHANGE

A few examples of infraspecific change within the Mollusca and Ostracoda further illustrate the potential of the two groups. Kerney (1963) noted that Late-glacial specimens of *Pupilla muscorum* (L.) were larger and more parallel-sided than those from the Holocene. Currey & Cain (1968) on banding morphs in *Cepaea nemoralis* noted that there was "a significant decrease of unbandeds from pre-Iron Age samples to the corresponding ones for the present day" in southern England, an observation that has been repeated by Preece (1980a). Cain (1971) on colour in *C. nemoralis*, showed browns to have increased greatly in frequency from Neolithic to later prehistoric times in an area of north Wiltshire; changes in frequencies of apex colour and of white lip were also discussed. Thomas (1978) discussed shell form in *C.*

nemoralis and, in particular, differences in relative growth rates between populations; this study, although entirely on modern material, was undertaken with the analysis of samples from archaeological contexts in mind. Preece (1980a) showed a decrease in height and breadth of the snail *Pomatias elegans* (Müller) from the mid-Holocene to the present day at a tufa site in Dorset, and Burleigh & Kerney (1982) showed the same from the Neolithic in Kent. Preece (1980b) showed an increase in height and breadth and a change in banding frequency of *Cernuella virgata* (da Costa) in the last 2000 years on the Isle of Wight. For the Pleistocene there is an interesting study of the freshwater species *Gyraulus laevis* (Alder) from Upton Warren (Coope, Shotton & Strachan, 1961) in which in a comparison of three assemblages, the shells in one were stunted and dwarfed, perhaps in a response to increased salinity.

The best molluscan species for these sorts of exercises are those that are present in quantity throughout entire sequences and which are phenotypically labile, for example *Trichia hispida, Anisus leucostoma* (Millet) and *Lymnaea peregra*.

The same is true of ostracod genera such as *Candona, Cyclocypris* and possibly *Cypridopsis. Candona neglecta* Sars is a particularly widespread and variable taxon, in which various forms have been accorded specific status by both neo- and palaeontologists (e.g. Petkovski, 1969; Caraion, 1973; Scharf, 1983; Krstic, 1988). Some of these 'species' are said to be linked to specific biotypes. There has been little work on microevolutionary change in freshwater Ostracoda, although vicariant divergence in more conservative marine lineages has been used to great advantage in some studies (e.g. Cronin, 1987; Cronin & Ikeya, 1990). In marine contexts, Cronin & Ikeya (1990) have shown that *K*-selected taxa would be expected to speciate more readily in response to tectonic and vicariant events than *r*-selected species. Just as marine faunas are predominantly composed of such *K*-selected species, the same is true of those inhabiting large and ancient lakes. It is certainly the case that such water bodies often host highly endemic faunas, e.g. Lake Baikal (Mazepova, 1990), Lake Ohrid (Petkovski, 1960, 1969) and Lake Tanganyika (Wouters, 1978; Martens, 1985; Wouters, Martens & De Deckker, 1989). In geologically permanent environments there appears to be some convergence of the types of selection pressures operating upon taxa, including a shift from physical to biological factors (Forester, 1991). Unlike many other groups within the Quaternary, for example, Coleoptera (Buckland & Coope, 1991), ostracods are evolutionarily dynamic with many extinct specific and generic taxa known from interglacial deposits (e.g. Diebel & Pietrzeniuk, 1975; Fuhrmann & Pietrzeniuk, 1990). Various modern species have also been shown to exhibit high levels of allozymic variation (Chaplin & Ayre, 1989; Havel, Hebert & Delorme, 1990a,b; Rossi & Menozzi, 1990). In addition, vicariant speciation has been reported from within the Holocene in the lakes of East Africa (Martens, 1990a,b).

THE PROPOSAL

Assemblages of land and freshwater Mollusca and freshwater Ostracoda can be extracted from certain Quaternary sedimentary contexts and these may closely reflect once-living populations. Within single species, intraspecific variation (of both adults and juveniles) and age structures can be identified. In particularly favourable circumstances, such parameters may be traced continuously in a single species for several hundreds or thousands of years from a particular location through the examination of loess and lake deposits.

By analysing various chemical and physical properties of the deposits, one can obtain an independent assessment of the amount of environmental change or stability as the deposits accumulated (Engstrom & Wright, 1984; Oldfield & Appleby, 1984). One can also do this from other biological material, not under direct study, particularly by the conventional palaeoenvironmental analysis of biological assemblages.

Dating, either by ^{14}C or, far better, by varve counting, allows rates to be added, so that an assessment of timescale can be obtained.

There is the potential to take a single species and examine its situation at different geographical scales, within local or site-specific populations, between local populations within a given geographical region, and by comparison between geographical regions at the 'race' or 'subspecific' level. "The relative magnitudes of genetic and environmental contributions to body size (or other characters) may well differ among these different levels" (Perry & Arthur, 1991).

Overall, there are possibilities of examining change in various aspects of a species through time and of judging to what extent they are related to environmental change.

An assessment of the relative contributions of ecophenotypic response and genuine genomic change is a major aim. How can this be achieved? After all, the problems are considerable, even with modern material. For example, varying claims have been made as to the relative importance of environment and genetics in forms of the aquatic snails *Lymnaea peregra* and *Lymnaea stagnalis* (L.) (Calow, 1981; Arthur, 1982; Lam & Calow, 1988), and the amount of environmental vs. genetic contribution to the striking forms of lake *Lymnaea peregra* from south-west Ireland, known since the last century, is still unresolved (Boycott, 1938). One approach is to select a situation where there is minimal gene flow between demes, e.g. non-dispersive benthic taxa with non-planktonic larvae. In lakes, isolation must follow dispersal from refugia along the edges of ice sheets or from ice-dammed, or proglacial waters (Segerstråle, 1957). This may lead to the rapid isolation of populations within the features of the new postglacial landscape. For benthic taxa, isolation may be compounded by a variety of interacting biotic and abiotic factors, not the least of which are the heterogeneity of the lacustrine benthos and site-specific peculiarities of mixis, hydrology and hydrochemistry. Although some studies of sub fossil material suggest minimal evolutionary change within the Pleistocene (e.g. Coope, 1987) the phenomenon of environmental tracking makes this difficult to reconcile with the conclusions derived from studies of the evolutionary genetics of modern species (e.g. Chapter 7). If Paterson (1985) is correct, then tight linkage between a species and its environment should be expected.

Another, and preferably complementary, approach would be the comparison of amplified DNA sequences through the use of the polymerase chain reaction (PCR). This approach is to be preferred to the use of allozyme studies as, particularly in the case of ostracods, it has proved difficult to obtain allozyme results from single individuals (Sywula *et al.*, 1985; Havel *et al.*, 1990a,b). Lake sediments, especially organic calcareous sediments such as marls, favour DNA survival (e.g. Golenburg *et al.*, 1990). Soft-part tissues are rarely preserved, but on occasions where the valves of the ostracod shell remain articulated and closed (common during periods of rapid sediment deposition) (Danielopol *et al.*, 1986) the survival of exoskeletal elements such as limb segments and the supporting framework of the genitalia is not uncommon

(Griffiths & Evans, unpublished). In Mollusca, it is not unusual to see the good preservation of the periostracum, the proteinaceous coating that covers the shell in life, in shells from organic, anaerobic deposits. It is certainly possible to extract DNA from modern ostracods, opening up at least the possibility of being able to compare mtDNA or rDNA sequences from modern and ancient material. The advantages of this are obvious; not only could genomic (as opposed to ecophenotypic) changes within lineages of a particular taxon be followed over thousands of years, but also accurate delimitation of species boundaries could be made in a time-transgressive record. One could also distinguish between invasive populations and *in situ* genetic change.

For these approaches, both palaeofaunal and molecular, lake deposits are the most suitable resource. Over northern Europe, from western Ireland across to northern Germany, and the North European Plain adjacent to the Baltic Sea and the Gulf of Finland, there are tens of thousands of lakes and infilled lake basins of varying sizes and in different phases of development. Many of these have long sequences of organic calcareous deposits in which both Mollusca and Ostracoda are abundant. These areas have been glaciated and the sequences begin at about 14 000 BP. Further south in Europe are areas outside the glacial region which have even longer sequences. The deposits are generally suitable for the preservation of organic material, which itself may be amenable to molecular study. Many lakes have been cored for pollen studies. ^{14}C and varve count data are often available (Usinger, 1975). Some species of mollusc such as *Lymnaea peregra, Valvata piscinalis* (Müller), *Gyraulus albus* (Müller), *Pisidium casertanum* (Poli) and *P. nitidum* are present in long sequences and in considerable abundance (e.g. Mouthon, 1987; Clerc, Magny & Mouthon, 1989; Magny & Mouthon, 1990). This is also true for ostracods such as *Candona, Ilyocypris*, and *Limnocythere* (Kempf & Scharf, 1980; Gunter, 1986; Evans & Griffiths, unpublished data). Furthermore, temporal resolution may be sufficiently good to provide information from annual deposits (e.g. Schmidt & Simola, 1991). At Le Grand Lemps, Bas Dauphiné, an infilled glacial lake, there is 10 m of deposit covering 5000 years, indicating a sedimentation rate of 2 mm year^{-1} (Clerc *et al.,* 1989). The potential of lake pulmonates for microevolutionary studies has also been discussed from a neontological angle by Russell-Hunter (1978), and although he is concerned with genetically controlled differences in bioenergetic partitioning which are impossible to detect directly in the fossil record, a proxy signal may be found in variations in size-frequency data, adult size, and in other morphometric data.

We therefore propose that Holocene material, and especially lake deposits, can provide data that could bridge the gap between the 'patterns in the fossil record' on the one hand, and the 'processes' of evolution on the other. We appreciate that we have presented no original research data of our own in this paper. What we are trying to convey is a viewpoint – a mind-set that allows interdisciplinary collaboration between workers in zoology, geology, environmental archaeology and genetics to tackle what remains the essential problem of biology: 'how are species generated?'.

ACKNOWLEDGEMENTS

We would like to express our thanks to the Science and Engineering Research Council for their financial support and to R.K. Butlin for his many helpful discussions.

REFERENCES

ABSOLON, A., 1973. Ostracoden aus einigen Profilen spät-und postglazialer Karbonatablagerungen in Mitteleuropa. *Mitteilungen der Bayerischen Staatssammlung für Paläontologie und historische Geologie, 13:* 47–94.

ALLISON, P.A., 1991. Variation in rates of decay and disarticulation of Echinodermata: implications for the application of actualistic data. *Palaios, 5:* 432–440.

ANDREWS, P., 1990. *Owls, Caves and Fossils.* London: British Museum (Natural History).

ARTHUR, W., 1982. Control of shell shape in *Lymnaea stagnalis. Heredity, 49:* 153–161.

BOSS, K.J., 1978. On the evolution of gastropods in ancient lakes. In V. Fretter & J. Peake (Eds), *Pulmonates, volume 2A, Systematics, Evolution and Ecology:* 385–428. London: Academic Press.

BOYCOTT, A.E., 1938. Experiments on the artificial breeding of *Limnaea involuta, Limnaea burnetti* and the other forms of *Limnaea peregra. Proceedings of the Malacological Society, 23:* 101–108.

BOYCOTT, A.E., OLDHAM, C. & WATERSTON, A.R., 1932. Notes on the lake *Lymnaea* of south-west Ireland. *Proceedings of the Malacological Society, 2:* 105–127.

BOYLAN, P.J., 1981. A new revision of the Pleistocene mammalian fauna of Kirkdale Cave, Yorkshire. *Proceedings of the Yorkshire Geological Society, 43:* 253–280.

BRADBURY, J.P., FORESTER, R.M., BRYANT, W.A. & COVICH, A.P., 1990. Paleolimnology of Laguna de Cocos, Albion Island, Rio Hondo, Belize. In M.D. Pohl (Ed.), *Ancient Maya Wetland Agriculture. Excavations on Albion Island, Northern Belize:* 119–154. Boulder: Westview Press.

BRIGGS, D.J., GILBERTSON, D.D. & HARRIS, A.L., 1990. Molluscan taphonomy in a braided river system and its implications for studies of Quaternary cold-stage river deposits. *Journal of Biogeography, 17:* 623–637.

BUCKLAND, P.C. & COOPE, G.R., 1991. *A Bibliography and Literature Review of Quaternary Entomology.* Sheffield: J.R. Collis Publications, University of Sheffield (Department of Archaeology and Prehistory).

BURLEIGH, R. & KERNEY, M.P., 1982. Some chronological implications of a fossil molluscan assemblage from a Neolithic site at Brook, Kent, England. *Journal of Archaeological Science, 9:* 29–38.

CAIN, A.J., 1971. Colour and banding morphs in subfossil samples of the snail *Cepaea.* In E.R. Creed (Ed.), *Ecological Genetics and Evolution:* 65–92. Oxford: Blackwell Scientific Publications.

CAIN, A.J., 1983. Ecology and ecogenetics of terrestrial molluscan populations. In W.D. Russell-Hunter (Ed.), *The Mollusca, volume 6, Ecology:* 597–647. London: Academic Press.

CALOW, P., 1981. Adaptive aspects of growth and reproduction in *Lymnaea peregra* (Gastropoda Pulmonata) from exposed and sheltered habitats. *Malacologia, 21:* 5–13.

CALOW, P., 1983. Life-cycle patterns and evolution. In W.D. Russell-Hunter (Ed.), *The Mollusca, volume 6, Ecology:* 649–678. London: Academic Press.

CARAION, F.E., 1973. Date noi privitoari la fauna de Candonine (Ostracoda-Cyprididae) din România. *Studii si cercetari de Biologie, Seria Zoologie, 25:* 17–23.

CHAPLIN, J.E. & AYRE, D.J., 1989. Genetic evidence of variation in the contributions of sexual and asexual reproduction to populations of the freshwater ostracod *Candonocypris novaezelandiae. Freshwater Biology, 22:* 275–284.

CLERC, J., MAGNY, M. & MOUTHON, J., 1989. Histoire d'un milieu lacustre du Bas-Dauphiné: Le Grand Lemps. Étude palynologique des remplissages Tardiglaciaires et Holocènes, et mise en évidence de fluctuations lacustres à l'aide d'analyses sédimentologiques et malacologiques. *Revue de Paléobiologie, 8:* 1–19.

CLUTTON-BROCK, J., 1987. *A Natural History of Domesticated Mammals.* London: British Museum (Natural History).

COHEN, A.S., 1984. Effect of zoobenthic standing crop on laminae preservation in tropical lake sediment, Lake Turkana, East Africa. *Journal of Paleontology, 58:* 499–510.

COOPE, G.R., 1987. The response of late Quaternary insect communities to sudden climatic change. In J.H.R. Gee & P.S. Giller (Eds), *Organisation of Communities. Past and Present. (27th Symposium of the British Ecological Society):* 421–438. Oxford: Blackwell Scientific Publications.

COOPE, G.R., SHOTTON, F.W. & STRACHAN, I., 1961. A late Pleistocene fauna and flora from Upton Warren, Worcestershire. *Philosophical Transactions of the Royal Society of London, B, 244:* 379–421.

CRONIN, T.M., 1987. Evolution, biogeography and systematics of *Puriana:* evolution and speciation in Ostracoda. *Journal of Paleontology (Memoir), 21:* 1–71.
CRONIN, T.M. & IKEYA, N., 1990. Tectonic events and climatic change: opportunities for speciation events in Cenozoic marine Ostracoda. In R.M. Ross & W.D. Allmon (Eds), *Causes of Evolution. A Palaeontological Perspective:* 210–248. Chicago: University of Chicago Press.
CURREY, J.D. & CAIN, A.J., 1968. Studies on *Cepaea.* IV. Climate and selection of banding morphs in *Cepaea* from the climatic optimum to the present day. *Philosophical Transactions of the Royal Society of London, B, 253:* 483–498.
DANIELOPOL, D.L., CASALE, L.M. & OLTEANU, R., 1986. On the preservation of carapaces of some limnic ostracodes; an exercise in actuopalaeontology. *Hydrobiologia, 143:* 143–157.
DE DECKKER, P., 1988. An account of the techniques using ostracodes in palaeolimnology in Australia. *Palaeogeography, Palaeoclimatology, Palaeoecology, 62:* 436–475.
DIEBEL, K. & PIETRZENIUK, E., 1975. Neue Ostracoden aus dem Pleistozän von Burgtonna (Bezirk Erfurt). *Tonnacypris* gen. n. und *Amplocypris tonnensis* sp. n. *Zeitschrift für Geologische Wissenschaften, 3:* 87–97.
ENGSTROM, D.R. & WRIGHT, H.E., 1984. Chemical stratigraphy of lake sediments as a record of environmental change. In E.Y. Haworth & J.W.G. Lund (Eds), *Lake Sediments and Environmental Change*: 11–67. Leicester: Leicester University Press.
EVANS, J.G., 1972. *Land Snails in Archaeology.* London: Seminar Press.
EVANS, J.G., 1991. An approach to the interpretation of dry-ground and wet-ground molluscan taxocenes from central-southern England. In D.R. Harris & K.D. Thomas (Eds), *Modelling Ecological Change:* 75–89. London: Institute of Archaeology.
EVANS, N.J., 1982. A note on collections of *Lymnaea peregra* (Müller) made in south-west Ireland. *The Irish Naturalists' Journal, 20:* 385–388.
FORESTER, R.M., 1991. Pliocene-climate history of the western United States derived from lacustrine ostracodes. *Quaternary Science Reviews, 10:* 133–146.
FREY, D.G., 1988. Littoral and offshore communities of diatoms, cladocerans and dipterous larvae, and their interpretation in paleolimnology. *Journal of Paleolimnology, 1:* 171–191.
FUHRMANN, R. & PIETRZENIUK, E., 1990. Die Ostrakoden Fauna des Interglazials von Grabschütz (Kreis Delitzsch). *Altenburger Naturwissenschaftliche Forschungen, 5:* 202–227.
GOLENBURG, E.M., GIANNASI, D.E., CLEGG, M.T., SMILEY, C.J., DURBIN, M., HENDERSON, D., & ZURAWSKI, G., 1990. Chloroplast DNA sequence from a Miocene *Magnolia* species. *Nature, 344:* 656–658.
GOODFRIEND, G.A., 1986. Variation in land-snail shell form and size and its causes: a review. *Systematic Zoology, 35:* 204–223.
GORDON, D. & ELLIS, C., 1985. Species composition parameters and life tables: their application to detect environmental change in fossil land molluscan assemblages. In N.R.J. Fieller, D.D. Gilbertson & N.G.A. Ralph (Eds), *Palaeoenvironmental Investigations: Research Design, Methods and Data Analysis:* 153–164. Oxford: British Archaeological Reports.
GUNTER, J., 1986. Ostracod fauna of Duvensee, an ancient lake in northern Germany. *Hydrobiologia, 143:* 411–416.
HAVEL, J.E., HEBERT, P.D.N. & DELORME, L.D., 1990a. Genetics of sexual Ostracoda from a low arctic site. *Journal of Evolutionary Biology, 3:* 65–84.
HAVEL, J.E., HEBERT, P.D.N. & DELORME, L.D., 1990b. Genotypic diversity of asexual Ostracoda from a low arctic site. *Journal of Evolutionary Biology, 3:* 391–410.
HOLOPAINEN, I.J. & JÓNASSON, P.M., 1989. Reproduction of *Pisidium* (Bivalvia, Sphaeriidae) at different depths in Lake Esrom, Denmark. *Archiv für Hydrobiologie, 116:* 85–95.
KEMPF, E.K., 1971. Ökologie, Taxonomie und Verbreitung der nichtmarinen Ostrakoden – Gattung *Scottia* im Quatär von Europa. *Eiszeitalter und Gegenwart, 22:* 43–63.
KEMPF, E.K. & SCHARF, B.W., 1980. Lebende und fossile Muschelkrebse (Crustacea: Ostracoda) vom Laacher See. *Mitteilungen Pollichia des Pfaelzischen Akademie der Wissenschaften, 60:* 205–236.
KERNEY, M.P., 1963. Late-glacial deposits on the Chalk of south-east England. *Philosophical Transactions of the Royal Society of London, B, 246:* 203–254.
KERNEY, M.P., 1977. A proposed zonation scheme for Late-glacial and postglacial deposits using land Mollusca. *Journal of Archaeological Science, 4:* 387–390.

KRSTIC, N., 1988. Some Quaternary ostracods of the Pannonian Basin, with a review of a few *neglectoida*. In T. Hanai, N. Ikeya & K. Ishizaki (Eds), *Evolutionary Biology of Ostracoda:* 1063–1072. Tokyo: Kodansha Ltd.

LAM, P.K.S. & CALOW, P., 1988. Differences in the shell shape of *Lymnaea peregra* (Müller) (Gastropoda: Pulmonata) from lotic and lentic habitats; environmental or genetic variance? *Journal of Molluscan Studies, 54:* 197–207.

LÖFFLER, H., 1969. Recent and subfossil distribution of *Cytherissa lacustris* (Ostracoda) in Lake Constance. *Mitteilungen der internationalen Vereinigung für theoretische und angewandte Limnologie, 17:* 240–251.

LOWE, J.J. & WALKER, M.J.C., 1984. *Reconstructing Quaternary Environments.* London: Longman.

LOŽEK, V., 1972. Le loess et les formations assimilées: corrélation entre l'Europe centrale et la France par la fauna de mollusques. In *Proceedings of the Second International INQUA Congress, Paris, 1969:* 597–606. Paris: INQUA.

MAGNY, M. & MOUTHON, J., 1990. Interprétation paléolimnimétrique d'une coup stratigraphique de la station 2 de Chalain (Jura): comparaison des approches sédimentologique et malacologique. *Archives des Sciences, Genève, 43:* 99–115.

MAGURRAN, A.E., 1988. *Ecological Diversity and its Measurement.* London: Croom Helm.

MARMONIER, P. & DANIELOPOL, D.L., 1988. Découverte de *Nannocandona faba* Eckman (Crustacea, Candoninae) en basse Autriche. Son origine et son adaptation au milieu interstitiel. *Vie et Milieu, 38:* 35–48.

MARTENS, K., 1985. *Tanganyikacypridopsis* gen. n. (Crustacea, Ostracoda) from Lake Tanganyika. *Zoologica Scripta, 14:* 221–230.

MARTENS, K., 1990a. Revision of African *Limnocythere s.s.* Brady, 1867 (Crustacea; Ostracoda), with special reference to the Rift Valley Lakes: morphology, taxonomy, evolution and (palaeo-) ecology. *Archiv für Hydrobiologie, 83(S):* 453–524.

MARTENS, K., 1990b. Speciation and evolution in the genus *Limnocythere* Brady, 1867 *sensu stricto* (Crustacea, Ostracoda) in the east African Gallo and Awassa Basins. *Courier Forschungsinstitut der Senckenberg, 123:* 87–95.

MAZEPOVA, G.F., 1990. *Pakyschovye Ratschki (Ostracoda) Baikala.* Sibirskoe Otdelenie Limnologicheskii Institut: Akademia Nauk SSSR.

MOUTHON, J., 1987. Malacocenoses des dépôts Tardiglaciares et Holocènes de Saint-Hilaire-du-Rosier (Dauphiné-France). *Revue de Paléobiologie, 6:* 257–260.

OLDFIELD, F. & APPLEBY, P.G., 1984. Empirical testing of ^{210}Pb-dating models for lake sediments. In E.Y. Haworth & J.W.G. Lund (Eds), *Lake Sediments and Environmental History:* 93–124. Leicester: Leicester University Press.

PATERSON, H.E.H., 1985. The recognition concept of species. In E.S. Vrba (Ed.), *Species and Speciation:* 21–29. Pretoria: Transvaal Museum.

PERRY, R. & ARTHUR, W., 1991. Environmental effects on adult shell size in *Cepaea hortensis. Biological Journal of the Linnean Society, 43:* 273–279.

PETKOVSKI, T.J., 1960. Zwei neue Ostracoden aus dem Ohrid- und Prespasee. *Izdanija Institut Pisciculture Macedoine, 3:* 57–60.

PETKOVSKI, T.J., 1969. Einige neue und bemerkenswerte Candoninae aus dem Ohridsee und einigen anderen Fundorten in Europa. *Acta Musei Macedonici Scientiarum Naturalium, 11, 5 (95):* 81–110.

PREECE, R.C., 1980a. The biostratigraphy and dating of the tufa deposit at the Mesolithic site at Blashenwell, Dorset, England. *Journal of Archaeological Science, 7:* 345–362.

PREECE, R.C., 1980b. The biostratigraphy and dating of a Postglacial slope deposit at Gore Cliff, near Blackgang, Isle of Wight. *Journal of Archaeological Science, 7:* 255–265.

ROSSI, V. & MENOZZI, P. 1990. The clonal ecology of *Heterocypris incongruens* (Ostracoda). *Oikos, 57:* 388–398.

RUSSELL-HUNTER, W.D., 1978. Ecology of freshwater pulmonates. In V. Fretter & J. Peake (Eds), *Pulmonates, volume 2A, Systematics, Evolution and Ecology:* 335–383. London: Academic Press.

SCHARF, B., 1983. Fossil (Quaternary) and living ostracods of the Meerfeld Maar (Germany). In R.F.

Maddocks (Ed.), *Applications of Ostracoda:* 255–262. Houston: University of Houston (Department of Geosciences).

SCHMIDT, R. & SIMOLA, H., 1991. Diatomeen-, pollen- und sediment mikrostratigraphische Untersuchungen zur anthropogenen Beeinflussung des Höllerer Sees (Oberösterreich). *Aquatic Science, 53:* 74–89.

SEGERSTRÅLE, S.G., 1957. On immigration of the glacial relicts of northern Europe, with remarks on their prehistory. *Commentationes Biologicae, 16 (16):* 1–117.

SUTCLIFFE, A.J., 1985. *On the Track of Ice-Age Mammals.* London: British Museum (Natural History).

SYWULA, T., SUOMALAINEN, E., SAURA, A. & LOKKI, J., 1985. Genetic structure of natural populations of Ostracoda. I. Electrophoretic methods. *Zeitschrift für zoologische Systematik und Evolutionsforschung, 23:* 194–198.

THOMAS, K.D., 1978. Population studies on molluscs in relation to environmental archaeology. In D.R. Brothwell, K.D. Thomas & J. Clutton-Brock (Eds), *Research Problems in Zooarchaeology:* 9–18. London: Institute of Archaeology.

USINGER, H., 1975. Pollenanalytische und stratigraphische Untersuchungen an zwei Spätglazial-Vorkommen in Schleswig-Holstein. *Mitteilungen der Arbeitsgemeinschaft Geobotanik in Schleswig-Holstein und Hamburg, 25:* 1–183.

WEIGELT, J., 1989. *Recent Vertebrate Carcasses and Their Paleobiological Interpretation.* Chicago: University of Chicago Press.

WOUTERS, K., 1978. *Kavalocythereis braconensis* gen. n., sp. n., a remarkable new cytheracean ostracod genus and species from Lake Tanganyika (Zaire). *Annales de Société royale Zoologique de Belgique, 108:* 177–186.

WOUTERS, K., MARTENS, K. & DE DECKKER, P., 1989. On the systematic position of *Tanganyikacypris* Kiss, 1961, with a description of *T. stapperse* sp. n. *Courier Forschungsinstitut der Senckenberg, 113:* 177–186.

CHAPTER

9

Stress, metabolic cost and evolutionary change: from living organisms to fossils

P.A. PARSONS

Introduction **140**

Stress extremes and morphological variability **141**

Energetic limitations **144**

Fitness, stress and reductionism **146**

Evolutionary change **147**
Selection and discontinuities 147
Level of stress: speciation and extinction 150

Acknowledgement **153**

References **153**

Keywords: Stress – asymmetry – metabolic rate – *Drosophila* – energetic limit – fossils – reductionism – anoxia – desiccation – extinctions.

Abstract

Phenotypic and genotypic variability tend to be high under severe environmental stress. A useful monitor of stress for morphological traits which appears applicable to the fossil record is fluctuating asymmetry (FA). Disturbance of normal development in a stressful environment is associated with increased metabolic energy requirement. Therefore, metabolic costs may be excessive for major range expansions at species borders even though variability may be high. Under stress, genotype–phenotype relationships become direct and potentially reducible to the gene level. Assuming that stress is the norm in natural populations, this reductionist approach to evolutionary change may have some generality. Furthermore, this 'simple' genetic architecture is permissive of rapid change and discontinuities in the fossil record. Even so, during periods of stasis, *Drosophila* selection experiments suggest that substantial physiological change is possible without significant morphological change. Maximum evolutionary

Evolutionary Patterns and Processes
ISBN 0-12-440895-8

rates may occur in moderately stressed regions where variability would be higher than in benign regions and metabolic costs not unduly restrictive. The level of physical stress, especially climatic, can be regarded as an environmental link between palaeontology, physiology, ecology and evolutionary genetics. In this way, it is possible to argue from observations on living organisms to fossils and vice versa. Because of the existence of stress-resistant and stress-sensitive species in closely related living taxa and the occurrence of stress evasion, examples of non-random extinctions emphasizing geographically restricted species are likely to increase during environmental catastrophes.

INTRODUCTION

Darwin (1859) regarded interactions between organisms and environments as central for an understanding of evolution. He argued for mainly a gradual process involving intricate and complex interactions among species, emphasizing the importance of competition. Even so, a continuum emerges whereby competition is exceedingly important in benign environments but of minor importance in extreme environments where the effects of environmental stress predominate (Hoffmann & Parsons, 1991). In parallel with Darwin, Fisher (1930) regarded the process of interactions among organisms as predominant, although when discussing the fission of species, he commented: "Any environmental heterogeneity which requires special adaptations, which are either irreconcilable or difficult to reconcile, will exert upon the cohesive power of the species a certain stress."

Then followed an increasing number of books devoted to the establishment and development of the synthetic theory of evolution. A particularly important landmark was Schmalhausen's (1949) book, *Factors of Evolution*. This contains extensive discussions of genotype by environment interactions including a consideration of extreme environments. For Schmalhausen, progressive evolution consisted of the continuous acquisition of new reaction norms especially under substantial environmental perturbations. The importance of climatic stress in mountainous, continental and polar regions is emphasized, and many examples given showing differences within species for stress resistance, especially for temperature extremes.

Early genetic studies on stress resistance traits in *Drosophila* are discussed in Parsons (1974). More recently, the focus has been on genetic variation at the protein and DNA levels. Gene frequencies at electrophoretic loci often tend to be correlated with temperature and other variables at the geographic level. Using electrophoretic variants, much work has been concerned with demonstrating genotypic differences under extreme conditions at the single locus level (Watt, 1985; Nevo, 1988). While this approach can reveal pathways from gene to phenotype, it cannot tell us the number of genes nor explore the interactions between genes with respect to a particular trait. Given the deficiencies in this approach, it is not surprising that the interest is now shifting to direct genetic studies of complex quantitative traits such as life history components. In some cases these components are compared under benign and stressful conditions (Hoffmann & Parsons, 1991).

In a summary of the diffuse literature on stress and its evolutionary consequences, Hoffmann & Parsons (1991) emphasize the extreme end of the stress gradient at the limits of resistance where non-reversible damage and mortality occur irrespective of the density of organisms. This is hard selection in the sense of Wallace (1981) as distinct from soft selection where the fate of an individual depends upon the phenotype or genotype of its immediate neighbours, and density and/or frequency dependence occurs from competitive interactions. The conclusion is that stress is important in many evolutionary and ecological processes, implying high selection intensities associated with rapid shifts in the frequencies of underlying genotypes. This is consistent with data on natural populations indicating that natural selection is often far more intense (Boag & Grant, 1981; Endler, 1986) and gene flow far more limited (Ehrlich & Raven, 1969) than envisaged by many of the earlier architects of the modern synthetic theory of evolution.

Here the aim is to extend the timescale for evolutionary change from that observed in living organisms to that of the fossil record. In this context, Eldredge & Gould (1988) consider that "it is palaeontologists more than any other cadre of evolutionary biologists who have realized that environmental change results in habitat shifting". In a book on the ecological history of life Vermeij (1987) argues for an approach based upon an "emphasis on the importance of demonstrating the presence and assessing the nature of environmental hazards as selective agencies". He considers that "enemies – competitors, predators, parasites, and agents of disease – are less important as selective agents than nonbiological hazards such as storms, floods, extreme heat and cold, scouring by ice and sand, fire, and heavy metal poisoning". Species should therefore adaptively track changes in the physical environment more predictably than changes in their biological surroundings. In considering the evolutionary consequences of stress at integration levels from the biogeographic to the molecular, Parsons (1987) and Hoffmann & Parsons (1991) came to parallel conclusions.

In studying adaptation in the fossil record, Vermeij (1987) regards a phenotypic approach to evolutionary change as mandatory. The organism is then necessarily the unit of selection in the framework of a modern approach to organism–environment interactions. In any case, gene frequencies are unknown and unknowable in the fossil record. However, as will be discussed in this paper, stress has the effect of magnifying genetic variability so that genotype–phenotype relationships become direct and in many cases reducible to the gene level (Parsons, 1992). In this way, convergences between the phenotype studied by the palaeontologist and the genes studied by evolutionary geneticists should emerge. In this paper, it will then be argued that conclusions from evolutionary studies on living organisms should be directly relatable to palaeontological change if, as argued by Hoffmann & Parsons (1991), extreme stress is a predominant environmental force underlying evolutionary processes.

STRESS EXTREMES AND MORPHOLOGICAL VARIABILITY

Substantial increases in variability are most likely at the extreme end of the stress gradient at ecological margins where just a small environmental perturbation could be lethal. Unfortunately, most data on stress and variability come from experiments with other primary aims, but there is now sufficient information to regard an association between stress and additive genetic variability as a working hypothesis (Hoffmann & Parsons, 1991). The implication is that any transfer to a novel environment is likely to increase variability; this includes the transfer of populations to toxin-stressed environments (Holloway, Sibly & Povey, 1990) and ecologically stressful environments generally (Parsons, 1987).

Recombination and mutation (including mutational events due to mobile genetic elements) also tend to be high under stress conditions. More generally, the existence of rapid reorganizations of the genome in response to environmental stress is now clearly documented, but the ways in which an organism perceives stress and responds to it are unknown (Cullis, 1990). The work on recombination under stress commenced in the first decade of genetical experiments on *Drosophila melanogaster*. Plough (1917) found an increase in recombination at both high and low temperatures especially under conditions close to threatening species continuity (Parsons, 1988). Other stresses that increase recombination include starvation in *D. melanogaster* (Neel, 1941) and behavioural stress from overcrowding in mice (Belyaev & Borodin, 1982).

In order to extrapolate to the fossil record, it is necessary to enquire about morphology. Taking recombination as a model, environmental stress should affect morphological variability in a parallel way. In the snake, *Natrix fasciatus*, vertebral number and the frequency of gross abnormalities increased at both high and low temperatures for experiments carried out over a developmental temperature range of 18–32°C (Osgood, 1978). In the sand lizard, *Lacerta agilis*, embryos were developed at a range of temperatures and developmental stability was estimated as the number of disturbances per individual (out of 204 forms of deviation from the normal pattern of pholidosis) for embryos developed at constant temperatures of 20, 25, 30 and 35°C. Figure 1 indicates that stability decreases at both low and high temperatures.

The epigenetic process that has attracted most attention recently at the population level is the fluctuating asymmetry (FA) of bilateral characters (Palmer & Strobeck,

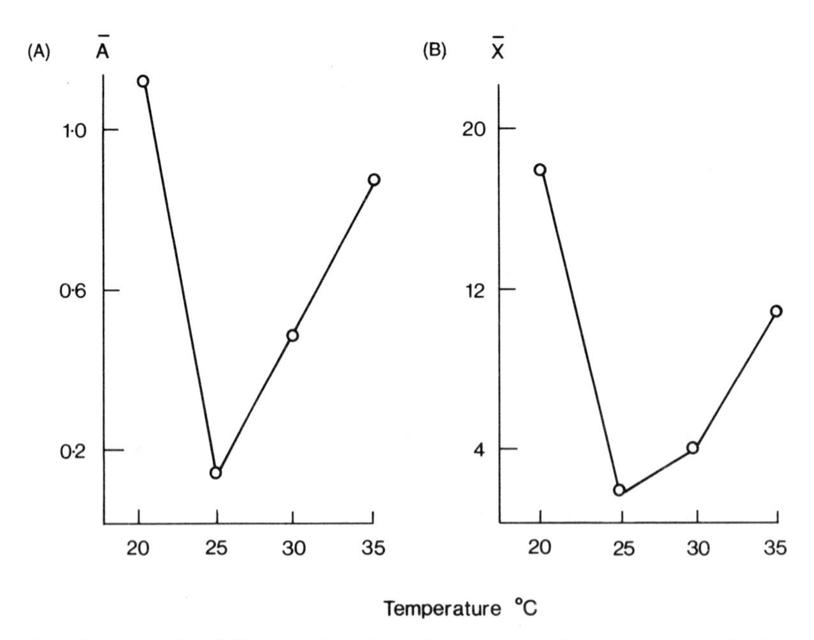

Figure 1 Developmental stability as a function of egg incubation temperature in the sand lizard, *Lacerta agilis*, measured as A: $\bar{A}$, fluctuating asymmetry and B: $\bar{X}$, mean number of disturbances in pholidosis characters (simplified from Zakharov, 1989).

1986; Zakharov, 1989; Parsons, 1990). For quantitative traits, FA increases away from optimum environments especially at stress levels approaching lethality. This is shown in Fig. 1A where FA increases away from 25°C in parallel with the frequency of disturbances in Fig. 1B. The predicted parallel with recombination is clear in these morphological studies.

In an Oregon R stock of *D. melanogaster* development was studied at 25°C and 30°C (which is close to lethality in terms of population continuity). Predictably, FA was substantially higher at 30°C than at 25°C (Fig. 2). Indeed at 30°C, there was much day-to-day variability which might be expected since the extreme nature of the stress means that slight changes in microenvironment have the potential for large effects. Furthermore, the increased developmental instability was manifested by the occurrence of flies with abnormalities, including crumpled wings (Parsons, 1961, 1962).

There is now a substantial literature on FA in a wide range of organisms. In the lizard, *L. agilis*, Zakharov (1989) found the highest FA at the ecological periphery of the range of the species, and provides examples of increased developmental instability in the three-spined stickleback, *Gasterosteus aculeatus*, when introduced into unusual water conditions. Even though there are many inconclusive reports in the literature, these and other examples (Zakharov, 1989; Parsons, 1990) suggest that there is an expectation in a wide range of taxa that FA should increase as the environment becomes increasingly stressful both in the laboratory and the field. This conclusion may be testable in the fossil record.

In summmary, an increase in FA suggests that normal development is upset at one or more of the molecular, chromosomal and epigenetic levels. There is now sufficient evidence to regard FA as a fitness measure whereby fitness is high when FA is low. In other words, the development of high FA in novel environments implies reduced fitness in these environments. However, *Drosophila* experiments have shown that

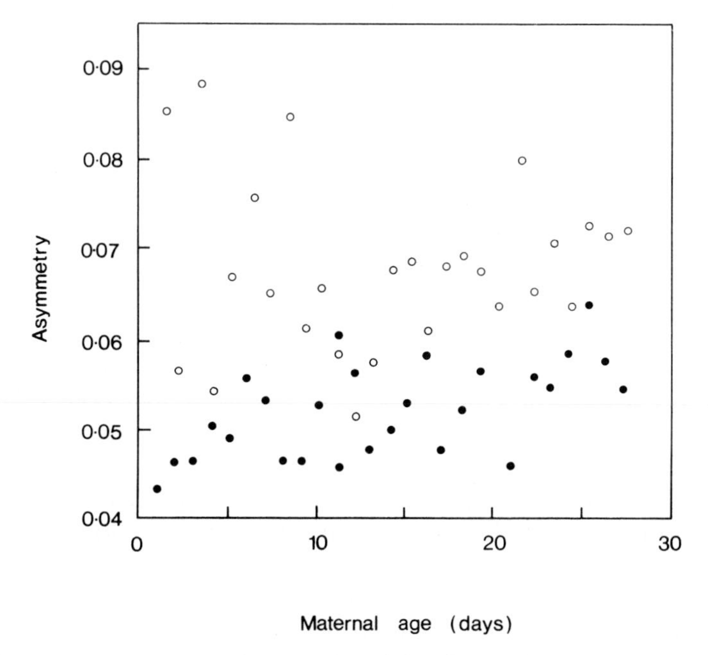

Figure 2 Fluctuating asymmetry of sternopleural bristle number of *D. melanogaster* plotted against maternal age for the Oregon-R stock grown at 25°C (●) and 30°C(○) (after Parsons, 1962).

progressive adaptation to novel environments can be associated with a slow decrease in FA (Thoday, 1958).

ENERGETIC LIMITATIONS

The underlying disturbance of normal development in a stressful environment may be associated with an increased metabolic energy requirement (Zotin, 1990). Organisms unable to respond to an environmental factor enter a stressed state where there is a diversion of energy from growth and reproduction to maintenance. This means that an organism may increase its resistance to stress by a reduction in metabolic rate. In *D. melanogaster*, a rapid selection response for desiccation was associated with reduced metabolic rate (Table 1). However, the cost was reduced fitness since both fecundity and behavioural activity levels were reduced. Since mechanisms increasing stress resistance may divert energy and other resources from growth and reproduction, resistance to environmental stresses should be negatively correlated with measures of many life-history traits, which implies a reduction in fitness in general terms. There are now sufficient data, especially in *Drosophila*, to regard this as a working model. However, such associations do depend upon severe stress to be fully manifested, since these are the conditions under which high additive genetic variability is most likely (Hoffmann & Parsons, 1991).

The association of stress resistance and reduced fitness is consistent with the view that species borders are regions where metabolic costs under stress are restrictive for range expansions (Root, 1988; Bozinovic & Rosenmann, 1989; Parsons, 1992). Energetic limitations at species borders would restrict normal physiological and behavioural processes, some of which are exceedingly expensive energetically, requiring almost all of the energy intake of an animal. Peterson, Nagy & Diamond (1990) analysed sustained metabolic rates (SusMR) which are time-averaged metabolic rates that are measured in free-ranging animals maintaining constant body mass over periods long enough that metabolism is fuelled by food intake rather than by depletion of energy reserves, and concluded that SusMR is between 1.5 and 7 times the resting metabolic rate, with most values being 5 and less. When these limits are consistently exceeded, the existence of the species is threatened. Examples of such energetically expensive activities include flight in birds, calling in frogs and insects, web-building in spiders and endothermy in insects (see Parsons, 1992).

In all cases, the high metabolic rates can only be maintained for short periods within the whole life cycle when activities that are crucial for the continuity of species tend to occur. At species borders, therefore, fitness would be reduced by restricting these activities consequent upon the substantial metabolic cost of elevated environmental perturbations in these habitats. Expressed in this way, other stresses such as diseases would interact to restrict further these energetically expensive activities. Indeed, the allocation of energy to stress resistance may increase immediate survival but at the cost of threatening activities necessary for continuity across generations.

Survival under extreme conditions is therefore associated with low metabolic rate, but these are circumstances where the chances of major adaptive change are low. Adaptive change and specialization are more likely when metabolic rates are above maintenance levels, even though reduced metabolic rate may evolve during the

Table 1 A comparison of lines of *D. melanogaster* selected for resistance to desiccation with control lines for metabolic rate, fecundity, activity, and various morphological traits.

	LT_{50} for resistance to desiccation	Metabolic rate (O_2 mg^{-1} h^{-1})	Fecundity*	Activity†	Head width (mm)	Wing width (mm)	Wing length (mm)	Wet weight ($g \times 10^{-3}$)
Selected lines 1	27.0	2.08	141.8	5.5	0.91	1.17	1.73	1.52
2	28.4	2.13	129.3	5.8	0.90	1.18	1.75	1.61
3	30.2	2.02	122.4	6.8	0.93	1.18	1.76	1.63
Control lines 1	17.8	2.74	158.5	9.7	0.93	1.20	1.77	1.49
2	20.1	2.45	162.3	9.2	0.93	1.18	1.73	1.57
3	17.8	2.65	166.3	8.2	0.90	1.19	1.73	1.52
Significance of difference between selected lines and control lines	<0.001	<0.01	<0.01	<0.01	n.s.	n.s.	n.s.	n.s.

Data derived from Hoffmann & Parsons, 1989a.
*Mean number of eggs laid per female over a 3-day period beginning the second day following eclosion.
†Mean number of female flies (out of 15) that moved in a 30-s interval.

period of change. For example, in adapting to the exceedingly high CO_2 levels found in cave environments, troglobitic arthropods show major changes compared with their epigean relatives, including reduced metabolic rates (Howarth, 1987). But if it is assumed that most extinctions in the fossil record are due to stress from weather-related catastrophes, organisms with high metabolic requirements could be the most susceptible to extinctions (Vermeij, 1987). Indeed, in *D. melanogaster,* high metabolic rate mutants such as the shaker mutants tend to be very sensitive to temperature and related stresses (Parsons, 1991a).

But these are circumstances where chances of major adaptive change are low. Organisms subjected to day-to-day and seasonal fluctuations should have some resistance to catastrophic mortality since continued exposure to a low but predictable level of stress would provide some protection. Provided that stress levels are not excessive, evolutionary change should be possible under these circumstances with a reduced threat of extinctions.

The two extremes of exceedingly high and low metabolic rates therefore appear restrictive, the former because of a high level of sensitivity to stress, and the latter because of a lack of metabolic energy. Evolutionary change therefore appears most likely between these extremes when there is some stress associated with sufficient available metabolic energy for some adaptive change (Parsons, 1991b).

FITNESS, STRESS AND REDUCTIONISM

Williams (1985) defined reductionism as "the seeking of explanations for complex systems entirely in terms of what is known of their component parts and processes". In this sense, to regard gene frequency as the essence of evolution is reductionist. This underlies the selfish gene view of evolutionary change (Dawkins, 1976). It contrasts with those who argue for the unity of the genotype and the importance of gene interactions and balanced gene complexes as units of evolution (Mayr, 1984). Such controversies have engaged many authors, covering the whole continuum of perspectives from genes to epistatic interactions (Lewontin, 1970; Brandon & Burian, 1984).

A consequence of stress is (1) high phenotypic and genotypic variability and (2) major fitness differences among phenotypes so that (3) genetic variation underlying fitness and associated traits should be high. Taking *D. melanogaster* as a case study, genetic correlations between stress resistance, low metabolic rate and life-history traits such as fecundity, development time and behavioural activity tend to be negative on the one hand (Table 1), while the reverse is true for traits involved with longevity on the other (Hoffmann & Parsons, 1993). Since these associations are difficult to perceive under benign conditions, stress can be regarded as an 'environmental probe' capable of revealing associations among an array of traits based upon an underlying dependence upon variations in metabolic rate. The efficiency of stress as an environmental probe follows from the high heritability of stress resistance, whereby contributions at the gene level are emphasized. Our understanding of specific genes for stress resistance is not great, however, a point applying even more to associated physiological traits.

The role of metabolic rate in this reductionist paradigm follows from the point that variations in metabolic rate can be reduced to the adenine nucleotide level, which is important in many reactions of intermediary metabolism. An involvement at this level incorporating stress has been suggested by Kohane (1988) in studies of larval fitness in *D. melanogaster*. The likelihood of generalizations via substances at the adenine nucleotide level has been emphasized by Morowitz (1989) in attempts at simplifying the matrix of biological information, and in descriptions of biological systems in terms of energetic costs (Hoffmann & Parsons, 1991).

In an invariant environment the genetic variance of fitness traits should approach zero in the process of maximizing fitness (Fisher, 1930). In natural populations this limit is not normally achieved. In a major survey of published data, Mousseau & Roff (1987) found that life-history traits measured under non-stressful conditions generally have lower heritabilities than do morphological traits, while those for behavioural and physiological traits fall between these extremes. Heritabilities for direct stress resistance traits may be very high, for example >0.60 for desiccation resistance in *D. melanogaster* (Hoffmann & Parsons, 1989a). This suggests that under stress, any life-history or fitness trait associated with desiccation resistance should show increased heritability and additive genetic variance compared with less stressful conditions. Mousseau & Roff (1987) point out that many explanations have been advanced to explain the maintenance of the genetic variance of fitness traits above zero. These include heterozygote advantage, frequency-dependent selection, variable selection in heterogeneous environments, diversifying selection, migration, and antagonistic pleiotropy. These need to be considered in relation to the effects of stressful environments, which may have more predictable consequences.

If environmental stress has a major role in determining evolutionary rates, it follows that the reductionist programme is a good approximation to the real world in many circumstances. Rather than viewing the reductionism of G.C. Williams and R. Dawkins on the one hand and the adaptive complexes of E. Mayr on the other as alternatives, it may, however, be preferable to consider these approaches as opposite ends of a continuum. Even so, the most usual model may be one where stress is an environmental probe magnifying genetic variability so that genotype–phenotype relationships become direct and in many cases reducible to the gene level. In this scenario, the organism remains the unit of selection, but following the principle adopted by Williams (1966) that adaptation should be attributed to no higher level of organization than is demanded by the evidence, reductionism is the expectation if extreme stress is a predominant environmental force underlying evolutionary processes.

EVOLUTIONARY CHANGE

Selection and discontinuities

Turner (1982) considered that periods of rapid evolutionary change involving mimicry occurred in *Heliconius* butterflies at a time when neotropical rainforests were fragmented and stressed during the cool dry periods of the Pleistocene. He argued that such 'jerky' evolution was underlain by major genes at the initial stages followed by modifiers improving resemblance until *Heliconius* closely resembled the new model. In other words, as for laboratory experiments studying FA (Thoday, 1958), there would be selection for increased fitness in the new situation following the initial major ecological and associated major genetic changes.

This postulated situation calls for experiments on the effect of ecological stress upon a range of traits, but such an approach is uncommon, even though a number of palaeontologists have realized that environmental change underlies many shifts in the fossil record (Eldrdge & Gould, 1988). However, this was a major aim of the desiccation selection experiments in Table 1. Three body measurements (head width, wing length, wing width) and wet weight were used to test for morphological changes associated with the selection reponse. No significant effects occurred, even though positive correlations between body size and desiccation resistance have been found among isofemale strains in natural populations (Parsons, 1974).

Stress response must presumably depend upon the primary target of selection in natural populations and how this is affected by intense stress. In the experiments in Table 1, the primary target could involve changes in the concentrations of energy carriers. Therefore, periods of morphological stasis in the fossil record could be periods of substantial change at the physiological level. Only when a reaction norm crosses certain critical thresholds will observable morphological change occur; this is likely to be rapid, giving 'jerky' evolution since a canalization zone in the sense of Waddington (1956) may be disrupted in the process. Clearly, it would be useful to monitor fitness shifts morphologically by FA in any transitions of this nature.

Under the 'punctuated equilibrium' model (Eldredge & Gould, 1972), evolution is a discontinuous process in which long periods of stability are interrupted by shorter periods of rapid morphological change. For example, in the Antarctic radiolarian, *Pseudocubus vema*, mean thoracic width increased in two major accelerated phases over a period of about 2.5 million years (Kellogg, 1975). Similarly, in a sequence of fossilized freshwater molluscs from the Turkana Basin in East Africa, intermediate forms existed relatively transiently between periods of environmental stability. In 13 lineages, the periods of rapid change coincided which suggests parallel environmental stresses affecting all lineages similarly (Williamson, 1981).

Assuming high genetic variability and reductionism to the gene level, discontinuities are likely in stressful situations. Unfortunately most laboratory selection experiments have been carried out on morphological traits. In Fig. 3, the response to selection for increased scutellar bristle number in *D. melanogaster* occurred in two periods of rapid response separated by a long period of stasis. Genetical experiments showed that the second period of rapid response was due to the recombination and selection of two genes of major effect segregating in the founder natural population (MacBean, McKenzie & Parsons, 1971). More generally, relatively few genes of major effect often appear to underlie responses to selection for morphological traits, and discontinuities as in Fig. 3 are not uncommon (Lee & Parsons, 1968). Recently, the localization of genes affecting quantitative traits (i.e. quantitative trait loci or QTLs) has become increasingly common. For example, using a systematic QTL mapping aproach, Paterson *et al.* (1988) resolved six quantitative traits of tomato into Mendelian factors.

Apart from providing a reductionist genetic interpretation of discontinuities, the selection experiment results are difficult to apply to the fossil record since continuous directional selection is unlikely in nature. Furthermore, optimal laboratory conditions are needed for the survival of extreme phenotypes which transcend the limits of variation normally occurring in a species in nature. The bizarre morphological characteristics of domesticated pigeon breeds noted by Darwin is a parallel situation since they occur in ecological situations created by man. Another example of rapid evolution is the process of 'domestication' in the production of corn from teosinte.

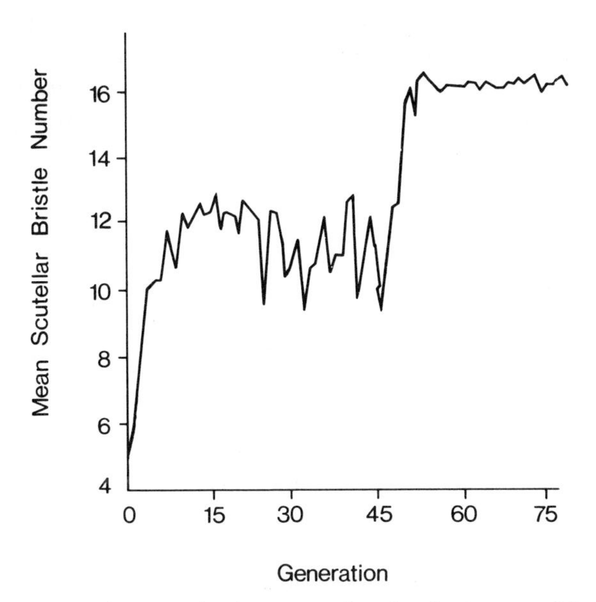

Figure 3 The response to directional selection in females for increased bristle number over 75 generations in *D. melanogaster*. Note the two accelerated responses lasting very few generations in each case (modified from MacBean *et al.*, 1971).

Recent molecular data suggest that as few as five genes can account for between 50 and 80% of this morphological change (Doebley *et al.,* 1990).

An example of rapid morphological change following a resource switch consequent upon drought is the observation of selection leading to large birds in the Galapagos finch (Boag & Grant, 1981). The average heritability combining seven characters came to 0.77 (Boag & Grant, 1978) which is compatible with very rapid responses to selection. However, the rainfall returned the next year indicating that periods of intense selection may only occur erratically in the wild. The observation of low rates of evolution in the fossil record may therefore greatly underestimate the rapidity with which populations can change, but over short periods of time.

The main exception to this conclusion would be in the colonization of empty islands or other unoccupied habitats, where organisms commonly diverge from their ancestors to the extent that they transcend previously established limits of form and behaviour. Under these conditions, as under domestication, normal ecological restrictions are reduced or absent, and there would almost certainly be a lack of stress from the usual complement of competitors and predators. Indeed, under these circumstances, new characters may evolve very rapidly so that small populations in geographically isolated regions could become species (Futuyma, 1979).

A laboratory experiment indicating change in an 'island' situation comes from body size studies in *D. pseudoobscura* in populations left in high and low temperature incubators for 12 years (Anderson, 1973). After 6 years there was a 7% difference in body size which evolved in a way consistent with the commonly observed body size clines of latitudinally widespread *Drosophila* species. Approximately 40% of the phenotypic variance could be ascribed to genetic variation. Evolutionary change in body size can therefore be effected by the physiological effect of temperature in an isolated situation.

However, the most substantial morphological changes are likely for genes acting late in a developmental pathway, because only a small subset of the larger developmental whole is affected. The more tightly integrated are the developmental subsets, the more exacting the functional interdependence among the structures so that a genetic change would almost certainly be deleterious. This is in accord with Futuyma (1989), who suggests that stasis and punctuation in the fossil record should be most commonly observed in 'specially' adaptive characters such as morphological variants of particular structures, by comparison with 'generally' adaptive characters such as body size. In other words, genetic changes conferring a general adaptation may be a small minority of those that are brought about by selection.

In summary, assuming that major genes normally underlie substantial responses to selection for quantitative traits, rapid changes can occur when ecological situations are permissive, especially in previously unoccupied habitats. The ultimate example is the Cambrian explosion which was a time of unparalleled opportunity when life was radiating into empty space following the evolution of multicellular life. This is shown in its most dramatic form in the Burgess shale (Conway Morris, 1979; Whittington, 1981; Gould, 1989). However, many changes in the fossil record are likely to be the result of an average level of selection in the direction of change, but which on a short-term basis may not be detectable because of major oscillations in selection pressures leading to rapid but transient responses in opposite directions.

Level of stress: speciation and extinction

Favourable conditions for substantial evolutionary change include the availability of metabolic energy beyond maintenance needs. At species boundaries, metabolic energy requirements become restrictive, since energy availability falls to maintenance levels, yet these are habitats where a consequence of stress is likely to be high genetic variability. Conversely, in more benign habitats of more central regions of the distribution of a species, there would be available metabolic energy permitting adaptive change, but genetic variability is low. On these arguments, evolutionary change appears most likely to occur in habitats between the extremes where there is some environmental stress and sufficient metabolic energy to accommodate adaptive change (Parsons, 1991b,c). Furthermore, these are the conditions under which phenotypic plasticity is likely to be maximized (Hoffmann, 1990; Hoffmann & Parsons, 1991). In addition, ecological systems appear to be the most diverse when subjected to a certain degree of stress (Moore, 1983). Indeed the greatest number of species is a feature of intermediate levels of disturbance (Petraitis, Latham & Neisenbaum, 1989).

These environmental circumstances appear to be a feature of the narrowly fluctuating tropics where highest speciation rates normally occur (Wallace, 1878). The diversity of species in the genus *Drosophila* in Australia is highest in the humid tropics. At more environmentally widely fluctuating and more heat-stressed forest edges, species diversity is lower (Parsons, 1991d).

Tropical species should be relatively sensitive to climate-related stresses. For example, comparative studies of 9 tropical and 11 mainly temperate *Drosophila* species for LT_{50} at -1°C gave a range of 2–37 h and 53–428 h for tropical and temperate species respectively with no overlap across zones, while LT_{50} at 0%RH at 25°C gave 4–11 h and 8–32 h for tropical and temperate species respectively with only *D. simulans* overlapping zones (Parsons, 1982, 1987). Tropical species therefore have a restricted range of stress resistance. This is especially true for species limited to rain forests of the humid tropics, as shown by studies on field-collected flies of fungus-feeding species of subgenus *Hirtodrosophila* (Parsons, 1985).

A two-way plot of resistance to cold and to heat/desiccation stress is instructive, since it indicates quite clearly a wide array of stress resistances in the genus *Drosophila* (Parsons, 1982) whereby the tropical fauna with the greatest species diversity is the most vulnerable to both stresses. Extending the taxonomic unit to *Scaptomyza,* another genus in the family Drosophilidae, the widespread *S. australis* appears to have broader physical limits for resource utilization than Australian endemic *Drosophila* species, enabling it to spread to both hot arid and cold regions where *Drosophila* are not found (Bock & Parsons, 1977).

Such data indicate that species distributions are closely associated with physical features of the environment emphasizing extremes. Stevens (1989), in a discussion of high species diversity in the tropics, suggested that "the greater annual range of climatic conditions to which individuals in high-latitude environments are exposed relative to what low-latitude organisms face has favored the escalation of broad climatic tolerances of high-latitude species". This is in accord with the wider latitudinal range of high-latitude *Drosophila* species and their greater resistances to stress than lower-latitude species. In any case, because of higher stress levels the energetic costs of adaptation would be expected to be higher in temperate species than in tropical species, which could be restrictive for major adaptive change and speciation in temperate species by comparison with tropical species.

In fossil Coleoptera of the late Cenozoic from the North Temperate zone, Coope (1979) found an extraordinarily low rate of evolution during a period subjected to rapid climatic change. Many species changed their ranges far more than could have been predicted from their modern biogeography. These changes follow an orderly pattern closely related to the large-scale fluctuations of the glacial–interglacial cycles. Within major cycles lesser oscillations occurred, each involving climatic changes exceeding those in historical time, and the geographical movements of species reflected these minor episodes with the same precision as the major ones.

Under conditions of severe climatic stress and where species distributions accurately track the climate, speciation rates therefore appear to be low both in the fossil and living biota. Differentiation within the lineage is more likely in regions with long periods of relative stability characterized by repeatable environmental changes compared with regions of major environmental perturbations. This appears to be in accord with the work of Sheldon (1987) on parallel gradualistic evolution of Ordovician trilobites where he suggests that gradual phyletic evolution can only be sustained by organisms living in or able to track narrowly fluctuating, slowly changing environments. In contrast, stasis seems to prevail in more widely fluctuating, rapidly changing environments.

In conclusion, an approach via observations on the living biota and fossils suggests that maximum evolutionary change is likely in habitats characterized by narrowly fluctuating environments implying moderate stress, moderate genetic variability of ecologically important traits, and not unduly restrictive metabolic costs (Parsons, 1991b,c). In this way, gaps between palaeontology, physiology, ecology and evolutionary genetics are being bridged whereby the level of physical stress, especially climatic, is the connecting environmental link. This association follows directly from the postulated association between stress resistance and metabolic rate. In organisms such as *D. melanogaster,* where experiments can be carried out under definable physical conditions, it should be possible to test some of the hypotheses advanced here and elsewhere to gain more understanding of the likely modes of evolutionary change in the fossil record.

For example, experimental work is needed to investigate the consequences of extreme stress in a range of organisms to ascertain underlying targets of selection and associated genetic changes. On the other hand, if rapid global warming occurs, the paradigm required may automatically appear in the environment at large. Can we predict what may happen under such circumstances? A model comes from studies of outlier populations and those at forest edges which suggest that extinctions of rare stress-sensitive rain forest *Drosophila* species are likely following quite small increases in temperature/desiccation stress. Biodiversity would therefore be reduced by eliminations of stress-sensitive species and replacements by stress-resistant generalist species (Parsons, 1991d). Resistant species would need to devote substantial metabolic energy to enable accommodation to stress, so that evolutionary change would be restricted. Such an energetic approach, as well as comparisons of tropical and temperate *Drosophila* species, suggests non-randomness concerning which species survive a period of mass extinction. Geographically restricted species are likely to be most heavily at risk, which concurs with palaeontological data (Jablonski, 1991).

Turning to the fossil record, priapulid worms which are relatively abundant in the Burgess shale, have survived through episodes of mass extinction until the present, although not now contributing much to organic diversity. Undoubtedly, their relative failure and yet ultimate survival derives from their occupation of unusually harsh and marginal environments (Conway Morris, 1979) including low oxygen levels, hydrogen sulphide, low or fluctuating salinity, and exposure to long periods of starvation. Experimental work in *D. melanogaster* has shown positive correlations between certain generalized stresses because of an underlying association with metabolic energy (Hoffmann & Parsons, 1989a,b, in press). These stresses include desiccation, starvation, heat, ionizing radiation, and toxic levels of ethanol and acetic acid. Although the experimental data are somewhat indirect, anoxia can be added to this list because both anoxia and desiccation resistance were found to be genetically correlated with body weight in a study of isofemale strains from a natural population of *D. melanogaster* (Parsons, 1974). This is relevant in the context of suggestions that anoxia stress is a feature of some mass extinctions (Donovan, 1989). In any case, adaptations to exploit environments with other than normal gas mixtures, especially abnormally high CO_2 concentrations, have evolved independently many times such as in cave and soil animals (Howarth & Stone, 1990). The implication of correlations among generalized stresses is that some stresses would exacerbate the effects of others, so that experimental work on combinations of stresses postulated to be of importance in mass extinctions would be of considerable interest.

Multiple stresses are emphasized in marine habitats in the terminal Cretaceous extinctions (Kaufman, 1984) and more generally in other mass extinctions throughout the palaeontological record (Donovan, 1989). Interestingly, diatoms came through the Cretaceous extinction relatively unscathed, apparently because of their ability to evade stress by forming a resting spore during periods of nutritional depletion (Kitchell, Clark & Gombos, 1986). Such stress evasion, whereby an organism's life-history is modified to minimize the time exposed to an unfavourable environment, is quite common (Hoffmann & Parsons, 1991). It has the effect of conserving energy since dormant stages have very low metabolic requirements.

In summary, as our detailed knowledge of mechanisms of stress resistance and evasion of various taxa increases, examples of non-random extinctions during environmental catastrophes are likely to increase. With the return of more benign conditions, the rate of speciation is expected to escalate but based upon a different constellation of species than before the catastrophe.

ACKNOWLEDGEMENT

I am grateful to Peter R. Sheldon for discussions and for pointing out that fluctuating asymmetry measures can be used in the fossil record.

REFERENCES

ANDERSON, W.W., 1973. Genetic divergence in body size among experimental populations of *Drosophila pseudoobscura* kept at different temperatures. *Evolution, 27*: 278–284.

BELYAEV, D.K., & BORODIN, P.M., 1982. The influence of stress on variation and its role in evolution. *Biologisches Zentralblatt, 100:* 705–714.

BOAG, P.T. & GRANT, P.R., 1978. Heritability of external morphology in Darwin's finches. *Nature, 274:* 793–794.

BOAG, P.T. & GRANT, P.R., 1981. Intense natural selection in a population of Darwin's finches (Geospizinae) in the Galapagos. *Science, 214:* 82–85.

BOCK, I.R., & PARSONS, P.A., 1977. Distributions of the dipteran genera *Drosophila* and *Scaptomyza* in Australia in relation to resource utilization. *Journal of Biogeography, 4:* 327–332.

BOZINOVIC, F. & ROSENMANN, M., 1989. Maximum metabolic rate of rodents: physiological and ecological consequences on distributional limits. *Functional Ecology, 3:* 173–181.

BRANDON, R.N. & BURIAN, R.M. (Eds), 1984. *Genes, Organisms and Populations: Controversies over the Units of Selection.* Cambridge, Massachusetts: The MIT Press.

CONWAY MORRIS, S., 1979. The Burgess shale (middle Cambrian) fauna. *Annual Review of Ecology and Systematics, 10:* 327–349.

COOPE, G.R., 1979. Late Cenozic fossil Coleoptera: evolution, biogeography, and ecology. *Annual Review of Ecology and Systematics, 10:* 247–267.

CULLIS, C.A., 1990. DNA rearrangements in response to environmental stress. *Advances in Genetics, 28:* 73–97.

DARWIN, C., 1859. *On the Origin of Species by Means of Natural Selection.* London: John Murray.

DAWKINS, R., 1976. *The Selfish Gene.* Oxford: Oxford University Press.

DOEBLEY, J., STEC, A., WENDEL, J. & EDWARDS, M., 1990. Genetic and morphological analysis of a maize-teosinte F_2 population: implications for the origin of maize. *Proceedings of the National Academy of Sciences, U.S.A. 87:* 9888–9892.

DONOVAN, S.K. (Ed.), 1989. *Mass Extinctions: Processes and Evidence.* London: Belhaven Press.

EHRLICH, P.R. & RAVEN, P.H., 1969. Differentiation of populations. *Science, 165:* 1228–1232.

ELDREDGE, N. & GOULD, S.J., 1972. Punctuated equilibria: an alternative to phyletic gradualism. In T.J.M. Schopf (Ed.), *Models in Palaeobiology:* 82–115. San Francisco: Freeman Cooper.

ELDREDGE, N. & GOULD, S.J., 1988. Punctuated equilibrium prevails. *Nature, 332:* 211–212.

ENDLER, J.A., 1986. *Natural Selection in the Wild.* Princeton: Princeton University Press.

FISHER, R.A., 1930. *The Genetical Theory of Natural Selection.* Oxford: Clarendon Press.

FUTUYMA, D.J., 1979. *Evolutionary Biology.* Sunderland, Massachusetts: Sinauer Associates.

FUTUYMA, D.J., 1989. Macroevolutionary consequences of speciation: inferences from phytophagous insects. In D. Otte & J.A. Endler (Eds), *Speciation and its Consequences,* 557–598. Sunderland, Massachusetts: Sinauer Associates.

GOULD, S.J., 1989. *Wonderful Life: The Burgess Shale and the Nature of History.* London: Penguin Books.

HOFFMANN, A.A., 1990. Acclimation for desiccation resistance in *Drosophila melanogaster* and the association between acclimation responses and genetic variation. *Journal of Insect Physiology, 36:* 385–391.

HOFFMANN, A.A. & PARSONS, P.A., 1989a. An integrated approach to environmental stress tolerance and life-history variation. Desiccation tolerance in *Drosophila. Biological Journal of the Linnean Society, 37:* 117–136.

HOFFMANN, A.A. & PARSONS, P.A., 1989b. Selection for increased desiccation resistance in *Drosophila melanogaster:* additive genetic control and correlated responses for other stresses. *Genetics, 122:* 837–845.

HOFFMANN, A.A. & PARSONS, P.A., 1991. *Evolutionary Genetics and Environmental Stress.* Oxford: Oxford University Press.

HOFFMANN, A.A. & PARSONS, P.A., 1993. Selection for adult desiccation resistance in *Drosophila melanogaster:* fitness components, larval resistance and stress correlations. *Biological Journal of the Linnean Society, 48:* in press.

HOLLOWAY, G.J., SIBLY, R.M. & POVEY, S.R., 1990. Evolution in toxin stressed environments. *Functional Ecology, 4:* 289–294.

HOWARTH, F.G., 1987. The evolution of non-relictual tropical troglobites. *International Journal of Speleology, 16:* 1–16.

HOWARTH, F.G. & STONE, F.D., 1990. Elevated carbon dioxide levels in Bayliss Cave, Australia: implications for the evolution of obligate cave species. *Pacific Science, 44:* 207–218.

JABLONSKI, D., 1991. Extinctions: a paleontological perspective. *Science, 253:* 754–757.

KAUFMAN, E.G., 1984. The fabric of Cretaceous marine extinctions. In W.A. Berggren & J.A. Couvering (Eds), *Catastrophes and Earth History: the New Uniformitarianism:* 151–246. Princeton: Princeton University Press.

KELLOGG, D.E., 1975. The role of phyletic change in the evolution of *Pseudocubus vema* (Radiolaria). *Palaeobiology, 1:* 359–370.

KITCHELL, J.A., CLARK, D.L. & GOMBOS, A.M., 1986. Biological selectivity of extinction: a link between background and mass extinction. *Palaios, 1:* 504–511.

KOHANE, M.J., 1988. Stress, altered energy availability and larval fitness in *Drosophila melanogaster. Heredity, 60*: 273–281.

LEE, B. T.O. & PARSONS, P.A., 1968. Selection, prediction and response. *Biological Reviews, 43:* 139–174.

LEWONTIN, R.C., 1970. The units of selection. *Annual Review of Ecology and Systematics, 1:* 1–18.

MacBEAN, I.T., McKENZIE, J.A. & PARSONS, P.A., 1971. A pair of closely linked genes controlling high scutellar number in *Drosophila. Theoretical and Applied Genetics, 41:* 227–235.

MAYR, E., 1984. The unity of the genotype. In R.N. Brandon & R.M. Burian (Eds), *Genes, Organisms, Populations: Controversies over the Units of Selection:* 69–84. Cambridge, Massachusetts: The MIT Press.

MOORE, P.D., 1983. Ecological diversity and stress. *Nature, 306:* 17.

MOROWITZ, H.J., 1989. Models, theory and the matrix of biological knowledge. *BioScience, 39:* 177–179.

MOUSSEAU, T.A. & ROFF, D.A., 1987. Natural selection and the heritability of fitness components. *Heredity, 59:* 181–197.

NEEL, J.V., 1941. A relation between larval nutrition and the frequency of crossing over in the third chromosome of *Drosophila melanogaster. Genetics, 26:* 506–516.

NEVO, E., 1988. Genetic diversity in nature: patterns and theory. *Evolutionary Biology, 23:* 217–246.

OSGOOD, D.W., 1978. Effects of temperature on the development of meristic characters in *Natrix fasciata. Copeia, 1*: 33–47.

PALMER, A.R., & STROBECK, C., 1986. Fluctuating asymmetry: measurement, analysis, patterns. *Annual Review of Ecology and Systematics, 17:* 391–421.

PARSONS, P.A., 1961. Fly size, emergence time and sternopleural chaeta number in *Drosophila. Heredity, 16*: 455–473.

PARSONS, P.A., 1962. Maternal age and developmental variability. *Journal of Experimental Biology, 39:* 251–260.

PARSONS, P.A., 1974. Genetics of resistance to environmental stresses in *Drosophila* populations. *Annual Reviews of Genetics, 7:* 239–265.
PARSONS, P.A., 1982. Evolutionary ecology of Australian *Drosophila:* a species analysis. *Evolutionary Biology, 14:* 297–350.
PARSONS, P.A., 1985. Tropical *Drosophila:* resistance to environmental stresses and species diversity. *Proceedings of the Ecological Society of Australia, 13:* 43–49.
PARSONS, P.A., 1987. Evolutionary rates under environmental stress. *Evolutionary Biology, 21:* 311–347.
PARSONS, P.A., 1988. Evolutionary rates: effects of stress upon recombination. *Biological Journal of the Linnean Society, 35:* 49–68.
PARSONS, P.A., 1990. Fluctuating asymmetry: an epigenetic measure of stress. *Biological Reviews, 63:* 131–145.
PARSONS, P.A., 1991a. Can atmospheric pollution be monitored from the longevity of stress-sensitive behavioural mutants in *Drosophila? Functional Ecology,* 5: 713–715.
PARSONS, P.A., 1991b. Evolutionary rates: stress and species boundaries. *Annual Review of Ecology and Systematics, 22:* 1–18.
PARSONS, P.A., 1991c. Stress and evolution. *Nature, 351:* 356–357.
PARSONS, P.A., 1991d. Biodiversity conservation under global climatic change: the insect *Drosophila* as a biological indicator? *Global Ecology and Biogeography Letters, 1:* 77–83.
PARSONS, P.A., 1992. Evolutionary adaptation and stress: the fitness gradient. *Evolutionary Biology, 26:* 191–223.
PATERSON, A.H., LANDER, E.S., HEWITT, J.D., PETERSON, S., LINCOLN, S.E. & TANKSLEY, H.D., 1988. Resolution of quantitative traits into Mendelian factors by using a complete RFLP linkage map. *Nature, 335:* 721–726.
PETERSON, C.C., NAGY, K.A. & DIAMOND, J., 1990. Sustained metabolic scope. *Proceedings of the National Academy of Sciences, U.S.A., 87:* 2324–2328.
PETRAITIS, P., LATHAM, R.E. & NEISENBAUM, R.A., 1989. The maintenance of species diversity by disturbance. *Quarterly Review of Biology, 64:* 393–418.
PLOUGH, H.H., 1917. The effect of temperature on crossing over in *Drosophila. Journal of Experimental Zoology, 24:* 148–209.
ROOT, T., 1988. Energy constraints on avian ranges. *Ecology, 69:* 330–339.
SCHMALHAUSEN, I.I., 1949. *Factors of Evolution.* Philadelphia: Blakiston.
SHELDON, P.R., 1987. Parallel gradualistic evolution of Ordovician trilobites. *Nature, 330:* 561–563.
STEVENS, G.C., 1989. The latitudinal gradient in geographical range: how so many species coexist in the tropics. *American Naturalist, 133:* 240–256.
THODAY, J.M., 1958. Homeostasis in a selection experiment. *Heredity, 12:* 401–415.
TURNER, J.R.G., 1982. Darwin's coffin and Doctor Pangloss – do adaptationist models explain mimicry? In B. Shorrocks (Ed.), *Evolutionary Ecology:* 313–361. Oxford: Blackwell Scientific Publications.
VERMEIJ, G.J., 1987. *Evolution and Escalation: an Ecological History of Life.* Princeton: Princeton University Press.
WADDINGTON, C.H., 1956. *Principles of Embryology.* London: George Allen & Unwin.
WALLACE, A.R., 1878. *Tropical Nature.* London: Macmillan.
WALLACE, B., 1981. *Basic Population Genetics.* New York: Columbia University Press.
WATT, W.B., 1985. Bioenergetics and evolutionary genetics: opportunities for a new synthesis. *American Naturalist, 125:* 118–143.
WHITTINGTON, H.B., 1981. Cambrian animals: their ancestors and descendents. *Proceedings of the Linnean Society of New South Wales, 105:* 79–87.
WILLIAMS, G.C., 1966. *Adaptation and Natural Selection.* Princeton: Princeton University Press.
WILLIAMS, G.C., 1985. A defence of reductionism in evolutionary biology. *Oxford Surveys in Evolutionary Biology, 2:* 1–27.
WILLIAMSON, P.G., 1981. Palaeontological documentation of speciation in Cenozoic molluscs from Turkana basin. *Nature, 293:* 437–443.
ZAKHAROV, V.M., 1989. Future prospects for population phenogenetics. *Soviet Science Reviews F, 4:* 1–79.

ZOTIN, A.I., 1990. *Thermodynamic Bases of Biological Processes: Physiological Reactions and Adaptations.* Berlin: Walter de Gruyter.

CHAPTER

10

The importance of rare events in polyploid evolution

CLIVE A. STACE

Introduction 158

Hybridization and polyploidy in plants 158

The genera *Festuca* and *Vulpia* and their hybrids 161

Fertility and chromosome pairing in × *Festulpia* 163

Evolution of polyploid complexes 164

Euploid gametes as the key step 166

Acknowledgements 167

References 169

Keywords: Polyploidy – introgression – hybridization – *Festuca* – *Vulpia* – chromosome banding – chromosome pairing.

Abstract

Speciation via hybridization and polyploidization is a well-known phenomenon in plants; it is possible that around 50% of all angiosperms might have arisen in this way. However, there are many other evolutionary routes at the polyploid level that deviate from the classical 'hybridization–polyploidy' cycle. Particular attention is paid to the grass genera *Festuca* and *Vulpia*, in which rapid introgression from the tetraploid to the hexaploid levels via highly sterile pentaploid intergeneric hybrids is occurring. The key step in this process is the rare production of euploid (in the latter case triploid) gametes. The increasing number of similar examples that can be cited from a wide range of angiosperm families leads to the conclusion that this is a significant evolutionary process.

Evolutionary Patterns and Processes
ISBN 0-12-440895-8

INTRODUCTION

An understanding of evolutionary pathways is made difficult for the obvious reason that we cannot go back in time, but instead must attempt a posteriori reconstruction as the next best thing. Three major approaches to this problem present themselves, all of them much used today, often to good effect. One is the sifting and analysis of phenetic evidence that is presented by the organisms themselves, whether it is in the form of morphological/anatomical characters, cytological behaviour, or DNA sequences. This is the easiest sort of evidence to obtain and, partly for that reason, it has been the most effective. A second approach is the study of fossils, which ideally would provide the best and most direct evidence, but all too often provides us with only a tantalizingly incomplete story due to their fragmentary nature. Thirdly, we can observe contemporary processes that produce evolutionary change and, by extrapolation, work out how these processes must have, in the past, produced the organisms that we see today.

The third approach is based upon the belief that evolutionary processes today are likely to be similar to those that operated in the past. It is what Stebbins (1972) called the theory of genetic uniformitarianism – the idea that evolutionary processes have remained essentially the same over evolutionary time, and are essentially the same at all taxonomic levels. In the absence of a good reason to doubt this to be true, it is a very reasonable hypothesis. At very least, even if it does not reconstruct phylogenetic pathways, it should indicate those pathways that are most feasible. It is the approach that will be primarily discussed in this paper, without forgetting that it would be foolish to attempt to reconstruct phylogenies by neglecting *any* of the three lines of evidence outlined above. The greatest attraction of the theory of genetic uniformitarianism to the writer is that it permits experiments on living organisms and in natural conditions.

Nevertheless, it must remain a matter for debate as to how far genetic uniformitarianism can be taken. For example, even if the same events do operate today as operated formerly, is it necessarily the case that the most important ones today were always the most important ones? There is some circumstantial evidence that suggests that this is not wholly so. Moreover, the main thesis of this paper is that the most common events are not necessarily the most important evolutionarily or, to put it more positively, that relatively rare events can be extremely significant phylogenetically. These two points – the long-term temporal change in significance of certain events, and the disproportionate importance of rare events – will now be taken further with reference to higher plants.

HYBRIDIZATION AND POLYPLOIDY IN PLANTS

It is an inescapable fact, and one far too frequently glossed over in the interests of a unified general evolutionary theory, that hybridization and polyploidy have been, and still are, of major phylogenetic significance in plant evolution, yet are relatively insignificant phenomena in animal evolution. Nevertheless, there is still a general belief that interspecific hybridization is a rare or unusual phenomenon. We tend to be conditioned to this view by our more intimate knowledge of higher animals, which in general cannot produce interspecific hybrids. *Homo sapiens* is a prime example, as almost equally are most of the higher animals with which we are familiar. Some obvious exceptions are well-known – the mule, the 'liger' and various groups of birds and fishes. But these *are* exceptions, and moreover the hybrids in nearly all these cases are quite sterile. The existence of the mule is completely inconsequential in terms of evolution, unless it could be shown that the reproductive rate of the donkey has thereby suffered a reduction.

Interspecific hybridization in plants, on the other hand, is a frequent occurrence. Although most interspecific hybrids are rare and highly sterile, many are common and most, conceivably all, show *some* degree of fertility; a good number are highly fertile. Even hybrids that are rare and totally sterile can have enormous evolutionary consequences. Some justification for these claims is needed.

Detailed figures of the number of hybrids are available, probably uniquely, for the flora of the British Isles (Stace, 1975). Since that survey about 100 extra interspecific hybrid combinations have been discovered, giving a present total of approximately 730 hybrids in a vascular plant flora of about 2500 species (ignoring agamospecies but including *c.* 1000 naturalized aliens). In addition, over 100 further records are not sufficiently well-documented to be acceptable at present, although a proportion of these must surely be well based, and over 400 additional combinations between two British taxa have been recorded abroad but not yet in the British Isles, again indicating a possible significantly greater real total. Since the figure of 2500 species is almost exactly 1% of the world's flora, it could be reasonably estimated that there might be about 73 000 interspecific hybrid combinations in nature throughout the world. Another fact, illustrating a vastly greater potential total, is that numerous artificial hybrids have been produced that have not so far been found in nature – there are over 70 000 of these in the Orchidaceae alone. Although there are some sizeable British genera that do not, so far as we know, produce hybrids (e.g. *Sedum, Trifolium, Vicia*), a list of such genera would be far shorter than one of sizeable genera that do produce hybrids.

Although it is true that sterility can be reliably used as an indication of hybridity in most groups, there are numerous examples across the whole range of the angiosperms (but less so in the pteridophytes) of highly to fully fertile hybrids. Well-known British genera in widely separated families that illustrate this are *Geum* (Rosaceae), *Linaria* (Scrophulariaceae), *Quercus* (Fagaceae), *Hyacinthoides* (Liliaceae), *Dactylorhiza* (Orchidaceae) and *Lolium* (Poaceae). In many further such genera, for example, *Rosa, Epilobium, Mimulus* and *Salix*, the existence of fertile hybrids has led to the natural formation of hybrids between three or even more species. Artificial hybridization has resulted in the production of plants with as many as 13 species of *Salix* or 6 genera of orchids in their parentage (Stace, 1986).

Most hybrids that seem to be quite sterile reveal some degree of fertility upon detailed study. Even where chromosome pairing is absent, or occurs at low frequencies, there is a possibility that some viable spores (and hence gametes) will be produced (e.g. Stace & Ainscough, 1984); the chances of this occurring are easily calculable. In other cases certain genotypes of the parental taxa produce hybrids that are more fertile than do most (e.g. Blackstock & Roberts, 1986).

The above facts taken together indicate that, in vascular plants, hybridization is not unusual or inconsequential; rather it is part of the normal genetic pattern of most oligo- or polytypic genera, in contrast to the situation in nearly all animal groups. This contrast amply explains the quite different emphasis placed by plant and animal taxonomists upon the use of breeding behaviour for species delimitation. In most groups of animals (notably the insects) the limits of the breeding unit are, with justification, considered synonymous with those of the species. In plants, however, breeding data are of value in species delimitation only as a comparative, not as an absolute, criterion (Stace, 1986).

The incidence of polyploidy in plants is more difficult to determine. We can be certain that it is very frequent. In the genus *Festuca* (x=7), for example, diploids, tetraploids, hexaploids, octoploids and decaploids all exist, and clearly all but the diploids ($2n$=14) are polyploids. There are numerous similar examples. But in the genus *Petasites* all 18 species have $2n$=58 (Toman, 1972); are all these diploids or all polyploids? It is not possible to recognize plants as diploids wherever n= a prime number, because many cases (e.g. in *Brassica*) are known of derived polyploids where this situation prevails. The known range of chromosome numbers in vascular plants ranges from $2n$=4 to $2n$=1440; the latter example is a race of the fern *Ophioglossum reticulatum* L., which is believed to be 96-ploid (Khandelwal, 1990). The highest known number in the dicots appears to be *Sedum suaveolens* Kimnach ($2n$=*c*. 640, ?*c*80-ploid) (Uhl, 1978) and in the monocots *Voanioala gerardii* Drans. ($2n$=*c*.596, ?*c*.50-ploid) (Johnson *et al*., 1989). According to the highest base number accepted as being primitively diploid (x=11 or 13 are frequently chosen), the proportion of polyploid species in the angiosperms has been calculated variously between about 35% and 70%. The percentage is probably higher in the pteridophytes. These figures refer only to diploidized or ancestral polyploids, completely ignoring true autoploids which probably occur occasionally by accident in all diploid species but which are rarely of evolutionary importance.

It is clear from the above that polyploidy is, like hybridization, an integral part of the normal range of the variation pattern of higher plants. It assumes even greater significance to man when one learns that many important crops, e.g. wheat, oil-seed rape, potato, tobacco, and weeds, e.g. *Poa annua* L., *Senecio vulgaris* L., *Taraxacum officinale* Wigg., *Urtica dioica* L., *Fallopia japonica* (Houtt).) Ronse Decraene, are polyploids.

Hybridization and polyploidy are, of course, intimately connected, since the former followed by the latter is a very well-understood and frequent route of evolution in higher plants. Perhaps its most intriguing feature is that there is, in general, an inverse relationship between the fertility of an F_1 hybrid and the fertility of the latter's derived polyploid. Hence a completely sterile F_1, in which there is no chromosome pairing, produces, upon chromosome number doubling, a completely fertile, diploid-like polyploid (amphidiploid). An increase in fertility (chromosome pairing) in the F_1 often produces a reduced fertility of the derived polyploid due to multivalent formation at meiosis. From the figures given earlier it can be presumed that around half of all vascular plants originated in this way, a route scarcely, if ever, exploited by animals.

Before devoting the rest of this paper to one particular aspect of polyploid evolution, I wish to enlarge upon three points that arise from the previous paragraph. First, the process of polyploidization has clearly proceeded in all the major taxonomic groups of vascular plants, for example from *Ranunculus* and *Rosa* to *Dactylis* and *Dactylorhiza* in the angiosperms, yet in all cases there are still diploids in existence alongside the polyploids. Unless we can recognize a major and very common mechanism for the regaining of the diploid chromosome number from the polyploid state (and we cannot at present), then this widespread co-existence of diploids and polyploids suggests that diploidy preceded polyploidy, and that the differentiation of the major groups of vascular plants took place at the diploid level. In that case (the original major differentiation being at the diploid level but subsequently about half the taxa appearing via polyploidization), there surely *is* a danger of overstating the theory of genetic uniformitarianism, because speciation will have operated via different pathways at different times.

Secondly, one could possibly solve this conundrum by suggesting that the major groups each evolved at both diploid and polyploid levels contemporaneously. This might suggest that the groups arose if not polyphyletically then at least polytopically and repeatedly. This might be near the truth, for polyploids are certainly sometimes of polytopic origin. The cases of *Tragopogon mirus* Ownbey and *T. miscellus* Ownbey (Ownbey & McCollum, 1954) are well-documented, and more recently good evidence has been gathered for *Senecio cambrensis* Rosser (Ashton & Abbott, 1992). The variability of many other polyploids, such as *Poa annua, Galeopsis tetrahit* L. and *Ranunculus penicillatus* (Dumort.) Bab., suggests similar parallel or successive origins.

Thirdly, several of the examples already given illustrate that hybridization between certain parental taxa might be a very rare event, but that such rarity is not necessarily reflected in the rarity of the products of hybridization. There is no good evidence that hybridization between *Spartina alterniflora* Lois. and *S. maritima* (Curtis) Fern. has occurred more than twice, yet its hybrid derivative *S. anglica* C. E. Hubb. is extremely successful and abundant in suitable habitats. The almost ubiquitous *Poa annua* apparently arose from hybridization between *P. infirma* Kunth and *P. supina* Schrader, which are sympatic only in very limited parts of the Mediterranean basin.

A survey of all the evidence, only a fraction of which has been summarized here, demonstrates unequivocally that the evolutionarily most significant hybridization events have been rare occurrences giving rise to highly sterile hybrids. This theme is continued in the following section.

THE GENERA *FESTUCA* AND *VULPIA* AND THEIR HYBRIDS

Festuca L. and *Vulpia* C. Gmelin are genera of the tribe Poeae of the Poaceae. *Festuca* consists of at least 300 species widely distributed in temperate regions. They are all perennials, but exhibit a wide range of habit: from densely tufted to extensively rhizomatous; from 2 m to a few centimetres tall; and with broad, flat leaves to narrow, tightly inrolled leaves. The species that concerns us here, *F. rubra* L., belongs to the type section of the genus, and is itself very variable and divisible into at least eight subspecies. It shows the full generic range of rhizomatousness and may possess flat or unrolled leaves, but in overall size it occupies the middle region of the total range. It exists as hexaploids ($2n=42$), octoploids ($2n=56$) and decaploids ($2n=70$); diploids and tetraploids were reported in the older literature but have not been confirmed recently, although they occur elsewhere in the genus.

Vulpia contains fewer than 20 species, again fairly widely distributed in temperate regions, but concentrated in the Mediterranean basin. The genus is closely related to *Festuca* and sometimes included within it, and indeed the two cannot be separated by any single character. The majority of the species, however, differ from *Festuca* in being annual and self-compatible (not perennial and self-incompatible), in having cleistogamous or semi-cleistogamous flowers with only one stamen (not chasmogamous flowers with three stamens), in having overlapping (not tubular) leaf-sheaths, and in having markedly (not slightly) unequal glumes. All these characters typical of *Vulpia* are applicable to *V. fasciculata* (Forsskål) Fritsch, the species with which we are concerned here. *V. fasciculata* is a tetraploid ($2n=28$) belonging to section *Monachne* Dumort., one of five sections in the genus. The genus contains diploids, tetraploids and hexaploids.

Natural hybrids (× *Festulpia* Melderis ex Stace & Cotton) occur between three species of *Vulpia*, representing all three ploidy levels and two sections (*V. bromoides* (L.) Gray and *V. myuros* (L.) C. Gmelin in section *Vulpia*; *V. fasciculata* in section *Monachne*) and two species of *Festuca* (*F. rubra*, mentioned above, and its close octoploid relative *F. arenaria* Osbeck) in four combinations, as shown in Table 1. Artificial hybrids have been made in two of these combinations, and in addition between *F. rubra* (both hexaploid and octoploid variants) and *V. sicula* (C. Presl) Link, a diploid, perennial, self-incompatible member of section *Loretia* (Duval-Jouve) Boiss. (Table 1). Extensive details of the characteristics and behaviour of these hybrids have been presented by Barker & Stace (1982, 1984, 1986).

The natural hybrid *F. rubra* × *V. fasciculata* (× *Festulpia hubbardii* Stace & Cotton) has been found on coastal sand-dunes with the two parents in south and south-west Britain from east Kent to south Lancashire, and in the Channel Isles (Ainscough, Barker & Stace, 1986). The plants are easily recognized by the intermediacy of their inflorescence characters, but they are perennials and in vegetative characters closely resemble some variants of *F. rubra*. They are highly sterile, showing pollen stainability of 0–2.5% and very rarely producing seed. There is circumstantial evidence that the plants do not survive for many years in the wild, and it is known that they do not survive for long in cultivation. No differences have been detected between wild and synthetic hybrids. Since *V. fasciculata* is a semi-cleistogamous inbreeder producing no or little aerial pollen, it is to be expected that the wild hybrids are produced as seed on *V. fasciculata* plants pollinated by the outcrossing *F. rubra*. In fact, in experimental crosses (Stace & Cotton, 1974; Barker & Stace, 1982) it was only crosses in that direction that succeeded, crosses using *F. rubra* as female producing no seed. Hence one can be reasonably certain of the male and female parentage of wild hybrids.

Table 1 Naturally occurring (N) and artificial (A) hybrids between *Festuca* L. and *Vulpia* C.C. Gmelin.

Festuca rubra (6*x*)	× *Vulpia bromoides* (2*x*)	– N (England)
Festuca rubra (6*x*)	× *Vulpia myuros* (6*x*)	– N (England, Wales, Holland), A
Festuca rubra (6*x*)	× *Vulpia fasciculata* (4*x*)	– N (England, Wales, Channel Isles), A
Festuca arenaria (8*x*)	× *Vulpia fasciculata* (4*x*)	– N (England)
Festuca rubra (6*x*)	× *Vulpia sicula* (2*x*)	– A
Vulpia sicula (2*x*)	× *Festuca rubra* (6*x*)	– A
Vulpia sicula (2*x*)	× *Festuca rubra* (8*x*)	– A

FERTILITY AND CHROMOSOME PAIRING IN × *FESTULPIA*

Meiosis in both wild and synthetic × *F. hubbardii* has been followed in some detail (Stace & Cotton, 1974; Barker & Stace, 1986). The hybrids are pentaploids ($2n$=35) with two genomes from the *Vulpia* parent and three from the *Festuca* parent (FFFVV). The maximum number of bivalents that might be expected is, therefore, 14, although in view of the high sterility of the hybrids a much lower number would usually be more likely. In fact considerable variation exists. The number of bivalents per pollen mother cell varied from 1 to 14 (mean 7.4); there were some multivalents and numerous irregularities (bridges, etc.), but 5.3% of cells had 14 bivalents. The similarity of karyotype morphology between the *Festuca* and *Vulpia* genomes as revealed by Feulgen or orcein staining (Fig. 1) is such that it is not possible to ascertain at meiosis using these stains whether the pairing observed is homogenetic (V–V and/or F–F) or heterogenetic (V–F). The presence of diploidizing genes (multivalent suppressors) in *F. rubra* has been demonstrated by Jauhar (1975), who found that, in crosses between hexaploid *Festuca* and diploid *Lolium* (FFFL), 10–14 bivalents were formed; any number of these over seven must have been F–F bivalents.

Despite the high degree of sterility of × *Festulpia* plants, Willis (1975, and personal communication), found that some seeds were formed and seedlings from them could be raised, though they died before maturity. Barker (1980) also raised one seedling from a natural × *F. hubbardii* plant (originating from Berrow, north Somerset, England); this was grown to maturity and was studied by Stace & Ainscough (1984). Surprisingly, it was found to be hexaploid or near-hexaploid ($2n$=41–42), to exhibit a high degree of pairing at meiosis (19–21 bivalents), and to produce a greatly enhanced pollen stainability (>70%) and seed-set, although many of the spikelets became pseudoviviparous instead of forming seeds (a common tendency in hybrid *Festuca*). This second-generation plant more closely resembles *F. rubra* than × *Festulpia*, and since the F_1 × *Festulpia* parent was grown in open conditions it is almost certain that it was pollinated by *F. rubra* pollen, which is abundant in the atmosphere of the experimental plots where the plants are grown. In order to produce hexaploid progeny following pollination by triploid pollen, the pentaploid × *Festulpia* must have produced a triploid embryo sac as the female parent of the cross. The genetic constitution of the triploid embryo sac would depend upon the

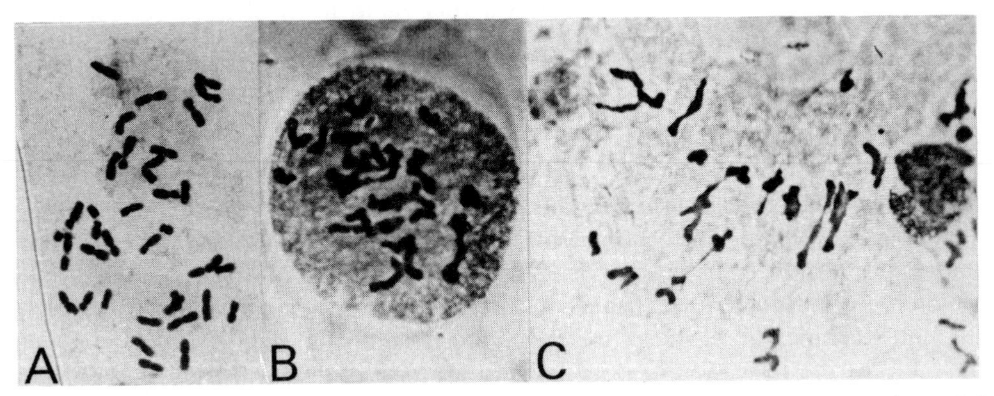

Figure 1 Orcein-stained chromosomes of × *Festulpia hubbardii*. A: Mitotic C-metaphase; B,C: meiotic metaphase-I with *c.* seven bivalents.

chromosome pairing behaviour and segregation in the F_1, especially upon whether the bivalents were V–V, V–F or F–F in nature. In fact, Stace & Ainscough (1984) showed that, whatever the pairing behaviour in the F_1, both homogenetic and heterogenetic pairing must have occurred in the pentaploid F_1 and/or in the hexaploid derivative (Fig. 3). This indirect demonstration of homogenetic F–F pairing supports the similar evidence of Jauhar (1975) and confirms the existence of diploidizing genes in *F. rubra.*

More recently Bailey & Stace (1992) have obtained for the first time direct visual evidence of both homogenetic and heterogenetic pairing in the F_1 × *Festulpia*. Giemsa C-banding studies of the two parents showed remarkable differences between the banding patterns of these two species – all 21 chromosomes of *F. rubra* possess strong terminal bands at one or both ends, while those of *V. fasciculata* possess no terminal bands but weak centromeric or interstitial bands are present on some of the 14 chromosomes (Fig. 2A,B). Every chromosome in squashes of F_1 × *Festulpia* mitotic cells could be unequivocally identified as originating either from the *Festuca* or from the *Vulpia* parent (Fig. 2C).

Banding was also carried out on meiotic pollen mother cells of × *F. hubbardii*, and this revealed the true nature of pairing. Clear examples of both homogenetic (V–V and F–F) and heterogenetic (V–F) pairing were found; 77.6% of bivalents that could be unequivocally scored were of the former sort (Fig. 2D,E).

Mitotic banding of the hexaploid derivative of the F_1 × *Festulpia* revealed a karyotype very similar to that of *F. rubra* (Fig. 2F). No *Vulpia*-type chromosomes are apparent, and the total number of terminal C-bands is close to if not identical with that in *F. rubra*. It is possible that some interstitial portions of *Vulpia* chromosomes exist. Certainly the hexaploid derivative shows some hybrid characteristics not typical of *F. rubra*, but it is conceivable that the Vulpioid characteristics are mainly or purely cytoplasmic effects (*Vulpia* was the female parent in the original cross and × *Festulpia* was the female parent in the cross producing the hexaploid derivative). It could be that viable backcrosses to a *Festuca* male parent can only be obtained using female F_1 gametes containing no whole (or large parts of) *Vulpia* chromosomes.

These possibilities are at present under investigation, but whatever the resolution of the problem it is clear that a reasonably fertile hexaploid derivative can arise in one generation from a highly sterile pentaploid hybrid that produces rare triploid female gametes containing chromosomes wholly or largely derived from one of the original parents. This process represents a very rapid form of introgression at the polyploid level (in this case from tetraploid to hexaploid).

EVOLUTION OF POLYPLOID COMPLEXES

The *Festuca rubra* aggregate represents a 'declining polyploid complex' in the sense of Stebbins (1971), in which polyploids exist at several levels of ploidy but related diploids are no longer to be found. Such a complex is the fourth in a series of five that Stebbins postulated as 'stages of maturity', ranging from initial to relictual. Although examples of all these five stages are not difficult to find, it is less easy to establish that the formation of polyploids in a group of taxa will necessarily lead to their becoming more common than diploids. It is still less certain that the diploids will become extinct in due course, and there seems no reason to doubt that the

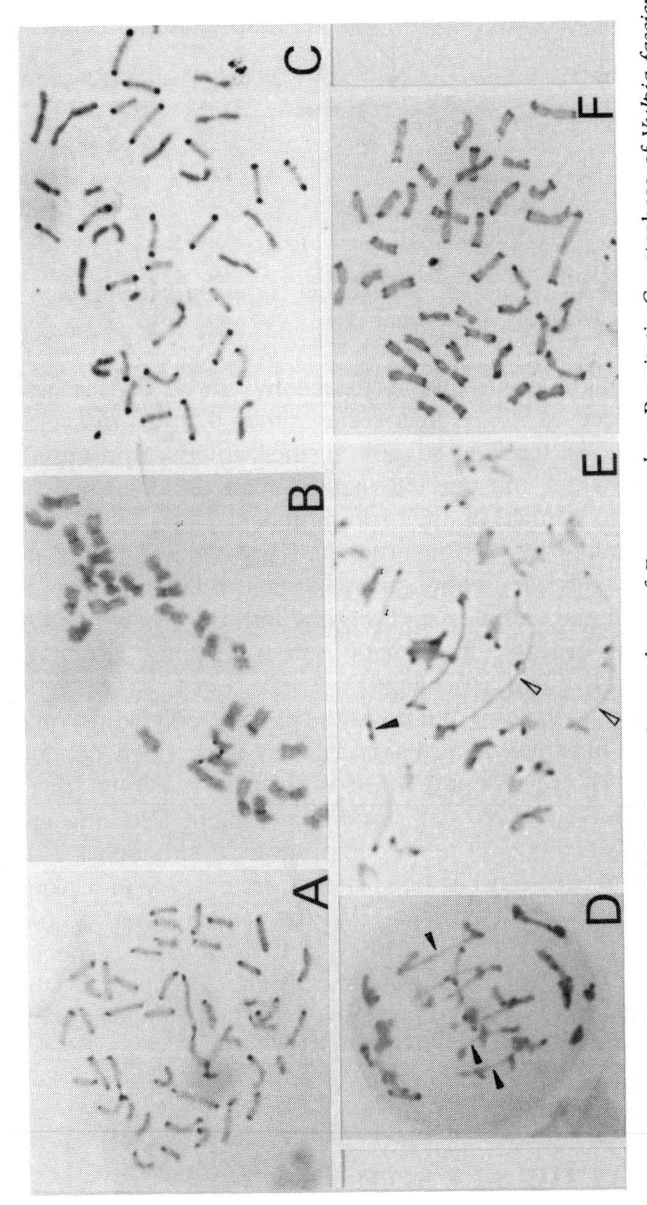

Figure 2 Giemsa C-banded chromosomes of: A: mitotic C-metaphase of *Festuca rubra*; B: mitotic C-metaphase of *Vulpia fasciculata*; C: mitotic C-metaphase of × *F. hubbardii*; D: meiotic late metaphase of × *F. hubbardii* (V-V bivalents arrowed); E: meiotic late metaphase of × *F. hubbardii* (F-F bivalent with solid arrow, F-V bivalents with hollow arrows); F: mitotic C-metaphase of hexaploid backcross derivative of × *Festulpia*.

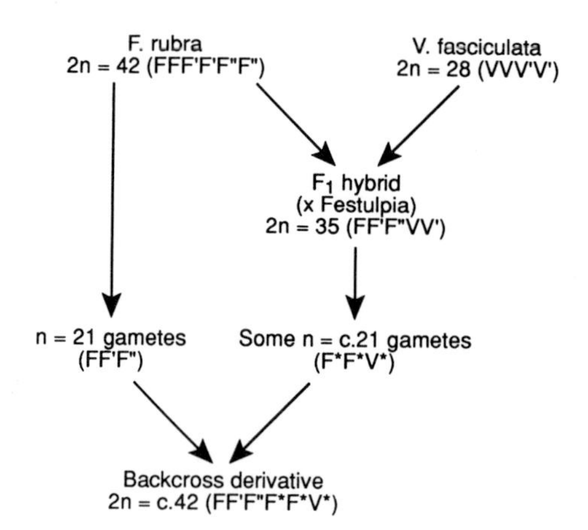

Figure 3 Hypothetical scheme for production of hexaploid backcross derivative of × *Festulpia hubbardii*, assuming homogenetic pairing in the latter.

extinction of polyploids might occur just as frequently. These five 'stages' might, therefore, be better regarded as nodes in a series rather than as a developmental sequence, as the latter concept tends to suggest an inevitability of progress towards diploid extinction. Nevertheless, the gradual extinction of diploids in a polyploid complex can be detected in a range of angiosperm genera.

Use of the word 'declining' by Stebbins signified his concept that a polyploid complex remains a dynamic evolving entity only so long as related diploids exist that can, by hybridization, enter the complex and thereby introduce new genomes to it. Once there are no diploids available the genetic system becomes 'closed' or finite, with mutation being the only means of change.

The introduction of new variation from a quite separate polyploid complex via rare, highly sterile hybrids, illustrates a means of maintaining an 'open' genetic system even where diploid relatives are absent. (There is also the possibility that diploid species of *Vulpia*, such as *V. sicula* or *V. bromoides*, might also introgress into polyploid *F. rubra*, but this has not yet been demonstrated.) Although it is true that in the present example introgression has been demonstrated only in a plant grown in cultivation, there is circumstantial evidence that the same process occurs in the wild. Plants of *F. rubra* with some characteristics of *V. fasciculata* (especially with long awns) are frequently found in coastal areas of Britain occupied by these two species. Their fertile pollen and hexaploid chromosome number has hitherto led to their being identified as variants of *F. rubra*, but the evidence presented above suggests that they might indeed have *V. fasciculata* in their ancestry.

EUPLOID GAMETES AS THE KEY STEP

The key step in this process of polyploid introgression via a highly sterile pentaploid F_1 hybrid is the production by the latter of a small number of triploid embryo sacs. For any given set of chromosome pairing data, the chances of obtaining an FFF (or near-FFF) pollen grain (and therefore male gamete) by random metaphase orientation and anaphase segregation can be calculated. Such a result might occur at extremely low frequency, but in nature millions of attempts are possible and hence such events do occur with significant frequency. A belief that this process might represent an important evolutionary route is supported by examples from several widely differing genera. A few examples will be mentioned briefly here.

In *Ranunculus* L. subgenus *Batrachium* (DC). A. Gray diploids, tetraploids and hexaploids occur. *R. penicillatus* is a hexaploid that is common in rivers in Britain, and its origin seems likely to be similar to the hexaploid backcross derivative of × *Festulpia hubbardii* (Cook, 1966). Apart from its fertility it is often not easy to distinguish from the sterile pentaploid hybrid between the tetraploid *R. fluitans* Lam. and the hexaploid *R. aquatilis* L., from which it is likely to have arisen by the production of triploid gametes fertilized by normal triploid *R. aquatilis* gametes (Fig. 4).

In *Centaurium* Hill, the two tetraploids *C. erythraea* Rafn and *C. littorale* (Turner ex Sm.) Gilmour form hybrids with low fertility in certain areas of western Britain. In Lancashire backcrossing to *C. erythraea* has produced fertile tetraploids that appear as introgressed *C. erythraea* (Ubsdell, 1976), to which the name *C. intermedium* (Wheldon) Druce has been applied (Fig. 5).

In *Euphrasia* L., diploids and tetraploids form triploid hybrids, as well as the diploids and tetraploids each hybridizing among themselves. In some cases the highly sterile triploids have produced haploid gametes in backcrosses to the diploid parent, producing a fully fertile diploid resembling but different from the diploid parent (Yeo, 1956). The three species mentioned in Fig. 6 represent the best-documented example.

In *Pilosella officinarum* F. Schultz & Schultz-Bip., all ploidy levels from diploid to heptaploid are known. The odd-number polyploids are mostly apomictic, but sexual plants occur; the pentaploids produce mostly diploid or triploid pollen, and the heptaploids produce mostly triploid pollen. In crosses between female tetraploids and male pentaploids 75% of the offspring were tetraploid and 20% pentaploid. In crosses between female tetraploids and male heptaploids 94% of the offspring were pentaploid (Gadella, 1991).

In the case of *Pilosella* it is male (rather than female, as in × *Festulpia*) sporogenesis that has selectively produced euploid spores (and hence gametes); in the other three examples the sexual direction of the backcross is unknown, as is the incidence of reciprocal backcrossing of this type.

There is enough circumstantial evidence, backed up by a reasonable amount of experimental work, to indicate that introgression between polyploids via rare, highly sterile hybrids might be a powerful evolutionary force. Its significance can be quantified only by far more experimental work on wild populations of polyploid complexes. It would be revealing to determine what constitutes a viable gamete; in particular the importance of euploidy rather than aneuploidy, and the role of the cytoplasmic environment, seem significant.

ACKNOWLEDGEMENTS

I am most grateful to Dr R. Cotton, Dr C. Barker and Dr J. Bailey for collaboration on *Festuca/Vulpia* research over many years, and for providing the photographs in Figs 1 and 2. Dr Bailey prepared the C-banded squashes in Fig. 2.

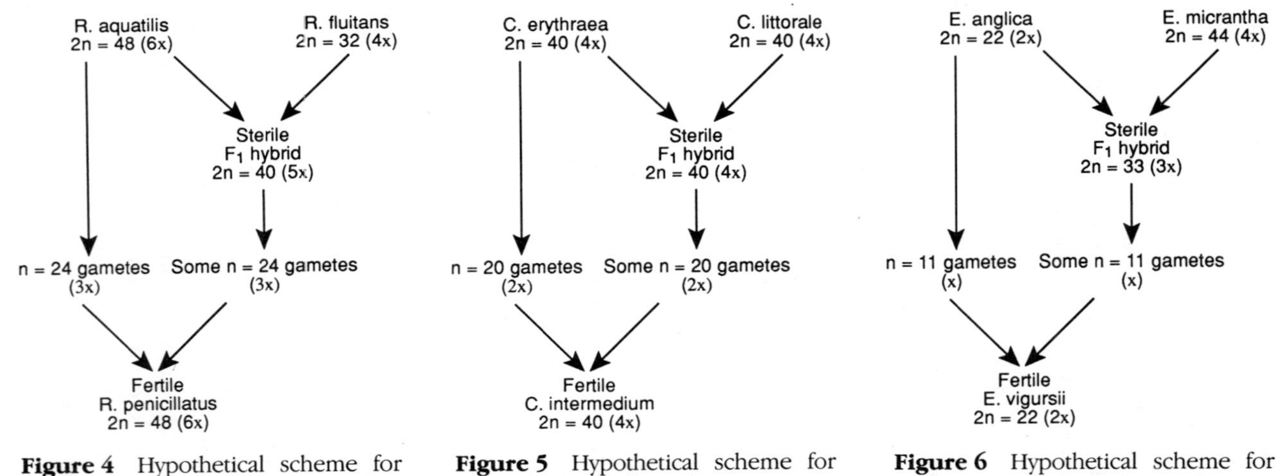

Figure 4 Hypothetical scheme for production of fertile hexaploid *Ranunculus penicillatus* from the sterile pentaploid *R. aquatilis* × *R. fluitans*.

Figure 5 Hypothetical scheme for production of fertile tetraploid *Centaurium intermedium* from sterile tetraploid *C. erythraea* × *C. littorale*.

Figure 6 Hypothetical scheme for production of fertile diploid *Euphrasia vigursii* from sterile triploid *E. anglica* × *E. micrantha*.

REFERENCES

AINSCOUGH, M.M., BARKER, C.M. & STACE, C.A., 1986. Natural hybrids between *Festuca* and species of *Vulpia* section *Vulpia*. *Watsonia, 16:* 143–151.

ASHTON, P.A. & ABBOTT, R.J., 1992. Multiple origins and genetic diversity in the newly arisen allopolyploid species *Senecio cambrensis* Rosser (Compositae). *Heredity, 68:* 25–32.

BAILEY, J.P. & STACE, C.A., 1992. Chromosome banding and pairing behaviour in *Festuca* and *Vulpia* (Poaceae:Pooideae). *Plant Systematics and Evolution, 182:* 21–28.

BARKER, C.M., 1980. 'Investigation into the relationships and ancestry of *Vulpia* and *Festuca*'. Unpublished Ph.D. thesis, University of Leicester, U.K.

BARKER, C.M. & STACE, C.A., 1982. Hybridisation in the genera *Vulpia* and *Festuca:* the production of artificial F_1 plants. *Nordic Journal of Botany, 2:* 435–444.

BARKER, C.M. & STACE, C.A., 1984. Hybridization in the genera *Vulpia* and *Festuca* (Poaceae): the characteristics of artificial hybrids. *Nordic Journal of Botany, 4:* 289–302.

BARKER, C.M. & STACE, C.A., 1986. Hybridization in the genera *Vulpia* and *Festuca* (Poaceae): meiotic behaviour of artificial hybrids. *Nordic Journal of Botany, 6:* 1–10.

BLACKSTOCK, T.H. & ROBERTS, R.H., 1986. Observations on the morphology and fertility of *Juncus* × *surrejanus* Druce ex Stace & Lambinon in north-western Wales. *Watsonia, 16:* 55–63.

COOK, C.D.K., 1966. A monographic study of *Ranunculus* subgenus *Batrachum* (DC.) A. Gray. *Mitteilungen der botanischen Staatssammlung München, 6:* 47–237.

GADELLA, T.W.J., 1991. Variation, hybridization and reproductive biology of *Hieracium pilosella* L. *Proceedings. Koninklijke Nederlandse Akademie van Wetenschappen, Amsterdam, 94:* 455–488.

JAUHAR, P.P., 1975. Genetic regulation of diploid-like chromosome pairing in the hexaploid species *Festuca arundinacea* Schreb. and *F. rubra* L. (Gramineae). *Chromosoma (Berlin), 52:* 363–382.

JOHNSON, M.A.T., KENTON, A.Y., BENNETT, M.D. & BRANDHAM, P.E., 1989. *Voanioala gerardii* has the highest known chromosome number in the monocotyledons. *Genome, 32:* 328–333.

KHANDELWAL, S., 1990. Chromosome evolution in the genus *Ophioglossum* L. *Botanical Journal of the Linnean Society, 102:* 205–217.

OWNBEY, M. & McCOLLUM, G.D., 1954. The chromosomes of *Tragopogon*. *Rhodora, 56:* 7–21.

STACE, C.A., 1975. Wild hybrids in the British flora. In S.M. Walters (Ed.), *European Floristic and Taxonomic Studies:* 111–125. London: Botanical Society of the British Isles.

STACE, C.A., 1986. Hybridisation and plant taxonomy. In B. Jonsell & L. Jonsell (Eds), *Biosystematics in the Nordic Flora. Symbolae Botanicae Upsalienses:* 27 (2).

STACE, C. A. & AINSCOUGH, M. M., 1984. Continuing addition to the gene-pool of the *Festuca rubra* aggregate (Poaceae:Poeae). *Plant Systematics and Evolution, 147:* 227–236.

STACE, C.A. & COTTON, R., 1974. Hybrids between *Festuca rubra* L. *sensu lato* and *Vulpia membranacea* (L.) Dum. *Watsonia, 10:* 119–138.

STEBBINS, G.L., 1971. *Chromosomal Evolution in Higher Plants.* London:Edward Arnold.

STEBBINS, G.L., 1972. Ecological distribution of centres of major adaptive radiation in angiosperms. In D.H. Valentine (Ed.), *Taxonomy, Phytogeography and Evolution:* 7–34. London: Academic Press.

TOMAN, J., 1972. A taxonomic survey of the genera *Petasites* and *Endocellion. Folia Geobotanica et Phytotaxonomica, Praha, 7:* 381–406.

UBSDELL, R.A.E., 1976. Studies on variation and evolution in *Centaurium erythraea* Rafn and *C. littorale* (D. Turner) Gilmour in the British Isles, 2. Cytology. *Watsonia, 11*: 33–43.

UHL, C.H., 1978. Chromosomes of Mexican *Sedum*, II. Section *Pachysedum*. *Rhodora, 80:* 491–512.

WILLIS, A.J., 1975. *Festuca* L. × *Vulpia* C.C. Gmel. = × *Festulpia* Melderis ex Stace & Cotton. In C.A. Stace (Ed.), *Hybridisation and the Flora of the British Isles:* 552–554. London: Academic Press.

YEO, P.F., 1956. Hybridisation between diploid and tetraploid species of *Euphrasia*. *Watsonia, 3:* 253–269.

CHAPTER

11

Rapid evolution in plant populations

M.S. DAVIES

Introduction	**172**
Potential for evolution	**172**
Evolutionary divergence over short distances	**173**
Reproductive isolation	**176**
Rapid evolution in time	**178**
Rapid evolution in crop domestication	**179**
Variability within populations	**181**
Conclusions	**184**
References	**184**

Keywords: Genetic variability – rapid evolutionary divergence – reproductive isolation – domestication of cereals.

Abstract

Evolutionary divergence between populations within plant species can occur over extremely short distances in the face of considerable gene flow and can also occur over very short timescales (< 10 years). There is field evidence which indicates the early stages of the evolution of reproductive isolating mechanisms between closely adjacent populations. Each field situation is an unique circumstance in which the evolutionary response of a given population to particular selection pressures is largely limited by the amount of heritable variation in that founder population. Considerable small-scale spatial heterogeneity can occur in the field and the genetic variability within clearly defined plant populations is related to this habitat variability. The domestication of early crops from their wild-type ancestors is an example of accelerated evolution and, in the case of primitive cereals such as the diploid wheats, this process may have occurred extremely rapidly.

Evolutionary Patterns and Processes
ISBN 0-12-440895-8

INTRODUCTION

Many plant species occur in a wide range of environments. Such wide ecological amplitude is generally achieved by the action of natural selection acting on genetically determined variation in morphological, physiological and reproductive characters, leading to the evolution of genetically differentiated populations, each of which is specifically adapted to its local environmental conditions (Heslop-Harrison, 1964).

Marked differences in the environment can occur over extremely short distances and can also be generated over relatively short periods of time; such small-scale spatial and temporal environmental heterogeneity often results from man's activities and presents the opportunity to study evolutionary processes over short distances, often over well-defined timescales and generally in the absence of reproductive isolation between diverging populations. This chapter considers briefly some examples of such evolutionary events in plant populations.

POTENTIAL FOR EVOLUTION

For any adaptation to evolve by natural selection there must be heritable variation for the character in question. There is a wealth of evidence that plant species are considerable reservoirs of genetic variation, thus allowing rapid evolutionary responses to particular directional selection pressures. For example, there are a number of experimental studies which have shown that populations of plant species, sampled from uncontaminated sites, contain heavy-metal-tolerant individuals at low frequencies; when seed of these 'normal' populations is sown on contaminated soil, metal-tolerant survivors can be selected within a single generation. In nature these individuals would act as founders for metal-tolerant populations at the contaminated site. However, genes for metal tolerance are not universally present in all populations of plant species. Some species lack genetic variability for tolerance to specific metals, thus resulting in their exclusion from soils contaminated by that metal (Gartside & McNeilly, 1974; Ingram, cited in Bradshaw, 1984). However, caution must be exercised in extrapolating the findings of studies of single populations of a species to the species as a whole. Symeonidis, McNeilly & Bradshaw (1985) found that zinc tolerance, although generally very common in the grass species *Agrostis capillaris*, was absent from certain populations. Al-Hiyaly, McNeilly & Bradshaw (1990) reported quite contrasting patterns in the occurrence of zinc-tolerant populations in five grass species on zinc-contaminated soil adjacent to the legs of galvanized electricity pylons in North Wales and Lancashire. Despite the fact that all species had evolved zinc-tolerant populations under at least some pylons, tolerant individuals of all species were not found beneath all pylons, even when the species was present in the surrounding vegetation. For example, the grasses *Anthoxanthum odoratum* and *Festuca ovina* were abundant in the surrounding vegetation of virtually all the pylons examined, yet individuals of these species occurred on the zinc-contaminated soils beneath just two and three pylons respectively. In contrast *Deschampsia caespitosa* failed to colonize the soil around the pylon legs only when it was absent from the surrounding vegetation. The result for *A. odoratum* parallels the finding that this species, despite having a wide distribution in relation to soil type, has produced zinc-tolerant populations on only one of the many zinc-contaminated mine spoils in North Wales.

There were considerable differences in the frequency of zinc-tolerant individuals between populations of *A. capillaris* sampled from nine pylons in North Wales (Al-Hiyaly, McNeilly & Bradshaw, 1988); this was interpreted as being due, at least in part, to chance variation in the availability of heritable variation in zinc-tolerance between the founding populations in the region of each pylon. Thus each pylon represents an unique evolutionary situation in which the magnitude of evolutionary change in the plant population will depend partly upon the availability of heritable variation in the founder population in the locality prior to the erection of the pylon, as well as upon the magnitude of selection pressures caused by the zinc-contamination.

Metal tolerance is a character which can be rapidly and easily measured using a rooting test (Wilkins, 1978). Consequently, much more work has been done on the occurrence of metal-tolerant genotypes in the 'gene pools' of plant populations than for other characters which are more difficult to evaluate. However, it is likely that there is similar variation between natural populations of plant species in the extent of heritable variation for other characters, which may be a reflection of mutation rate, population size, life history, breeding system and reproductive biology and also the past and present selective regimes to which the population has been subjected. The evolutionary responses of each population in respect of these characters to particular selection pressures will reflect these inherent differences in genetic diversity. Some populations fail to show evolutionary responses to selection because of lack of heritable variation for the selected character, thus exhibiting the phenomenon of 'genostasis' (Bradshaw, 1984). There are, however, very few studies where the evolutionary responses of plant populations to specific selection pressures in the field has been examined in 'replicated situations' as described above.

EVOLUTIONARY DIVERGENCE OVER SHORT DISTANCES

Marked differences in environmental factors can occur over very short distances. Examples of such localized environmental heterogeneity include metal-contaminated soils resulting from mining activity and which are surrounded by uncontaminated soils, localized variation in soil chemistry resulting from differential application of lime and fertilizers, woodland and adjacent unshaded habitats and variation in soil flooding resulting from small-scale variation in topography and hydrology. At the outset such localized environmental heterogeneity would impose disruptive selection pressures on the source population inhabiting the area, followed by directional selection for genotypes with increased fitness in the divergent habitats and perhaps finally stabilizing selection operating in each microhabitat if it remained unchanged with time. Such localized variation in the environment presents ideal situations in which to study evolutionary processes since the plant populations in the different habitats are, in many instances, exchanging genes and thus it is possible to examine the relative influences of differential selection pressures and gene flow on evolutionary divergence.

Mine spoil heaps, contaminated by heavy metals, have been used for several such studies (Jain & Bradshaw, 1966). Where sharp boundaries occur between the mine spoil and adjacent uncontaminated soil, differentiation of plant populations within a species can occur over a matter of metres, resulting in a steep cline in metal tolerance of adult plants across the boundary (e.g. Jain & Bradshaw, 1966; Antonovics &

Bradshaw, 1970). The steepness of the cline will depend mainly on two factors: the extent of gene flow between the diverging populations and the intensity of selection operating on genotypes in the contrasting habitats. Antonovics & Bradshaw (1970) showed that some characters (e.g. zinc-tolerance) of the grass *A. odoratum* showed a steep clinal change across the boundary between the Trelogan zinc/lead mine in North Wales and normal pasture; other characters showed a more gradual clinal change while others showed no change at all. All characters were measured on the same plants and thus were subjected to the same amount of gene flow between mine and pasture plants. The steepness of the cline for a particular character was a reflection of the intensity of the selection operating on that character in the various sites along the transect.

The interplay between selection and gene flow on the pattern of divergence at mine–pasture boundaries was nicely demonstrated in *Agrostis capillaris* on a small copper mine in Snowdonia (McNeilly, 1968). The mine is situated in a steep valley and the wind direction is highly polarized for most of the flowering season. McNeilly compared two transects spanning the mine–pasture boundary, one on the upwind and the other on the downwind side of the mine. In each transect, both adult plants and seed were collected in each sampling zone; the former had been subjected to selection while the latter had been influenced by gene flow via pollen transfer but had not yet been exposed to selection. Comparisons of seed and adult populations showed that considerable gene flow occurred from pasture to mine plants in the upwind transect and from mine to pasture plants on the downwind transect. However, the cline in increasing copper tolerance in adult plants was extremely steep in the upwind side of the mine, indicating that selection against plants with low copper tolerance on the mine was intense and able to counteract the effects of gene flow. On the downwind side of the mine, however, the cline showing decreased copper tolerance from mine to pasture plants was much shallower, reflecting the weaker selection against tolerant plants on the pasture. The effects of gene flow from mine plants was still evident 160 m downwind from the mine boundary. Selection pressures against metal-tolerant phenotypes on normal soil seem not to be directed against the tolerant character itself but rather against the poorer growth and competitive ability of mine phenotypes which has arisen as a result of simultaneous selection pressures on the mine for adaptation to the poor nutrient status of most mine spoils (Hickey & McNeilly, 1975; Macnair & Watkins, 1983; Nicholls & McNeilly, 1985).

The Park Grass Experiment at Rothamsted is another area which has proved to be an ideal field laboratory for the study of evolutionary divergence of plant populations over short distances. Various fertilizer treatments have been imposed annually since 1856 and differential liming treatments have been applied at 4-year intervals since 1903. Apart from these the plots, which vary in size from 25 × 15 m to 40 × 30 m, have been managed identically (Brenchley & Warrington, 1958; Thurston, 1969). The vegetation is cut for hay in June each year and cut again between September and October. The fertilizer and liming treatments have resulted in very large differences in soil chemical composition which is reflected in vegetation height and yield and the botanical composition of the vegetation (Thurston, 1969; Snaydon & Davies, 1972). The area as a whole is now a mosaic of contrasting soil and aerial environments with very sharp boundaries between them. The grass *A. odoratum* occurs on most of the plots; populations collected from contrasting plots differ in a number of morphological and physiological characters which are correlated with analagous measures of the vegetation and soil chemical environment of their source plots (Snaydon & Davies, 1972; Davies & Snaydon, 1973a,b, 1974; Davies, 1975). Evolutionary divergence between populations of this short-lived outbreeding species has occurred over distances of less than 30 m and within 60 years, with each population being specifically adapted to the environment of its source plot. In some instances differences between populations from adjacent plots in physiological characters are as large as those between quite ecologically contrasting plant species (Davies & Snaydon, 1973a).

Snaydon & Davies (1976) sampled plants of *A. odoratum*, both as adults and as seed, along a 4-m transect across a very sharp boundary between two plots of the Park Grass Experiment which differed in nitrogen and potassium fertilizer application. There was marked differentiation between adult populations in a number of morphological characters. For example, the dry weight yield and panicle height differed significantly between adult populations sampled only 10 cm apart at the boundary region (Fig. 1A,B) but there was less differentiation between plants derived from the seed collection. Comparison of the seed and adult populations allows estimates of the direction and magnitude of gene flow. Appreciable effects of gene flow were apparent up to 0.5 m on the downwind (left hand) side of the boundary and detectable amounts up to 2 m downwind (Fig. 1C). Despite this gene exchange,

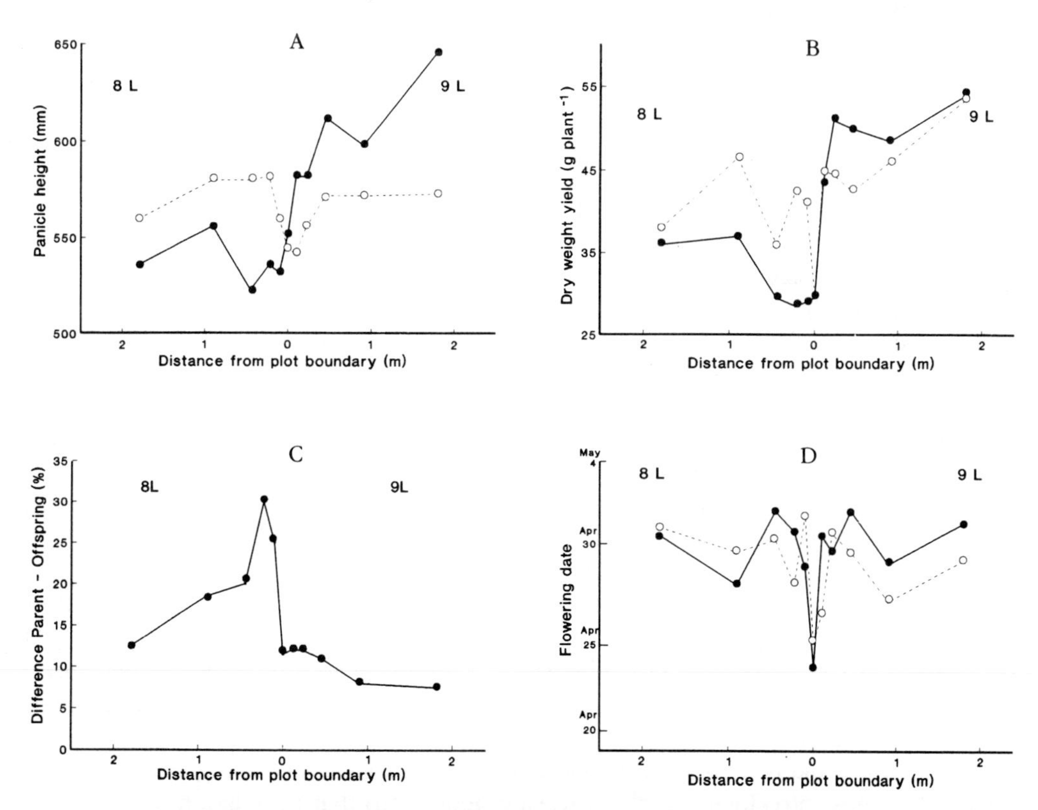

Figure 1 Characters measured on populations of *Anthoxanthum odoratum* collected at various distances from the boundary along a transect across the boundary between plots 8 limed and 9 limed of the Park Grass Experiment, Rothamsted, when grown in a spaced plant trial. A: Panicle height; B: dry weight yield per plant; C: the mean percentage difference between seedling (progeny) populations and tiller (parent) populations, calculated over 11 attributes; D: flowering date. ○ Populations collected as seed (progeny); ● populations collected as tillers (parent).

sharp differences have been maintained between adult populations growing *in situ,* illustrating the capacity of selection in the various sampling zones to negate the effects of this gene flow.

REPRODUCTIVE ISOLATION

Although evidence from field situations, such as those described above, indicates that evolutionary divergence between closely adjacent populations can occur in the face of considerable gene flow, it might be expected that any mechanism which would reduce the amount of gene exchange between diverging populations in the contrasting environments would be an advantage. For example, if the hybrids between the two divergent populations are at a disadvantage in either habitat, selection might favour assortative mating (Endler, 1977).

The power of selection in bringing about rapid reproductive isolation is well-documented in artificial selection experiments. Selection against hybrids between lines of *Drosophila* resulted in partial reproductive isolation of the lines (Thoday & Gibson, 1962). Paterniani (1969) reduced the degree of natural outcrossing between the two lines of maize from 40% to less than 5% within five generations by selection of individuals showing the least outcrossing. This resulted from both a divergence in time of flowering and a reduction in pollen compatibility between the two lines.

There is evidence from field situations of the early stages of the evolution of reproductive isolation between closely adjacent populations. McNeilly & Antonovics (1968) showed that mine plants of *A. odoratum* and *A. capillaris* flowered about a week earlier than adjacent pasture populations; the differences in flowering time were maintained in culture and appeared to be genetically determined. The flowering time differences in *Agrostis* were heritable with dominance in the direction of early flowering (Nicholls, 1979, cited in Macnair, 1981). Ducousso *et al.* (1990) found that populations of *Arrhenatherum elatius* from mine sites flowered on average 10–15 days later than populations from normal soils in the same area; the mine population continued flowering well into the autumn whereas normal populations generally completed flowering by mid-July. These late flowering mine plants would be effectively isolated from the non-tolerant plants off the mine.

It is possible that the differences in flowering time have arisen in response to mechanisms other than by direct selection for reproductive isolation (Wallace effect). The soils of mines and pastures differ in characters other than metal content, for example, in soil moisture status and soil nutrient status; these factors may well affect the growth performance of plants and perhaps also flowering time. Stam (1983) proposed a model illustrating how phenotypic differences in time of flowering caused by the growth environment could eventually lead to genetically controlled differences in flowering time; for example, plants which flowered early for environmental reasons would be more likely to mate with individuals whose early flowering is determined genetically, thus introducing early flowering genes into that population.

In the Park Grass transect (see above) there were no differences in time of flowering in uniform culture between the majority of populations collected along the transect. However, the adult population sampled in a 100-mm-wide strip along the boundary between the two plots flowered 6 days earlier than populations either side of it (Fig. 1D) and completed the flowering process more quickly (Snaydon, 1973). This pattern occurred in two contrasting years of monitoring and thus is unlikely to be due to environmental effects. The seed progeny population in this sampling zone also showed the earlier flowering trait. There was considerable similarity between adult plants and their seed progeny in this population for a number of morphological characters (Snaydon & Davies, 1976). This boundary population seems to be temporally isolated from its adjacent neighbours. A similar phenomenon was observed in *Anthoxanthum* at another plot boundary on Park Grass (R.W. Snaydon, personal communication). However, the evolutionary significance of the earlier flowering of the plants occupying the narrow zone of the boundary is not clear. Although this population may itself be reproductively isolated from its neighbours it does not eliminate gene flow across the boundary since appreciable gene flow was detected in the population 0.5 m downwind from the boundary (see above).

Antonovics (1968) demonstrated that in the outcrossing species, *A. capillaris* and *A. odoratum*, mine populations had a greater degree of self-fertility than adjacent pasture populations; the differences in self-fertility were heritable and were interpreted as a mechanism for reducing gene flow between mine and pasture plants. However, the mean population self-fertility was low ($< 10\%$) and the increase in population self-fertility was due to a few highly self-fertile individuals in the tolerant population. Mine populations of the grass *A. elatius* have nearly three times the seed set under artificially imposed selfing than populations from normal soils and a lower seed set than normal populations under outcrossing (Lefèbvre & Vernet, 1990); some mine individuals had potential self-fertilities of greater than 50%. Similarly, in *Armeria maritima*, high seed set under artificial selfing was recorded in individuals from mine soils (Lefèbvre & Vernet, 1990). Arctic populations of this species are highly self-compatible. However, studies on natural populations from mine soils in Europe failed to detect a significant degree of selfing in the wild. Thus although a greater capacity for self-fertility can sometimes be demonstrated experimentally in mine plants, there is no conclusive evidence that a significant degree of self-fertilization occurs in the field in populations of outcrossing species on mine spoils. There is clearly a need for more extensive work in this area using morphological, enzymic, and perhaps molecular markers to investigate the breeding system of closely adjacent plant populations in the field.

Crosses between mine and pasture plants produce as many viable seeds as crosses within population types in *A. odoratum, A. capillaris* and *A. maritima* (McNeilly & Antonovics, 1968; Lefèbvre & Vernet, 1990). In *A. elatius,* however, the germination of pollen from mine and pasture plants was significantly lower on the stigma of the opposite population type than on stigmas of the same population type (Lefèbvre & Vernet, 1990); crosses within types produced greater seed set than crosses between types.

Macnair & Christie (1983) have reported a gene conferring post-mating isolation in *Mimulus guttatus;* this is closely linked to or pleiotropic to the gene conferring copper tolerance. Crosses between plants from a copper mine and non-tolerant genotypes revealed a post-mating barrier resulting in the death of some F_1 offspring. If strong selection on the mine area favours the isolating gene directly, or because it is linked to or pleiotropic to the gene for tolerance, it would be fixed rapidly in the population. This is one of the very few examples where an isolating barrier is closely associated with a major gene which is favoured by selection. Such a system might promote very rapid divergence between populations and perhaps accelerate greatly the process of speciation.

The above observations are based upon short-term studies. Long-term studies are required to determine whether reproductive isolation between diverging populations increases with time. However, it is already clear that such reproductive isolation is not a prerequisite for evolutionary divergence, which has occurred in the face of considerable gene flow, but it may nevertheless reinforce and accelerate the effects of differential selection. There are many examples of situations where interfertile species co-exist in mosaic environments, yet maintain their discrete identities (Heslop-Harrison, 1956; Briggs, 1962). Many species have no intrinsic isolating mechanisms and may hybridize freely, or do under certain environmental conditions (see Chapter 10). In view of the fact that reproductive isolation can evolve rapidly under selection (Thoday & Gibson, 1962; Paterniani, 1969), it is perhaps surprising that intrinsic isolation is not the general rule. Several authors have concluded that isolation mechanisms may be the result, rather than the cause, of evolutionary divergence and speciation (Lewis, 1966; Ehrlich & Raven, 1969). Long-term studies on adjacent plant populations could be used to test these ideas.

RAPID EVOLUTION IN TIME

There are well-documented examples of very rapid changes in the structure of plant populations. Rapid shifts in the genotypic composition of mixtures of cultivars of crop species can occur both in annual crops (Harlan & Martini, 1938; Blijenburg & Sneep, 1975) and in perennial crops (Charles, 1961, 1964; Brougham & Harris, 1967). Rapid responses to artificial selection for herbicide resistance have been demonstrated (Holliday & Putwain, 1977) and variation between wild populations in simazine resistance is correlated with the duration of simazine treatment in their source habitats (Holliday & Putwain, 1980). Widespread herbicide resistance occurs in weed populations in North America, following continuous application of herbicides over the past few decades (Ryan, 1970; Bandeen *et al.*, 1979).

Rapid evolution of resistance to sulphur dioxide pollution has also been demonstrated (Taylor & Murdy, 1975; Bell & Mudd, 1976). Wilson & Bell (1985) showed a rapid increase in the frequency of SO_2-tolerant genotypes over a 3–4-year period in sown plots of several grass species at a polluted site, but the tolerance of the populations later disappeared following a decline in ambient SO_2 levels. They also examined lawns of different ages in a polluted area and demonstrated an evolutionary series of increasing tolerance to acute SO_2 injury with increasing age of the lawn.

Marked divergence between populations of *Poa annua* from greens, fairways and roughs of a golf course in temperature-induced seed dormancy has been shown to occur within 20 years (Wu, Till-Bottraud & Torres, 1987).

Contamination of soils by toxic metals presents situations where evolutionary processes can be studied over known timescales since the history of contamination is often well-defined. Thus Al-Hiyaly *et al.* (1988) showed the evolution of zinc tolerance in populations of *A. capillaris* under galvanized electricity pylons which were between 21 and 31 years old. In *D. caespitosa* and *A. odoratum*, zinc-tolerant populations have evolved beneath pylons within 17 years (Al-Hiyaly *et al.*, 1990). Snaydon (cited in Bradshaw, McNeilly & Gregory, 1965) found that evolution of zinc tolerance had occurred in a population of *Agrostis canina* under a galvanized wire fence within 25 years.

Even more remarkable rates of evolution have been reported in response to aerial pollution by toxic metals. Zinc-tolerant populations have evolved in *A. capillaris* within 3 years after the beginning of emissions from an adjacent zinc smelter in Germany (Ernst, 1976; Ernst, Verkleij & Vooijs, 1983). Similarly, rapid evolution of copper tolerance has occurred in *Agrostis stolonifera* in the vicinity of a copper refinery at Prescot, Merseyside (Wu, Bradshaw & Thurman, 1975). Different aged lawns at this site represent a historical series of the evolution of increased copper tolerance over 70 years. Even in a lawn which was only 4 years old, there was a marked shift towards an increased frequency of copper-tolerant individuals compared with populations from unpolluted sites. The high copper tolerance of the populations from the older lawns was not the result of vegetative spread of a few copper-tolerant individuals, since these populations consisted of a large number of different genotypes. Evolution of copper tolerance in these populations has arisen as the result of rapid selection of copper-tolerant genotypes that occur at low frequency in normal populations.

The rapid evolution of metal tolerance in plant populations reflects the intensity of selection pressures operating in metal-contaminated sites (Jain & Bradshaw, 1966). All studies to date show that metal tolerance also has a high heritability. Only a small number of genes are generally involved in metal tolerance (Macnair, 1990), although the character is polygenically controlled in some species. The direction of dominance is normally towards tolerance. Characters with high heritability and showing dominance would be expected to respond more rapidly to selection than those with low heritability or which are recessive.

The marked morphological and physiological divergence of populations of *A. odoratum* on the Park Grass Experiment (Snaydon & Davies, 1972; Davies & Snaydon, 1973a,b, 1974; Davies, 1975) have arisen within 60 years. However, in 1965 new liming treatments were imposed, with the aim of producing the same range of soil pH on each fertilizer plot (Warren, Johnston & Cooke, 1965). These have resulted in marked changes in soil chemical composition, yield of herbage and botanical composition (Johnston, 1972; Williams, 1974, 1978). Snaydon & Davies (1982) found that genetic divergence in a number of morphological characters had occurred in *A. odoratum* in response to the new liming treatments only 6 years after their imposition. The rapidity of these changes rivals that resulting from metal contamination and is a reflection of the intense selection pressures that occur in what might be regarded as 'normal' environments (Davies & Snaydon, 1976).

RAPID EVOLUTION IN CROP DOMESTICATION

The change of early man from hunter-gatherer to farmer resulted in marked shifts in the selection pressures which operated on the ancestors of our modern crops, particularly the cereals. This change in practice eventually brought about the domestication of crop plants. Domestication was first recognized as an example of accelerated evolution by Darwin (1859, 1868) and de Candolle (1886) but it was

Vavilov (1926) who first postulated specific evolutionary pathways for the domestication of cereals such as wheat and barley. The narrow classical definition of domestication covers the process which occurs under cultivation in populations of the early wild-type crops, which were sown originally from seed gathered from wild stands. The process advantages rare mutants in the wild-type population and continues until these mutants dominate the crop population and the original wild-type phenotypes are eliminated or maintained at very low frequency. In many instances, because of the lack of wild-type adaptive features, the domestic type may not survive outside cultivation. There has been considerable speculation in the literature about the timescale required to achieve domestication; the estimates vary widely, from a few years to over a millennium!

In cereals such as wheat, even the most primitive domesticated forms differ from their wild-type progenitors in a number of polygenically determined characters such as awn robustness, grain size, number of fertile florets, tillering characteristics, uniformity of grain ripening, etc. However, it is generally accepted that the most critical adaptive changes occurring through domestication are the loss of wild-type seed dormancy and the loss of rachis fragility and only the latter is readily apparent morphologically. In the wild-type diploid wheats (*Triticum monococcum* subsp. *boeoticum*) the mature rachis disarticulates between each of the fertile spikelets allowing spontaneous shedding. The arrow-like morphology of the spikelets, their long springy awns and backward pointing barbs and hairs ensures that they penetrate the surface litter and wedge themselves in cracks in the ground where they are better protected against rodents and seed-eating ants. However, even in the most primitive of the domesticated diploid wheats *T. monococcum* subsp. *monococcum*), the rachis fails to disarticulate spontaneously and the ears remain intact until the crop is harvested and threshed and is termed 'semi-tough'. These domestic forms will not survive in the wild since their spikelets are not efficiently disseminated and are not protected from predation. In modern diploid wheats the rachis fragility character is controlled by two unlinked loci (Sharma & Waines, 1980) but in the initial stages of domestication mutation to the recessive semi-tough rachis character at only one locus was likely to have been involved.

In order to estimate the possible timescale over which domestication of the diploid wheats might have occurred, it is necessary to estimate the relative advantages of the fragile-rachis and semi-tough-rachis phenotypes under various harvesting methods. Prior to cultivation, hunter-gatherers probably harvested grain from wild stands by beating or using a 'swinging basket', thus gathering the fragile spikelets into a suitable container. Following the change to cultivation, the early farmers would probably have harvested their crops by uprooting or by sickle-harvesting, both of which would favour the semi-tough-rachis homozygotes (Hillman & Davies, 1990).

In a field trial in Turkey (Hillman & Davies, 1990), harvesting by beating clearly favoured the fragile-rachis wild-type form but harvesting by either sickles or uprooting gave estimates of selection coefficients of *c.* 0.6 against the fragile-rachis type, compared with 0 for the semi-tough-rachis domestic type. Assuming that the semi-tough-rachis allele was present only in heterozygotes in wild populations at a low frequency of 10^{-6} (homozygotes would be lethal in the wild) a simple computer model was used to predict the increase in frequency of semi-tough-rachis homozygotes under cultivation at a range of selective values against the fragile-rachis wild-type. The model also incorporated various levels of selfing although primitive and

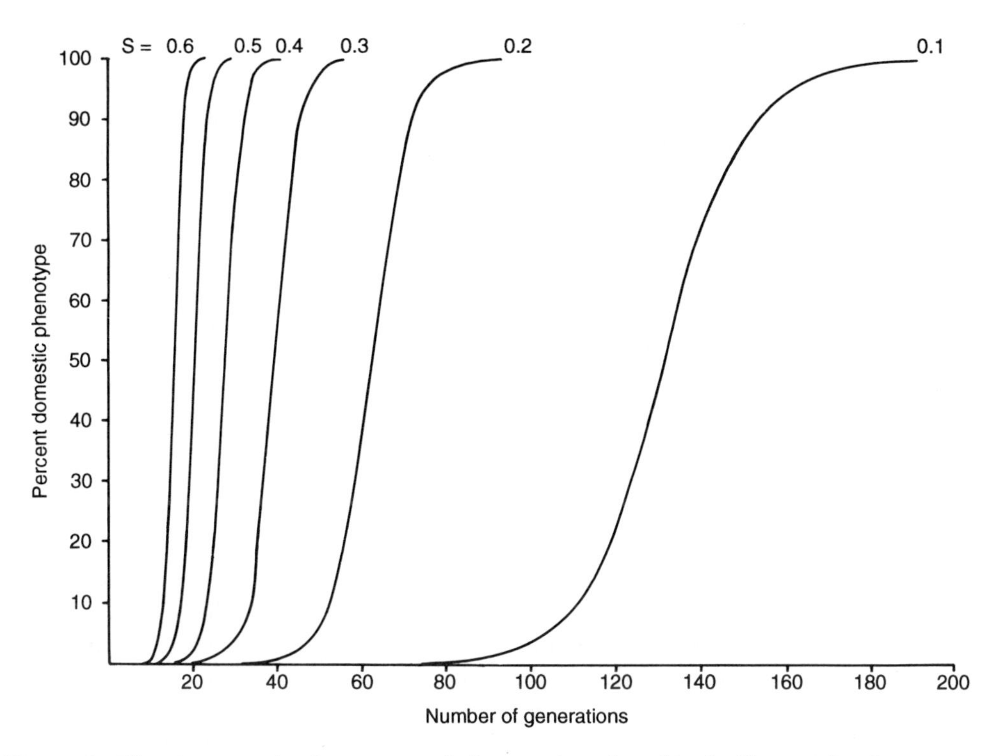

Figure 2 The increase in frequency of the semi-tough-rachised (domestic) phenotype in populations of initially fragile-rachised wild-type einkorn wheat (*Triticum monococcum* subsp. *boeoticum*) under a range of selective intensities against the wild-type, assuming 100% inbreeding and an initial frequency of the semi-tough-rachis allele of 10^{-6} (with annual sowing and one plant generation per annum).

domesticated diploid wheats are almost entirely inbreeding (Hillman & Davies, 1990) and assumed that farmers retained part of their harvest as seed corn for the following year's crop.

Figure 2 shows the predicted increase in frequency of the domesticated form at various levels of selection against the wild-type and assuming 100% selfing. At the selection coefficients estimated in the Turkish field trial (0.6), the semi-tough-rachis phenotype reaches fixation in only 20 generations. However, it is probably realistic to assume lower selective values against the wild-type; even at a selection coefficient of 0.2 against the brittle-rachis wild-type, which might occur if farmers harvested the crop when not fully ripe, domestication is still complete in *c.* 70 generations.

This would be an extremely rapid evolutionary change resulting from a change in farming practice and, in archaeological terms, is virtually instantaneous. Thus the process of domestication would stand little chance of being recognized as a clinal process in plant remains recovered from archaeological sites.

VARIABILITY WITHIN POPULATIONS

Many studies on natural plant populations have demonstrated the existence of considerable variability between genotypes within populations. The factors controlling the degree of genetic variability are numerous and complex and include the breeding system, life-cycle, population size, mutation rate, recombination rate and the nature and intensity of selection. Several theoretical models have predicted that genetic variability should be greater in heterogeneous environments (e.g. Maynard Smith, 1962; Levins & MacArthur, 1966; Maynard Smith & Hoekstra, 1980) and there have been a number of laboratory studies aimed at testing these ideas (e.g. McDonald & Ayala, 1974; Minawa & Birley, 1978). The topic has been reviewed by Hedrick (1986) and for plants by Ennos (1983).

There are, however, few studies which attempt to relate genetic diversity within clearly defined populations in the field to measures of environmental heterogeneity on a microscale. The Park Grass Experiment allows us to examine these relationships in populations from the contrasting liming and fertilizer plots. The genetic variation within each population of *A. odoratum* in relative growth response to acid and calcareous soils is closely correlated with the spatial variability in soil pH on the source plot of that population (Fig. 3A). Correlation between genotypic variability within populations and measures of environmental variability can be demonstrated on an even smaller scale in the population samples collected in various zones along the transect across the boundary between plot 9 limed and 8 limed (see above); measures of genotypic variability within the subpopulations are closely correlated with measures of spatial variability in the environment of their sampling zone (Fig. 3B).

The clear inference from these correlations is that particular genotypes are associated with particular microhabitats in the area. However, this has not been tested experimentally. Harberd (1961) and Silander (1979) have shown that the spatial distribution of individual clones of vegetatively reproducing grasses is related to small-scale variation in soil factors. Rocovich & West (1975) found a highly significant correlation between the arsenic tolerance of individual genotypes of *Andropogon scoparius* sampled from an arsenic mine and the arsenic content

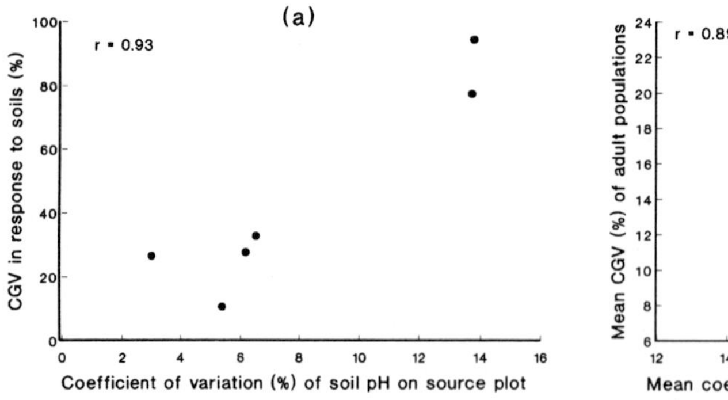

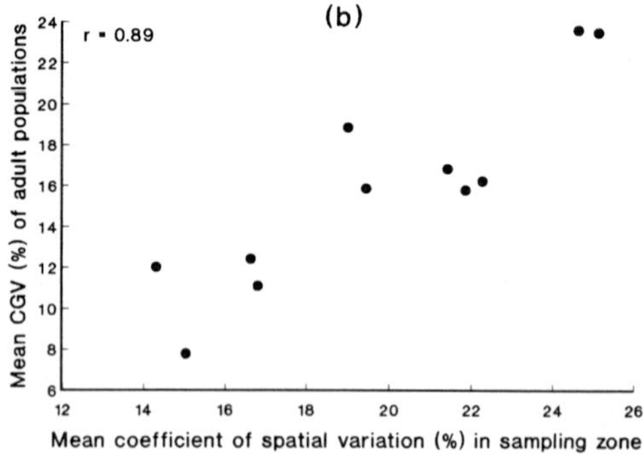

Figure 3 Relationship between genetic variation within tiller populations of *Anthoxanthum odoratum* from the Park Grass Experiment, Rothamsted and measures of the spatial variability in the environment from which they originated. A: Coefficient of genetic variation in relative response to calcareous and acidic soil in relation to the coefficient of variation in soil pH on the source plot. B: Mean coefficient of genetic variation for 10 morphological characters, measured in a spaced plant trial, within tiller populations sampled along a transect across the boundary between plots 8 limed and 9 limed, in relation to the mean coefficient of spatial variability (vegetation height and soil pH) in the sampling zone in which they were collected.

of the soil in the precise area in which the plant was growing. Selection acts upon individuals in the immediate environment in which they germinate and establish. Most studies to date have related the mean morphological and physiological characteristics of populations, defined and delimited from each other in arbitrary ways, to mean measures of the environment in their habitat. There is clearly a need to narrow the scale of microevolutionary studies to include investigations of specific features of individual genotypes and relate these to their specific microhabitat.

However, spatial heterogeneity in the environment is more diverse than may be at first apparent. In addition to spatial variation in measurable physical and chemical characters, the environment will vary in 'biotic' factors, e.g. the influence of other genotypes of the same and different species and the complex competitive interactions which can thereby arise. The influence of pests and disease will also increase the environmental complexity and diversity. Thus it is likely that individual genotypes are subject to selection not only in terms of the physical and chemical aspects of their immediate locality but also for co-existence with members of the same and other species in that microenvironment. There is evidence that 'co-adaptation' can occur among genotypes of *Drosophila* (Seaton & Antonovics, 1967) and genotypes of cereals (Allard & Adams, 1969) but evidence from field situations is more limited. Turkington & Harper (1979) concluded that micropopulations of *Trifolium repens* had evolved within a single pasture, each of which was adapted to co-exist with the specific grass species in its precise area of origin in the field, that is, each showed 'biotic differentiation'. However, the micro-habitats also varied edaphically and this, together with other limitations of the experimental system, precludes conclusive interpretation of the data. Populations of four grass species collected from various plots of the Park Grass Experiment were grown together in two-species mixtures (Snaydon, 1978). Mixtures of species-populations from the same plot yielded more when grown together than did combinations from different plots, both on fertilized and unfertilized soil, and also formed more stable associations under experimental conditions than those from different plots. Similarly, Evans *et al.* (1985) found that populations of white clover, from geographically separate sites in Europe, performed better under experimental conditions when grown with the perennial ryegrass population collected from the same site than with an 'alien' population. The grass population also generally grew better when combined with the co-existing clover population. Thus there is preliminary evidence of co-adaptation between species-populations in natural environments but the data are scanty and much more experimental work is required in this area.

Individual genotypes may also cope with a degree of environmental heterogeneity in both time and space by phenotypic plasticity; there are genetically determined differences between individuals in their capacity to respond phenotypically to the environment (e.g. Winn & Evans, 1991). Plastic responses of the phenotype to environmental factors and the breadth of tolerance of individual genotypes to a range of environments, are characters which are themselves subject to natural selection and evolutionary processes (for reviews see Bradshaw, 1965; Schlichting, 1986; Sultan, 1987; Bradshaw & Hardwick, 1989).

CONCLUSIONS

There is now ample field evidence that evolutionary divergence can occur in plant populations over extremely short distances, often in the face of considerable gene flow. Similarly, divergence can occur over very short periods of time following a change in the environment. A necessary prerequisite for such divergence is the availability of sufficient heritable variation in the population of the species at the outset. Each field situation is an unique evolutionary event in which the extent of evolutionary change will be a reflection of the genetic structure of the source population as well as the nature and magnitude of selection pressures. There are very few studies where evolutionary responses within natural populations have been studied in replicated situations.

The extent of population divergence and its rapidity indicates that selection pressures in the field are very strong. Experimental studies using reciprocal transplants of contrasting genotypes in different environments (e.g. Davies & Snaydon, 1976; Lovett Doust, 1981; McGraw & Antonovics, 1983; Ashmun & Pitelka, 1985; Davies & Long, 1991) have confirmed the existence of intense selection pressures in 'ordinary' field situations, not just in extreme environments.

Work to date on evolutionary divergence in closely adjacent plant populations has generally involved short-term studies (3–5 years). There is need for much longer term investigations, or subsequent follow-up work to previous studies, particularly in relation to the extent of development of reproductive isolation mechanisms between the diverging populations.

Most studies to date have centred on 'the population'. There is need to recognize the spatial and temporal heterogeneity and ecological complexity of natural habitats and to narrow the scale of evolutionary studies to incorporate investigations of the responses of individual genotypes in relation to the physical, chemical and biotic factors of their microhabitat.

REFERENCES

ALLARD, R.W. & ADAMS, J., 1969. Population studies of predominantly self pollinating species. XIII. Intergenotypic competition and population structure in barley and wheat. *American Naturalist, 103:* 621–645.

AL-HIYALY, S.A.K., McNEILLY, T. & BRADSHAW, A.D., 1988. The effect of zinc contamination from electricity pylons – evolution in a replicated situation. *New Phytologist, 110:* 571–580.

AL-HIYALY, S.A.K., McNEILLY, T. & BRADSHAW, A.D., 1990. The effect of zinc contamination from electricity pylons. Contrasting patterns of evolution in five grass species. *New Phytologist, 114:* 183–190.

ANTONOVICS, J., 1968. Evolution in closely adjacent plant populations. V. Evolution of self fertility. *Heredity, 23:* 219–238.

ANTONOVICS, J. & BRADSHAW, A.D., 1970. Evolution in closely adjacent plant populations. VIII. Clinal patterns at a mine boundary. *Heredity, 25:* 349–362.

ASHMUN, J.W. & PITELKA, L.F., 1985. Population biology of *Clintonia borealis*. II. Survival and growth of transplanted ramets in different environments. *Journal of Ecology, 73:* 185–199.

BANDEEN, J.D., PAROCHETTI, J.V., RYAN, G.F., MALTAIS, B. & PEABODY, D.V., 1979. Discovery and distribution of triazine resistant weeds in North America. *Abstract of the Meeting of the Weed Science Society of America, 179:* 108–109.

BELL, J.N.B. & MUDD, C.H., 1976. Sulphur dioxide resistance in plants: a case study of *Lolium perenne.* In T.A. Mansfield (Ed.), *Effects of Air Pollutants on Plants:* 87–103. Cambridge: Cambridge University Press.

BLIJENBURG, J.G. & SNEEP, J., 1975. Natural selection in a mixture of eight barley varieties grown in six successive years. I. Competition between the varieties. *Euphytica, 24:* 305–315.

BRADSHAW, A.D., 1965. Evolutionary significance of phenotypic plasticity in plants. *Advances in Genetics, 13*: 115–155.

BRADSHAW, A.D., 1984. The importance of evolutionary ideas in ecology and vice versa. In B. Shorrocks (Ed.), *Evolutionary Ecology:* 1–25. Oxford: Blackwells.

BRADSHAW, A.D. & HARDWICK, K., 1989. Evolution and Stress – genotypic and phenotypic components. *Biological Journal of the Linnean Society, 37:* 137–155.

BRADSHAW, A.D., McNEILLY, T. & GREGORY, R., 1965. Industrialisation, evolution and the development of heavy metal tolerance in plants. In G.T. Goodman, R.W. Edwards & J.M. Lambert (Eds), *Ecology and the Industrial Society:* 327–343. Oxford: Blackwell Scientific Publications.

BRENCHLEY, W.E. & WARRINGTON, K., 1958. *The Park Grass Plots at Rothamsted, 1856–1949.* Rothamsted: Rothamsted Experimental Station.

BRIGGS, B.G., 1962. Interspecific hybridization in the *Ranunculus lappaceus* group. *Evolution, 16:* 372–390.

BROUGHAM, R.W. & HARRIS, W., 1967. Rapidity and extent of changes in genotype structure induced by grazing in a ryegrass population. *New Zealand Journal of Agricultural Research, 10:* 56–65.

CHARLES, A.H., 1961. Differential survival of cultivars of *Lolium, Dactylis* and *Phleum. Journal of the British Grassland Society, 16:* 69–75.

CHARLES, A.H., 1964. Differential survival of plant types in swards. *Journal of the British Grassland Society, 19:* 198–204.

DARWIN, C., 1859. *On the Origin of Species by Means of Natural Selection.* London: John Murray.

DARWIN, C., 1868. *The Variation of Animals and Plants Under Domestication, volume 1.* London: John Murray.

DE CANDOLLE, A., 1886. *Origin of Cultivated Plants* (English translation of the 2nd edition. New York & London: Hafner, 1967).

DAVIES, M.S., 1975. Physiological differences among populations of *Anthoxanthum odoratum* L. collected from the Park Grass Experiment, Rothamsted. IV. Response to potassium and magnesium. *Journal of Applied Ecology, 12:* 953–964.

DAVIES, M.S. & LONG, G.L., 1991. Performance of two contrasting morphs of *Brachypodium sylvaticum* transplanted into shaded and unshaded sites. *Journal of Ecology, 79:* 505–517.

DAVIES, M.S. & SNAYDON, R.W., 1973a. Physiological differences among populations of *Anthoxanthum odoratum* L. collected from the Park Grass Experiment, Rothamsted. I. Response to calcium. *Journal of Applied Ecology, 10:* 33–45.

DAVIES, M.S. & SNAYDON, R.W., 1973b. Physiological differences among populations of *Anthoxanthum odoratum* L. collected from the Park Grass Experiment, Rothamsted. II. Response to aluminium. *Journal of Applied Ecology, 10:* 47–55.

DAVIES, M.S. & SNAYDON, R.W., 1974. Physiological differences among populations of *Anthoxanthum odoratum* L. collected from the Park Grass Experiment, Rothamsted, III. Response to phosphate. *Journal of Applied Ecology, 11:* 669–707.

DAVIES, M.S. & SNAYDON, R.W., 1976. Rapid population differentiation in a mosaic environment. III. Measures of selection pressures. *Heredity, 36:* 59–66.

DUCOUSSO, A., PETIT, D., VALERO, M. & VERNET, P., 1990. Genetic variation between and within populations of a perennial grass *Arrhenatherum elatius. Heredity, 65:* 179–188.

EHRLICH, P.R. & RAVEN, P.H., 1969. Differentiation of populations. *Science, 165:* 1228–1232.

ENDLER, J.A., 1977. *Geographical Variation, Species and Clines.* Princeton: Princeton University Press.

ENNOS, R.A., 1983. Maintenance of genetic variation in plant populations. *Evolutionary Biology, 16:* 129–155.

ERNST, W.H.O., 1976. Physiological and biochemical aspects of metal tolerance. In T.A. Mansfield (Ed.), *Effects of Air Pollutants on Plants:* 115–133. Cambridge: Cambridge University Press.

ERNST, W.H.O., VERKLEIJ, J.A.C. & VOOIJS, R., 1983. Bioindication of a surplus of heavy metals in terrestrial ecosystems. *Environmental Monitoring and Assessment, 3:* 297–305.

EVANS, D.R., HILL, J., WILLIAMS, T.A. & RHODES, I. (1985). Effects of coexistence on the performance of white clover-perennial ryegrass mixtures. *Oecologia, 66:* 536–539.

GARTSIDE, D.W. & McNEILLY, T., 1974. The potential for evolution of heavy metal tolerance in plants. III. Copper tolerance in normal populations of different species. *Heredity, 32:* 335–348.
HARBERD, D.J., 1961. Observations on the population structure and longevity of *Festuca rubra* L. *New Phytologist, 60:* 184–206.
HARLAN, H.V. & MARTINI, M.L., 1938. The effect of natural selection in a mixture of barley varieties. *Journal of Agricultural Research, 57:* 189–201.
HEDRICK, P.W., 1986. Genetic polymorphism in heterogeneous environments: a decade later. *Annual Review of Ecology and Systematics, 17:* 535–566.
HESLOP-HARRISON, J., 1956. Some observations on *Dactylorchis incarnata* L. *Proceedings of the Linnean Society of London, 166:* 51–83.
HESLOP-HARRISON, J., 1964. Forty years of genecology. *Advances in Ecological Research, 2:* 159–247.
HICKEY, D.A. & McNEILLY, T., 1975. Competition between metal tolerant and normal plant populations. A field experiment on normal soil. *Evolution, 29:* 458–464.
HILLMAN, G.C. & DAVIES, M.S., 1990. Domestication rate in wild-type wheats and barley under primitive cultivation. *Biological Journal of the Linnean Society, 39:* 39–78.
HOLLIDAY, R.J. & PUTWAIN, P.D., 1977. Evolution of resistance to simazine in *Senecio vulgaris* L. *Weed Research, 17:* 291–296.
HOLLIDAY, R.J. & PUTWAIN, P.D., 1980. Evolution of herbicide resistance in *Senecio vulgaris:* variation in susceptibility to simazine between and within populations. *Journal of Applied Ecology, 17:* 779–791.
JAIN, S.K. & BRADSHAW, A.D., 1966. Evolutionary divergence among adjacent plant populations. I. The evidence and its theoretical analysis. *Heredity, 21:* 407–441.
JOHNSTON, A.E., 1972. Change in soil properties caused by the new liming scheme on Park Grass. *Report Rothamsted Experimental Station, 1971(2):* 177–180.
LEFEBVRE, C. & VERNET, P., 1990. Microevolutionary processes on contaminated deposits. In A.J. Shaw (Ed.), *Heavy Metal Tolerance in Plants: Evolutionary Aspects:* 285–299. Boca Raton, Florida: CRC Press.
LEVINS, R. & MacARTHUR, R., 1966. The maintenance of genetic polymorphism in a spatially heterogeneous environment: variations on a theme by Howard Levene. *American Naturalist, 100:* 585–589.
LEWIS, H., 1966. Speciation in flowering plants. *Science, 152:* 167–172.
LOVETT DOUST, L., 1981. Population dynamics and local specialization in a clonal perennial (*Ranunculus repens*). II. The dynamics of leaves and a reciprocal transplant-replant experiment. *Journal of Ecology, 69:* 757–768.
McDONALD, J.F. & AYALA, F.J., 1974. Genetical response to environmental heterogeneity. *Nature, 250:* 572–574.
McGRAW, J.B. & ANTONOVICS, J., 1983. Experimental ecology of *Dryas octopetala* ecotypes. I. Ecotypic differentiation and life cycle stages of selection. *Journal of Ecology, 71:* 879–897.
McNEILLY, T., 1968. Evolution in closely adjacent plant populations. III. *Agrostis tenuis* on a small copper mine. *Heredity, 23:* 99–108.
McNEILLY, T. & ANTONOVICS, J., 1968. Evolution in closely adjacent plant populations. IV. Barriers to gene flow. *Heredity, 23:* 205–218.
MACNAIR, M.R., 1981. Tolerance of higher plants to toxic metals. In J.A. Bishop & L.M. Cook (Eds), *Genetic Consequence of Man Made Change:* 177–207. London: Academic Press.
MACNAIR, M.R., 1990. The genetics of metal tolerance in natural populations. In A.J. Shaw (Ed.), *Heavy Metal Tolerance in Plants: Evolutionary Aspects:* 235–253. Boca Raton. Florida: CRC Press.
MACNAIR, M.R. & CHRISTIE, P., 1983. Reproductive isolation as a pleiotropic effect of copper tolerance in *Mimulus guttatus. Heredity, 50:* 295–302.
MACNAIR, M.R. & WATKINS, A.D., 1983. The fitness of the copper tolerance gene of *Mimulus guttatus* in uncontaminated soil. *New Phytologist, 95:* 133–137.
MAYNARD SMITH, J., 1962. Disruptive selection, polymorphism and sympatric speciation. *Nature, 195:* 60–62.
MAYNARD SMITH, J. & HOEKSTRA, R., 1980. Polymorphism in a varied environment. How robust are the models? *Genetical Research, 35:* 45–57.
MINAWA, A. & BIRLEY, A.J., 1978. The genetical response to natural selection by varied environments. I. Short term observations. *Heredity, 40:* 39–50.

NICHOLLS, M.K. & McNEILLY, T., 1985. The performance of *Agrostis capillaris* L. genotypes, differing in copper tolerance, in ryegrass swards on normal soil. *New Phytologist, 101:* 207–217.

PATERNIANI, E., 1969. Selection for reproductive isolation between two populations of maize, *Zea mays* L. *Evolution, 23:* 534–547.

ROCOVICH, S.E. & WEST, D.A., 1975. Arsenic tolerance in a population of the grass *Andropogon scoparius* Michx. *Science, 188:* 263–264.

RYAN, G.F., 1970. Resistance of common groundsel to simazine and atrazine. *Weed Science, 18:* 614–616.

SCHLICHTING, C.D., 1986. Evolution of phenotypic plasticity in plants. *Annual Review of Ecology and Systematics, 17:* 667–693.

SEATON, A.J.P. & ANTONOVICS, J., 1967. Population interrelationships. I. Evolution in mixtures of *Drosphila* mutants. *Heredity, 22:* 19–33.

SHARMA, H. & WAINES, J.G., 1980. Inheritance of tough rachis in crosses of *Triticum monococcum* and *Triticum boeoticum*. *Journal of Heredity, 71:* 214–216.

SILANDER, J.A., 1979. Micro-evolution and clone structure in *Spartina patens*. *Science, 203:* 658–660.

SNAYDON, R.W., 1973. Ecological factors, genetic variation and speciation in plants. In V.H. Heywood (Ed.), *Taxonomy and Ecology:* 1–29. London: Academic Press.

SNAYDON, R.W., 1978. Genetic change in pasture populations. In J.R. Wilson (Ed.), *Plant Relations in Pastures:* 253–269. Melbourne: C.S.I.R.O.

SNAYDON, R.W. & DAVIES, M.S., 1972. Rapid population differentiation in a mosaic environment. II. Morphological variation in *Anthoxanthum odoratum*. *Evolution, 26:* 390–405.

SNAYDON, R.W. & DAVIES, M.S., 1976. Rapid population differentiation in a mosaic environment. IV. Populations of *Anthoxanthum odoratum* at sharp boundaries. *Heredity, 37:* 9–25.

SNAYDON, R.W. & DAVIES, T.M., 1982. Rapid divergence of plant populations in response to recent changes in soil conditions. *Evolution, 36:* 289–297.

STAM, P., 1983. The evolution of reproductive isolation in closely adjacent plant populations through differential flowering time. *Heredity, 50:* 105–118.

SULTAN, S.E., 1987. Evolutionary implications of phenotypic plasticity in plants. *Evolutionary Biology, 21:* 127–178.

SYMEONIDIS, L., McNEILLY, T. & BRADSHAW, A.D., 1985. Interpopulation variation in tolerance to cadmium, copper, lead, nickel and zinc in nine populations of *Agrostis capillaris* L. *New Phytologist, 101:* 317–324.

TURKINGTON, R. & HARPER, J.L., 1979. The growth, distribution and neighbour relationships of *Trifolium repens* in a permanent pasture. IV. Fine scale biotic differentiation. *Journal of Ecology, 67:* 245–254.

TAYLOR, G.E. & MURDY, W.H., 1975. Population differentiation of an annual plant species *Geranium carolinianum* in response to sulfur dioxide. *Botanical Gazette, 136:* 212–215.

THODAY, J.M. & GIBSON, J.B., 1962. Isolation by disruptive selection. *Nature, 193:* 1164–1166.

THURSTON, J.M., 1969. The effect of liming and fertilizers on the botanical composition of permanent grassland and the yield of hay. In I.H. Rorison (Ed.), *Ecological Aspects of the Mineral Nutrition of Plants:* 3–10. Oxford: Blackwell.

VAVILOV, N.I., 1926. *Studies on the Origin of Cultivated Plants.* Leningrad: Institute of Applied Botany and Plant Breeding.

WARREN, R.G., JOHNSTON, A.E. & COOKE, G.W., 1965. Changes in the Park Grass Experiment. *Report Rothamsted Experimental Station, 1964:* 224–228.

WILKINS, D.A., 1978. The measurement of tolerance to edaphic factors by means of root growth. *New Phytologist, 80:* 623–633.

WILLIAMS, E.D., 1974. Changes in yield and botanical composition caused by the new liming scheme on Park Grass. *Report Rothamsted Experimental Station, 1973 (2):* 67–73.

WILLIAMS, E.D., 1978. *Botanical Composition of the Park Grass Plots at Rothamsted 1856–1976.* Harpenden: Rothamsted Experimental Station.

WILSON, G.B. & BELL, J.N.B., 1985. Studies on the tolerance to SO_2 of grass populations in polluted areas. III. Investigations on the rate of development of tolerance. *New Phytologist, 100:* 63–67.

WINN, A.A. & EVANS, A.S., 1991. Variation among populations of *Prunella vulgaris* L. in plastic responses to light. *Functional Ecology, 5:* 562–571.

WU, L., BRADSHAW, A.D. & THURMAN, D.A., 1975. The potential for evolution of heavy metal tolerance in plants. III. The rapid evolution of copper tolerance in *Agrostis stolonifera*. *Heredity, 34:* 165–187.

WU, L., TILL-BOTTRAUD, I. & TORRES, A., 1987. Genetic differentiation in temperature – enforced seed dormancy among golf course populations of *Poa annua* L. *New Phytologist, 107:* 623–631.

CHAPTER

12

Does genome organization influence speciation? A reappraisal of karyotype studies in evolutionary biology

ANN KENTON, ALEXANDER S. PAROKONNY,
SIMON T. BENNETT & MICHAEL D. BENNETT

Chromosomal diversity and phylogeny **190**
Karyotype 190
Analysis of meiosis 192
Chromosome banding 193
Molecular cytogenetics 195
Genomic *in situ* hybridization (GISH) **196**
GISH determines the origin of two natural allopolyploids 197
GISH discriminates between identical karyotypes 199
GISH detects homology among species that will not hybridize 200
Conclusions: genome organization and speciation **202**
Acknowledgements **202**
References **203**

Keywords: Karyotype – molecular cytogenetics – genome evolution – speciation – genomic *in situ* hybridization (GISH) – phylogeny.

Abstract
The evolutionary significance of the ways in which DNA is organized into chromatin and chromosomes remains, in many cases, an enigma. While changes in gene expression are clearly important in creating phenotypic differences, these may relate as much to chromatin configuration as to the linear arrangement of sequences on the chromosomes. Additions, deletions and rearrangements of DNA by chromosome structural mutations often affect the phenotype remarkably little, but when sufficient

Evolutionary Patterns and Processes
ISBN 0-12-440895-8

to disrupt chromosome pairing, can influence speciation by initiating reproductive isolation between populations. The genomic changes associated with speciation may be cryptic because conventional cytogenetic methods are insufficiently sensitive to detect them. New molecular techniques in cytogenetics offer a greater degree of certainty in determining the origins of sequences, chromosomes and whole genomes. Some examples in higher plants are discussed with a view to assessing the validity of the karyotype as a phylogenetic marker.

CHROMOSOMAL DIVERSITY AND PHYLOGENY

Karyotype

In an attempt to understand the mechanisms involved in phylogenetic divergence, genome analyses have been made for a great many organisms. This approach has been regarded as fundamental to evolutionary problems, because it detects changes in the material directly concerned with heredity. However, the evolutionary significance of the ways in which DNA is organized into chromatin and chromosomes remains poorly understood. This is largely due to a paradoxical variation in chromosome number, shape and size among, and frequently within, taxonomic groupings. In angiosperms, for example, the amount of DNA in a 4C nucleus ranges from 1 pg or less (e.g. *Myriophyllum verticillatum,* Fig. 1; Baum & Kenton, unpublished) to over 500 pg in *Fritillaria assyriaca* Baker (Bennett & Smith, 1976, cf. Fig. 1C) but bears no overall relationship to evolutionary complexity. Chromosome numbers range from $2n = 4$, for example, in *Brachycome lineariloba* (DC.) Druce (Asteraceae, Fig. 1B; Smith-White, 1968) to $2n = c.\ 640$ in *Sedum suaveolens* Kimnach (Uhl, 1978). The palm *Voanioala gerardii* J. Dransf. has the highest number known in monocots ($2n = 596$), contrasting sharply with most other palm species (usually $2n = 32$ or $2n = 36$) and with no outstanding changes in phenotype (Johnson *et al.*, 1989).

In some plant groups, such as the tribe Aloineae in the Liliaceae, or the subtribe Tigridiinae in the Iridaceae, the number and morphology of the chromosomes is usually constant among species (Brandham, 1983; Molseed, 1970). Changes in chromosome size, and presumably in linear sequence, are incorporated into the same basic karyotype pattern that exemplifies all related taxa. Other groups show a remarkable diversity in chromosome number, shape and size between close relatives (e.g. *Ornithogalum* Rafin., Johnson, Garbari & Mathew, 1991; *Narcissus* L., Brandham & Kirton, 1987) and even within taxonomic species (e.g. *Crocus speciosus* M. Bieb., Brighton, Mathew & Rudall, 1983). Many others exhibit a predictable series of changes, such as whole-arm fusions and fissions (e.g. *Cymbispatha* Pichon, Jones, Kenton & Hunt, 1981), repeated interchanges (e.g. *Oenothera* L., Cleland, 1972), or a progressive increase in symmetry as genome size increases (e.g. Ranunculaceae, Rothfels, Sexsmith & Heimburger, 1966, and *Lathyrus* L., Narayan & Durrant, 1983).

Conserved types of organizational change are by no means unique to specific groups of organisms, and indeed may occur in very different ones. For example, bimodal karyotypes (having two size classes of chromosomes) occur widely in both animals and plants. In animals, they predominate in lizards (Bickham, 1984), fishes (Ohno *et al.*, 1969) and especially birds (Christidis, 1990). In plants, they are typical of many petaloid monocots, particularly *Agave* L. and its near relatives (Stebbins,

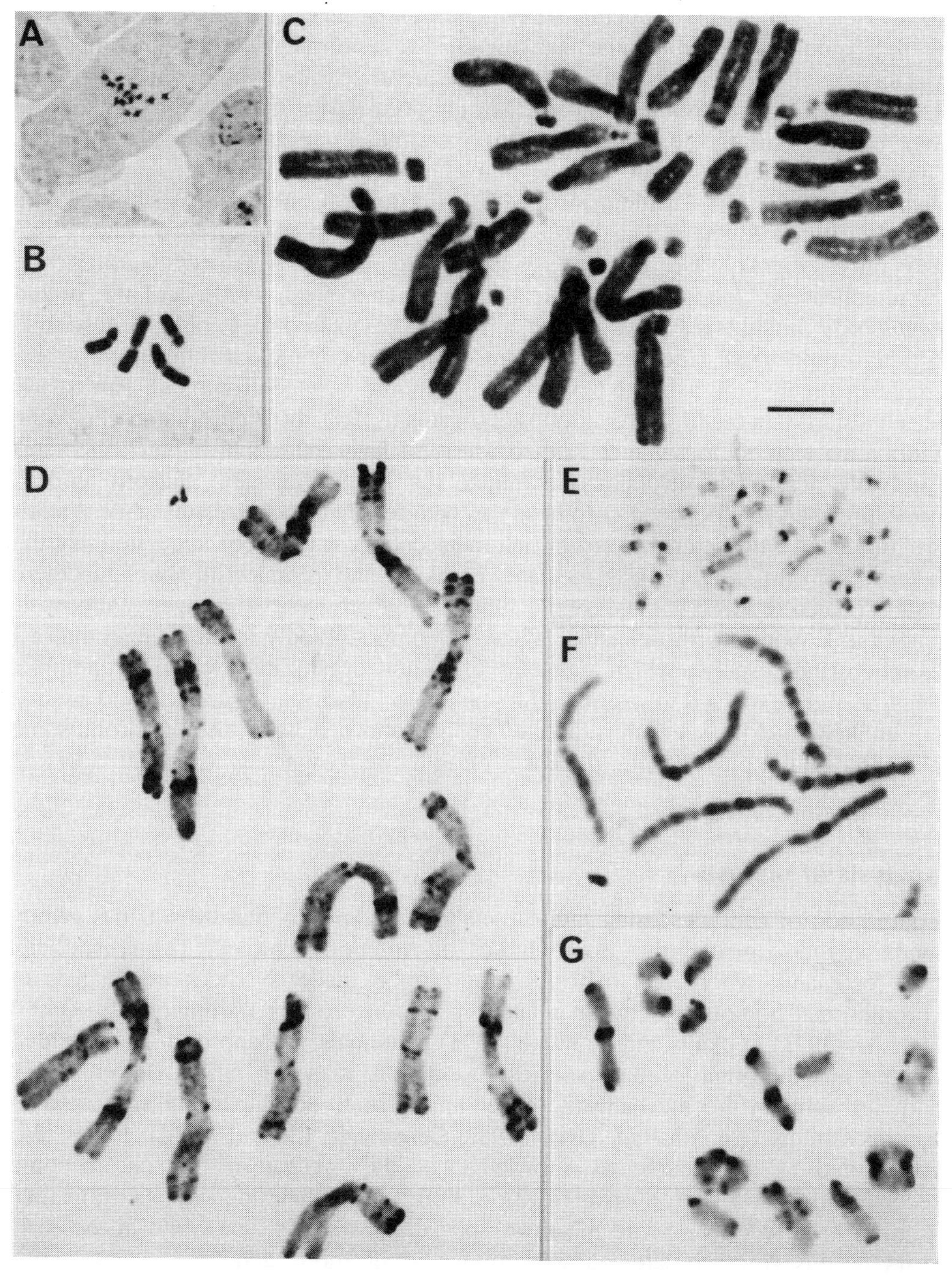

Figure 1 A: *Myriophyllum verticillatum, 2n* = 14, 4C = 1.04 pg.; B: *Brachycome lineariloba, 2n* = 4 (= *B. dichromosomatica* C.R. Carter); C: *Fritillaria* sp., *2n* = 24; D: *Gibasis pulchella, 2n = 15, C-banding; E: Tigridia meleagris, 2n* = 28, C-banding; F: part cell of *Tradescantia ohiense* (*2n* = 24) to show G-banding; G: *Gibasis karwinskyana* × *G. consobrina,* meiotic metaphase I, C-banding. Bar = 10 μm.

1971). Heterochromatin (tandemly repeated DNA sequences) also tends to accumulate in the same places on similar karyotypes. For example, karyotypes comprising metacentric and telocentric chromosomes frequently exhibit whole-arm fusions and fissions. Such chromosomes characteristically accumulate heterochromatin around their centromeres and in new short arms. Examples in animals include ants, whose chromosome numbers range from $n = 1$ (Imai & Taylor, 1989) to $n = 42$ (Imai, Crozier & Taylor, 1977); muntjac deer ($n = 3$ to $n = 40$, Brinkley *et al.*, 1984); gophers (Patton & Sherwood, 1982); kangaroo rats ($n = 26$–37, Hatch *et al.*, 1976) and mice (Pathak, Hsu & Arrighi, 1973) and in plants, *Cymbispatha* Pichon (Commelinaceae, Jones *et al.*, 1981; Kenton & Drakeford, 1990), and the orchid *Paphiopedilum* Pfitz (Karasawa & Tanaka, 1980). These karyotypes could be described as relatively primitive, since rearrangements have involved breaks and heterochromatin amplification in only a limited part of the genome (i.e. centromeres). Particularly detailed studies in the first three groups of organisms have shown that heterochromatic short arms tend to increase in size to a critical level, after which rearrangements (probably pericentric inversions) eliminate the heterochromatin. Fusions tend to occur preferentially between chromosomes with less heterochromatin, while fissions are associated with heterochromatin-rich metacentrics. It has been suggested that the heterochromatin in some way mediates breakage and re-union in the centromere region (Hatch *et al.,* 1976; Redi *et al.*, 1986). These progressive changes (collectively known as karyotype orthoselection, White, 1973) undoubtedly result from an internal 'homoeostatic' response which maintains an optimal organization within the genome. Thus, they are expected to maintain the status quo rather than altering it, and as such are unlikely to drive speciation directly. For the above reasons, use of chromosome morphology as a phylogenetic marker is often limited, especially at the suprageneric level.

Analysis of meiosis

Comparison of genomes using meiotic analysis of experimental hybrids has greatly increased the scope of chromosome studies for evolutionary biology. This is especially true for plants, where test hybridizations can be made to check the degree of chromosomal homology between related taxa. Pioneered by Darlington (1958) and Stebbins (1971) in plants and by White (1973) in animals, this approach has provided insights into the origin of many species important to man (e.g. wheat, Dewey, 1984), and cast light on the mechanisms of genomic change accompanying speciation in natural systems (e.g. *Clarkia,* Lewis, 1962; *Oenothera,* Cleland, 1972). It has also contributed to a fundamental knowledge of the mechanisms of chromosome recombination and reproductive isolation, both of which influence speciation (see John, 1990 for review). Some putatively specialized systems discovered in the early years of this research later turned out to be of more general importance for understanding genome organization and evolution (e.g. the discovery of transposable elements, McClintock, 1951).

One drawback of using meiotic chromosome pairing in hybrids for assessing genomic affinities is that it can be affected by genetic and environmental factors (Evans, 1988; Kimber, Alonso & Sallee, 1981), or even by insects or viral diseases (Kostoff, 1933). Detection of cryptic structural rearrangements is also dependent upon a chiasma involving the rearranged segment, so that small rearrangements may be missed. Finally, species that appear to be similar chromosomally, and/or phenotypically, often cannot be hybridized successfully, due either to specialized reproductive barriers (incongruence), or failure of hybrid seed to develop. While the latter can be overcome by embryo rescue techniques (Brink, Cooper & Auscherman, 1944), the former often prevents successful comparisons of genome homology. Protracted development times also make this method impractical in many plant groups (e.g. in many bulbs, cycads, and most trees).

Chromosome banding

In the 1970s, methods for banding chromosomes were developed. It was hoped that these would provide detailed chromosome maps which could be used to compare chromosome structural organization in related species. While this has been achieved to a large extent in many animals, it has been less successful in plants. G-bands, so-called because they were produced with the stain, Giemsa, were used to characterize every chromosome of the karyotype in man and other mammals (Schnedl, 1972; Evans, Buckland & Sumner, 1973), birds (Christidis, 1990) and fishes (Mayr, Kalat & Ràb, 1990). They are now being used with specific molecular probes to determine the chromosomal locations of genes in phylogenetic and clinical studies (Bhatt & McGee, 1990). Among mammals, primates closest to man have extremely similar G-banding patterns (de Grouchy *et al.*, 1972) which have been used to map structural rearrangements characteristic of different species. G-bands have also been used as a basis for cladistic assessments (e.g. in bats, Haiduk & Baker, 1982). Recent work has shown that these bands have a structural and a molecular component. The light bands are rich in short interspersed repeats (SINES), replicate early in nuclear DNA synthetic phase of the cell cycle, and contain housekeeping genes. The dark bands replicate late and contain tissue-specific genes (i.e. concerned with ontogeny), which are facultatively heterochromatinized. They are characterized at the molecular level by long interspersed elements, or LINES (Chen & Manuelidis, 1989). Holmquist (1989) proposed that a gene's function depends upon which of these domains it resides in. Thus there may, in mammals, be a direct connection between the organization of sequences in chromosomes, development and speciation.

In plants, G-like structural bands have been demonstrated in several groups, including gymnosperms, Liliaceae, Commelinaceae and legumes (Wang & Kao, 1988, and Figs 1F and 2G; M. Cornell, unpublished) using the methods originally developed for mammals. Bands can also be produced with some restriction enzymes, especially if recognition sites occur within heterochromatin (e.g. Schubert, 1990, and Fig. 2F; Kenton, unpublished). However, whether these structural bands have a molecular organization that is in any way comparable to that in mammalian chromosomes is as yet unknown. It has been suggested that plant chromosomes, like those of Amphibia, may have a different organization from those of mammals, birds and fishes (Greilhuber, 1977).

C-bands, unlike G-bands, have no clear function and usually do not contain genes, although they may contain regulatory elements and be important in the mechanics of chromosome behaviour. In plants, they are composed of constitutive heterochromatin which is condensed throughout the cell cycle and is (superficially at least) the same as that demonstrated in animals by similar methods (Pardue & Gall, 1970). In most mammals, with the notable exception of cetaceans (Árnason & Widegren, 1989), C-

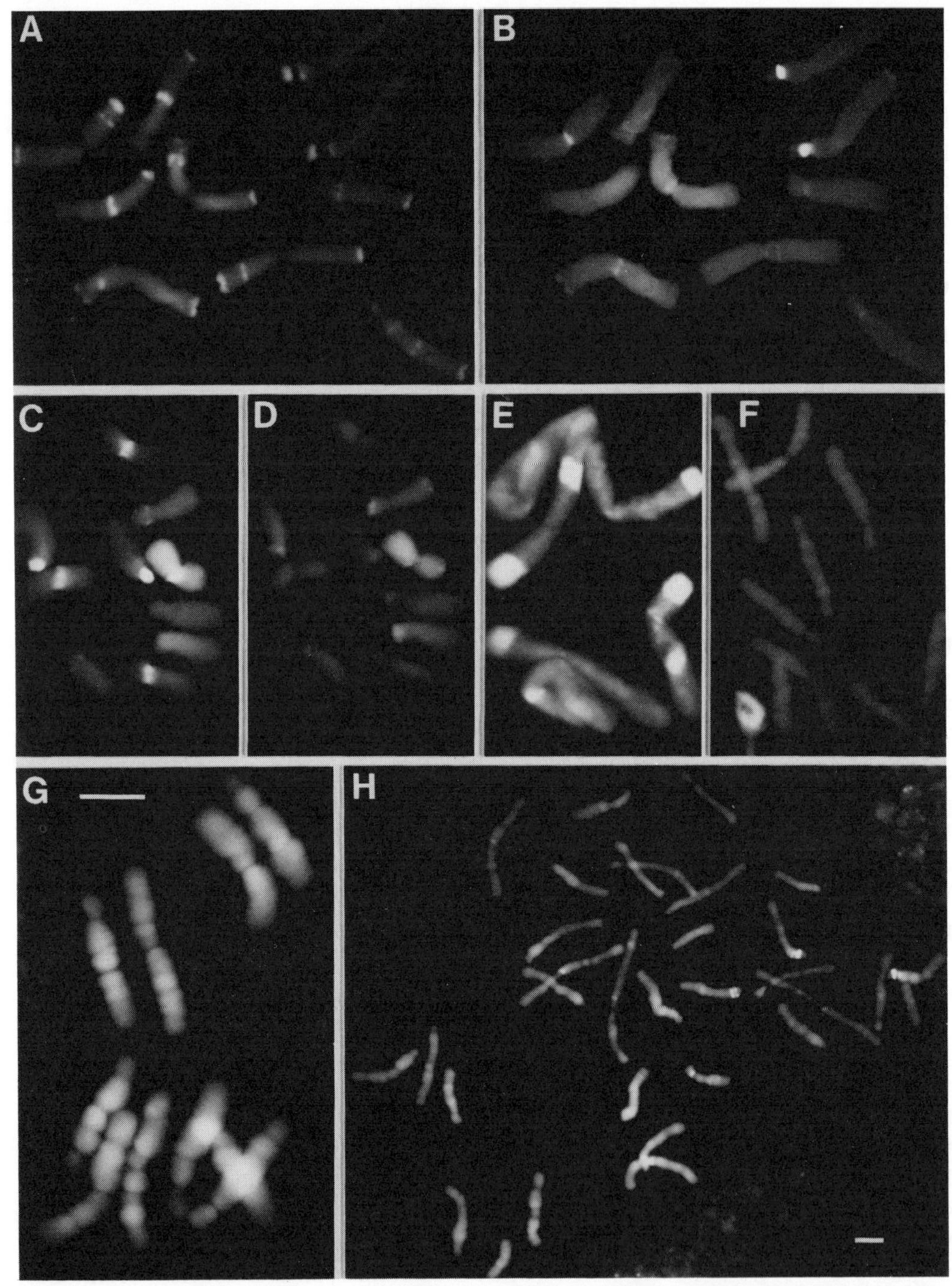

Figure 2 Fluorochrome banding. A: *Gibasis karwinskyana* subsp. *palmeri,* AT-rich bands with Hoechst 33258; B: same cell as A, GC-rich bands with CMA; C: *G. karwinskyana* subsp. *karwinskyana,* CMA; D: as C, Hoechst 33258; E: *Scilla sibirica,* pollen grain mitosis, CMA; F: *G. karwinskyana* subsp. *palmeri, in situ* digestion with Hae III, stained with propidium iodide; G: as F, G-banding; H: *Lilium candidum,* CMA. Bar = 10 μm.

bands are present only at the centromere (hence the term C). In plants, however, they can be very numerous and positionally diverse (e.g. Fig. 1D), and have been used in some groups as nuclear markers (e.g. in optical sections of pollen grains, White *et al.*, 1991), or to identify chromosome rearrangements in phylogeny (e.g. in complex interchange heterozygotes, Kenton, Davies & Jones, 1987), or for gene mapping (e.g. in *Hordeum* L., Linde-Laursen, 1982). C-bands can be differentiated into GC-rich and AT-rich components by applying base-specific fluorochromes (Schweizer, 1976). Both types can be visualized in the same cell by changing the excitation wavelength (Fig. 2A–D). Combined banding patterns produced in this way have been useful in detecting population divergence in incipient species (cf. *Gibasis karwinskyana* subsp. *palmeri*, Fig. 2A,B) and subsp. *karwinskyana* (Fig. 2C,D; Kenton, 1991) and affinities within aggregates (e.g. in *Scilla* L., Greilhuber & Speta, 1976). The molecular sequences present in the bands can be cloned, sequenced, and hybridized *in situ* to chromosomes of related species (e.g. Plate 1B*, and Schweizer, Strehl & Hagemann, 1990). The latter is a more sensitive means of detecting similarities and differences within and between closely related taxa. While common C-bands may contain a shared repeated sequence in some species groups (e.g. in *Scilla*, Deumling & Greilhuber, 1982), this is not universally true. Satellite DNA comprising C-bands can evolve relatively rapidly (Schweizer *et al.*, 1990) by concerted evolution of a new repeated sequence (see Dover & Flavell, 1984; Waye & Willard, 1989). Thus, a similar banding pattern is not necessarily a reliable indicator of phylogenetic relatedness, and each material must be considered independently.

Although banding patterns have proved very useful in selected plant groups, it now seems unlikely that they will, in the foreseeable future, attain in plants the resolution and sensitivity that has been possible in animal chromosomes.

Molecular cytogenetics

Molecular cytogenetics has allowed detailed gene mapping and cloning of breakpoints involved in chromosome rearrangements (Trask, 1991), and recent work is applying *in situ* gene mapping to evolutionary studies (e.g. Jung, Warter & Rumpler, 1991). For example, the consensus sequence of human telomere hybridizes to chromosomal locations in addition to the chromosome ends in some species of vertebrates. Such additional locations of telomere-like sequence may predispose karyotypes of some taxa to rearrangements in specific locations (Meyne *et al.*, 1990). In animals, man and mouse were model organisms for the application of molecular cytogenetic techniques, and much of the research stemmed from the need for clinical applications, for example, screening for chromosome abnormalities. A similar approach has been used for plants, where well-known crop species, whose cytogenetics has been intensively studied, have provided the models for exploring the uses of molecular techniques in genome analysis and plant breeding (reviewed by Heslop-Harrison, 1991). However, only relatively recently has molecular cytogenetics been applied to assessing evolutionary relationships in natural plant populations (Deumling & Greilhuber, 1982; Schweizer *et al.*, 1990; Bennett, Kenton & Bennett, 1992; Parokonny *et al.*, 1992). This is because in plants, molecular cytogenetics has mainly been conducted by molecular biologists who seek localization of sequences isolated from

* Plate 1 appears as a frontispiece to this volume.

plant genomes as final confirmation of their work. In animals, non-molecular specialists such as cytogeneticists, histologists and pathologists have adapted much more readily to incorporating *in situ* hybridization of DNA probes routinely into their investigations, perhaps because of the greater number of probes available for clinical work. However, the solid foundation provided by classical studies in plant chromosome evolution has provided an ideal and exciting base from which to explore the relevance of genome organization to evolution, using molecular probes.

GENOMIC *IN SITU* HYBRIDIZATION (GISH)

For most natural plant populations, insufficient evidence is available to determine the molecular and structural relationships of genomes. Data on higher order nuclear structure are even more limited. Studies that compare the distribution and homology of cloned sequences in different species are most informative but are time-consuming and require some experience with cloning. Also, they are informative about only one part of the genome, which may evolve differentially from the rest. A rapid, visual method which can give an idea of overall sequence homology is preferable. In the Jodrell Laboratory, we have recently been using genomic *in situ* hybridization (GISH) as a supplementary method of genome analysis in cases where we are unable to proceed further by light microscopy and experimental hybridization.

GISH involves isolating the total genomic DNA, labelling it with a reporter molecule and annealing it under specific conditions to denatured chromosomal DNA on microscope slides (Schwarzacher *et al.*, 1989; Anamthawat-Jónsson *et al.*, 1990). The method is outlined in Fig. 3. Effectively, it combines karyotype analysis with DNA–DNA hybridization. By varying the stringency of the post-hybridization washing, it is possible to discriminate between parental genomes even when they share more than 80% of sequence homology. We have used this method to test genomic homology in a

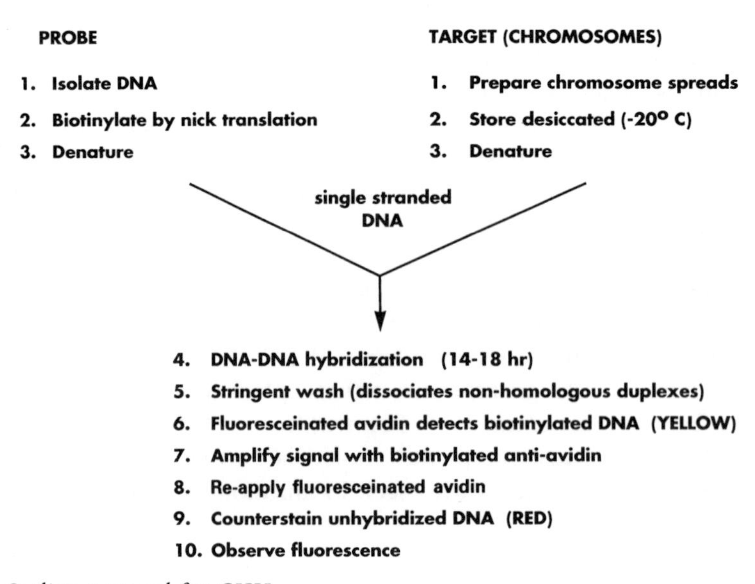

Figure 3 Outline protocol for GISH.

number of naturally occurring taxa where it was impossible to proceed further by conventional genome analysis. Since GISH is relatively costly, it may often be wise to carry conventional genome analysis as far as possible first. The probe homology can also be tested by Southern hybridization (Fig. 4), which will give an idea of the amount of homology but not its distribution. Below, we illustrate some specific examples where GISH has supplied further information about genomic relationships than was possible with conventional cytology.

GISH determines the origin of two natural allopolyploids

GISH was used to test the allopolyploid origin of the naturally occurring grass *Milium montianum* Parl. (Bennett *et al.*, 1992). This species, previously included in the *M. vernale* Bieb. aggregate, has a distinctive bimodal karyotype comprising 8 large (L) and 14 smaller (S) chromosomes. Owing to a marked similarity in chromosome morphology, Bennett & Thomas (1991) suggested that the $2n = 8$ cytotype of *M. vernale* was the ancestral diploid species that had contributed the L complement to *M. montianum*. A conventional analysis of genome relationships, using chromosome behaviour in experimental hybrids as an indication of chromosomal homology, was precluded since there was a strong reproductive barrier between these two species, thus rendering it impossible to make test hybrids. This hypothesis was therefore tested using GISH. Using the technique described in Fig. 3, biotinylated genomic DNA from *M. vernale* hybridized preferentially to the 8L chromosomes of *M. montianum* (Plate 1C). GISH therefore showed that the L complements of *M. montianum* and *M. vernale* genomes share a common origin, and that the S chromosomes certainly derive from a different, and as yet unknown taxon, thus confirming the allopolyploid origin of *M. montianum*.

Here, allopolyploidy is important in establishing a new species. *Milium montianum* constitutes a fertile amphidiploid between two genomes that are very distinct karyotypically, in molecular sequence, and presumably, genetically. Meiotic pairing

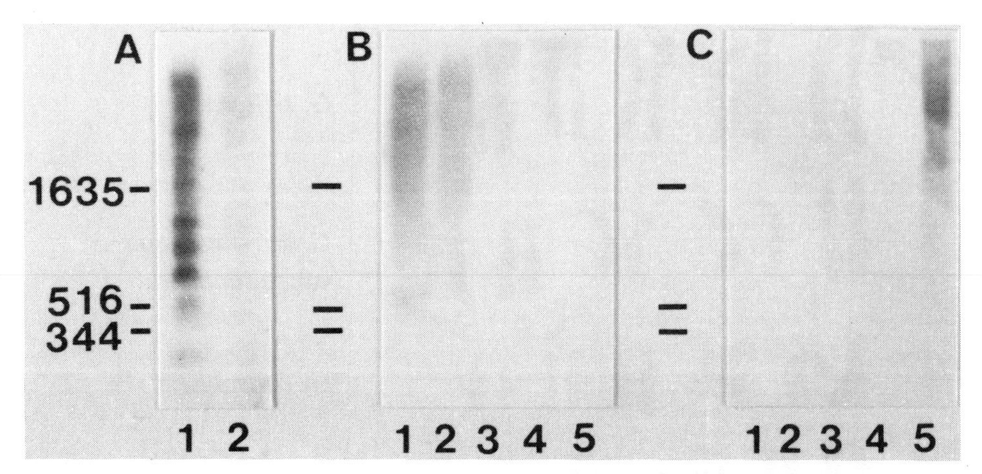

Figure 4 Southern hybridization of Hae III digests using total genomic DNA. A: Probed with *G. karwinskyana* DNA, washed at high stringency; B: probed with *G. karwinskyana* DNA, washed at low stringency; C: probed with *G. pulchella* DNA, washed at low stringency. Lanes: 1, *G. karwinskyana;* 2, *G. consobrina;* 3, *G. schiedeana;* 4, *G. triflora;* 5, *G. pulchella.* Size marker = 1 kb DNA ladder hybridized with biotinylated DNA from pBR322. (Numbers refer to size of fragments in kilobases).

between the two haploid sets from the same species will ensure that the new genotype breeds true, whereas any backcrosses to the parental species will be unstable. As with all allopolyploids, this increases heterozygosity and may increase fitness, but it is still not known how very differently organized genomes will be integrated into a common nucleus, and whether gene expression will change as a direct result. In muntjacs, species with 46 small and 6 large chromosomes, respectively, give rise to apparently normal (although sterile) F_1 hybrids, and it has been argued that the way in which the genes are packaged into chromosomes makes little difference to gene expression during development (John & Miklos, 1988). However, as yet, little is understood about how the chromosomes are arranged relative to one another in the new combination, and it may be this higher order structure which is more important for gene expression than linear sequence arrangement.

It is known that hybrids exhibit some 'adjustment' mechanisms which may compensate for the sudden addition of new DNA through amplification, hybridization or artificial manipulation. For example, in somatic hybrids between *Petunia* Juss. species, the chiasma frequency of the component species is markedly changed relative to the parents, although the overall chiasma frequency is similar (White & Rees, 1985). In *Brassica* L. allopolyploids, there is an adjustment in chromosome size, so that chromosomes in the allopolyploid are more contracted than in the respective diploids (Verma & Rees, 1974), while in Triticale (wheat × rye), there is an adjustment in the rate of replication fork movement relative to both parents. (Kidd, Francis & Bennett, 1992). Adjustments also occur in interspecific hybrids to allow chromosomes of different sizes to pair at meiosis (Jenkins, 1985), and to regulate the number of active ribosomal genes organizing the nucleolus (Flavell, 1989). So far, nothing is

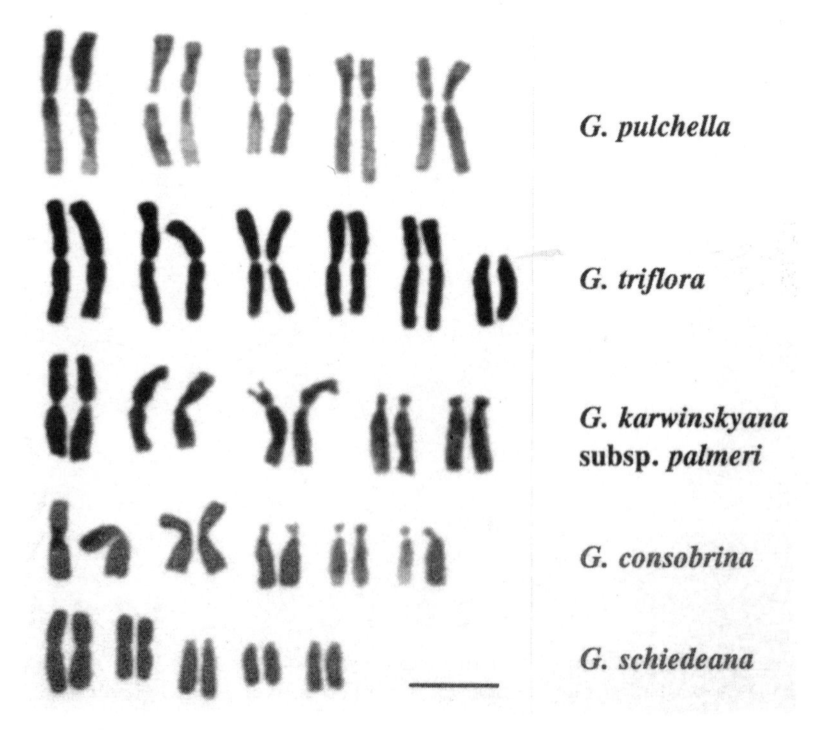

Figure 5 Karyotypes of five species of *Gibasis*. Bar = 10 μm.

known about whether or how genomes compensate for structural rearrangement of chromosomes.

In Plate 1C, there is a strong tendency for the ancestral genomes in *M. montianum* to lie apart. GISH therefore has another application in that it can be used to localize genetically dissimilar genomes in the nucleus, especially at interphase when chromosome morphology cannot be used for identification. The tendency of chromosomes to occupy specific domains in nuclei of many hybrids is now well-demonstrated (e.g. Finch, Smith & Bennett, 1981; Gleba *et al.*, 1987; Schwarzacher *et al.*, 1989; Leitch *et al.*, 1991; van Dekken *et al.*, 1989; Brandriff *et al.*, 1991). This has exciting possibilities for determining why genome reorganization takes place and how the genome may compensate for it.

Nicotiana tabacum L. is "the classic example of a natural amphiploid" (Gerstel, 1960). Based on genetical studies, karyotype morphology and chromosome pairing, its nearest extant relative is likely to be *N. sylvestris* Speg. & Comes (Goodspeed, 1954; Gerstel, 1960). However, Gerstel (1963) demonstrated pronounced differences between loci of the two species in genetically marked amphiploids between them, and concluded that some chromosomes from *N. tabacum* had remained completely homologous to *N. sylvestris,* while others had diverged.

Repeated sequences isolated from a *N. tabacum* genomic library are specific for *N. tomentosiformis* Goodsp. and *N. sylvestris,* respectively (Koukalová, Reich & Bezděk 1990; Kuhrová *et al.*, 1991), suggesting that these species are close to the ancestors of *N. tabacum.* However, *N. tomentosiformis*, in particular, may additionally be very closely allied to other species within the *tomentosae* (Goodspeed, 1954). Also, similarity in a repeated sequence does not necessarily imply similar overall homology, especially if the repeat can change rapidly by unequal crossing-over and gene conversion. GISH is therefore currently being used to localize overall homology between *N. tabacum* and its putative ancestors. Biotinylated DNA from *N. sylvestris* ($2n = 24$) labelled 24 chromosomes in *N. tabacum* ($2n = 48$; Plate 1D). The post-hybridization washing was highly stringent, so that the labelled chromosomes in *N. tabacum* were expected to have more than 85% homology with *N. sylvestris.* In the absence of spindle inhibitors (normally used to contract and spread the chromosomes), the *N. sylvestris* genome could be clearly located in *N. tabacum* nuclei at different stages of mitosis, and appeared, as in *Milium montianum,* to have a localized domain (Plate 1E,F). Interestingly, some chromosome arms were labelled less intensely than others with the *N. sylvestris* probe. Small cross-hybridizing segments were also apparent on six otherwise unlabelled chromosomes. Current work is exploring whether the '*N. sylvestris*' component of *N. tabacum* has a complex origin. Here, GISH has the advantage of speed over breeding experiments or molecular cloning for detecting an ancestral genome within a species, and for cytogeneticists, requires no additional special knowledge. Use of total genomic DNA may also be more reliable than cloned probes as an indicator of overall relatedness, since cross-hybridization of repeated-sequence DNA may be confined to a single conserved sequence.

GISH discriminates between identical karyotypes

The ability to discriminate rapidly and unequivocally between closely related species with almost identical karyotypes is a great advantage in evolutionary biology. Many classical studies have attempted it but failed, either because experimental hybridization is impossible, or because of a lack of chromosome markers.

Gibasis karwinskyana and *G. consobrina* D.R. Hunt are vicarial species, or microspecies, which occur north and south, respectively, of the trans-Mexican volcanic belt. They can be diploids ($2n = 10$) or cytological autopolyploids ($2n = 20, 30$) and have very similar karyotypes except that their C-banding patterns are different for some chromosomes and genotypes (Kenton, 1978), and they have different numbers of nucleolus organizing regions (NORs). While diploid *G. karwinskyana* has only two NORs per genome, *G. consobrina* has 4–6 (Jones, Papes & Hunt, 1975; Plate 1A). Artificial hybrids have been synthesized at the diploid and tetraploid levels. Diploid hybrids are weak and difficult to maintain. Meiotic pairing is very reduced and they are sterile. Pairing in the more vigorous tetraploid hybrids involves only chromosomes from the same parent (Fig. 1G; Kenton & Jones, 1985). A study of early interphase pairing showed that multivalents were formed in the tetraploid hybrid at zygotene, but these were later resolved into strictly homologous bivalents (Davies, Jenkins & Rees, 1990). So, was pairing failure in this hybrid caused by genetic controls limiting pairing to strictly homologous chromosomes (which could be the effect of a single gene), or was there a general lack of linear homology between morphologically similar homoeologues from the two species?

Biotinylated DNA from *G. karwinskyana* hybridized to Southern blots under different conditions suggested that the two species have between 70% and 80% sequence homology (Fig. 4A,B). GISH was carried out on both diploid and tetraploid hybrids. The chromosomes of the two parents were completely discriminated by probing with biotinylated DNA from one of them and washing at high stringency (Plate 1G–J). In the diploid hybrid (Plate 1G,H), one metacentric chromosome with a secondary constriction is a marker for *G. consobrina*. In the tetraploid hybrid, the *G. karwinskyana* chromosomes can be identified by their centromeric heterochromatin, which fluoresces more intensely (but non-specifically) than the rest of the chromosome. These features provided a check on the specificity of GISH for at least some chromosomes. Tetraploids have areas of clear cross-hybridization, probably representing conserved sequences, adjacent to secondary constrictions.

Here, GISH shows that the 20–30% difference in sequence homology beteween *G. consobrina* and *G. karwinskyana* is distributed throughout the karyotype, with only small areas of highly conserved sequences. Most of the DNA that is differentially labelled is likely to comprise middle-repetitive sequences. The highly repeated sequences that appear in chromosome bands are also differentiated both in Southerns and by GISH. Molecular differentiation of these microspecies is consistent with their geographical separation by the trans-Mexican volcanic belt. Their similar karyotypes belie an overall difference in sequence homology, which could well be the cause of pairing failure between homoeologues. This represents a classic case of allopatric speciation. Molecular divergence is almost certainly due to the lack of opportunity for homogenization of middle-repetitive sequences, which comprise the major part of large plant genomes (Flavell, O'Dell & Hutchinson, 1981).

GISH detects homology among species that will not hybridize

Because GISH allows a comparative study of homology without conventional genome analysis, it can be used, as in *Milium*, to examine the relationships of species that will not form experimental hybrids. Below is an example where the probe species was used as an internal control to examine the relationships of diploid species of *Gibasis.*

Figure 5 shows Feulgen-stained karyotypes of five species in the same section of *Gibasis. G. karwinskyana, G. consobrina* and *G. schiedeana* (Kunth) D.R. Hunt are clearly similar chromosomally. They have karyotypes with a basic number of $x = 5$, comprising 2 metacentrics and 3 acrocentrics. *G. pulchella* (Kunth) Rafin. ($2n = 10$) and *G. triflora* (Martens & Gal.) D.R. Hunt ($2n = 12$) display a remarkable similarity in karyotype except for the extra pair of subtelocentric chromosomes in *G. triflora. G. pulchella* has the largest genome – as a guide, 10 whole genomes of *Arabidopsis thaliana* ($4C = 4 \times 10^5$ kbp) would fit on to one of these chromosomes. Figure 4B shows a Southern blot using biotinylated total DNA from *G. karwinskyana* as a probe, with 3 μg of target DNA per lane and a low stringency wash. Differentiation between *G. karwinskyana* and *G. consobrina,* which will cross-fertilize and produce an F_1, is far less noticeable under these conditions than after the high-stringency washing and low target DNA concentration shown in Fig. 4A. *Gibasis pulchella* has no detectable homology to *G. karwinskyana,* even at the lower stringency level. When biotinylated *G. pulchella* DNA is used as a probe on the same five species, only lanes from *G. pulchella* and *G. triflora* are labelled (Fig. 4C). This shows that *G. triflora* has more molecular homology to *G. pulchella* than any of the other three species. Interestingly, *G. triflora* occurs in the same area in Oaxaca as *G. pulchella* and *G. schiedeana.* The three species grow at different altitudes, with *G. schiedeana* the lowest and *G. pulchella,* the highest, and may represent parts of an ancient cline (D.R. Hunt, personal communication). It will be interesting, therefore, to determine whether their genomic relationships reflect a common origin.

When experimental hybrids cannot be obtained, GISH can be performed using a mixture of cells from two different species on the same slide. Thus, material known to be 100% homologous to the probe (i.e. taken from the same individual plant) acts as an internal control. Root tip nuclei of *G. consobrina* and *G. schiedeana*, two species that will cross-fertilize but do not produce viable seed, are clearly distinguished in this way when biotinylated DNA from *G. consobrina* is used as a probe (Plate 1M, N). Regions of cross-hybridized chromatin are visible, as in the hybrid between *G. consobrina* and *G. karwinskyana.* The positions of regions on the chromosomes (Plate 1L) and around the nucleolus in interphase nuclei (Plate 1M) suggests that they are associated with ribosomal genes. There is also a low level of more general cross-hybridization, which is especially clear when no counterstain has been used (Plate 1M).

Gibasis pulchella and *G. karwinskyana* do not cross-fertilize either naturally or experimentally. Although some pollen tubes reach the ovary after cross-pollination, they are very abnormal. When a mixture of root tip nuclei from *G. karwinskyana* and *G. pulchella* was probed with biotinylated total DNA from *G. karwinskyana* and washed at low stringency, only the nuclei from *G. karwinskyana* were labelled (Plate 1K), as in the Southern blot. In this case, nuclei from *G. pulchella* could also be identified by their larger size. The apparently complete lack of cross-homology between *G. pulchella* and *G. karwinskyana* is despite a very similar banding pattern in diploids (not shown). The similarity in organization of heterochromatin in these disjunct species is still not understood. Either their genomes have differentiated subsequently to band positioning, or tandemly repeated sequences of a certain type (in this case, AT-rich) have tended to accumulate preferentially in a particular part of the chromosome.

These preliminary results suggest that the amount of genomic compatibility in terms of crossing potential and chromosome pairing (both involving recognition processes) is reflected in the success of GISH. In *Gibasis*, banding patterns do not seem to be a reliable marker of phylogenetic relationships among species, although within *G. karwinskyana*, they have been useful in identifying populations (Kenton, 1991). This is consistent with the idea that tandemly repeated sequences evolve more rapidly than the rest of the genome.

CONCLUSIONS: GENOME ORGANIZATION AND SPECIATION

We have reviewed evidence for and against the involvement of karyotype variation, heterochromatin and molecular divergence in the formation of new species. Although there are many examples where karyotype reorganization accompanies speciation, there are others where karyotype changes have no apparent effects and it is difficult to prove that the two factors are correlated. Karyotype morphology is probably most useful as an indicator of phylogenetic relationship at the generic level and below. G-banding patterns are a reliable indicator in animals (especially mammals), probably because they have ontogenetic significance, but have no certain parallel in plants. C-banding patterns are more useful at the lowest taxonomic rankings (e.g. subspecies and population), as C-bands may change both qualitatively and quantitatively in response to internal changes within the genome. Molecular cytogenetics will greatly improve and speed-up assessment of genome reorganization, especially in the absence of physical chromosome markers. In particular, GISH offers a reliable, fast and reproducible method for making broad comparisons of close and more distant relatives. Ultimately, it is phenotypic differentiation that decides whether a genotype is recognized as a new species. Usually, genome reorganization plays a secondary role in this, for example by inducing reproductive isolation and allowing mutations to accumulate. However, established evidence for 'catastrophic' reorganization of the genome, leading to the rapid generation of new species (Lewis, 1962), deserves exploration with molecular cytogenetics. Correlative studies, although useful in defining the range and distribution of variation, should be accompanied by experiments which seek to understand the complex machinery of the genome and how gene expression is controlled. Only in this way can we hope to understand the processes underlying evolution.

ACKNOWLEDGEMENTS

We thank the following colleagues for photographs: A. Baum for *Myriophyllum verticillatum;* M. Cornell for G-banding in *Tradescantia* and *Gibasis,* and CMA in *Lilium;* M. Johnson for *Brachycome lineariloba* and *Fritillaria* sp.; and L. Meredith for karyotypes of *Gibasis.* We are also grateful to colleagues at the Jodrell Laboratory, and to Professor K. Jones, for helpful discussions. The photograph of *A. blanda* chromosomes was taken during a collaborative study at the University of Vienna (with Professor D. Schweizer and Dr S. Hagemann).

REFERENCES

ANAMTHAWAT-JÓNSSON, K., SCHWARZACHER, T., LEITCH, A.R., BENNETT, M.D. & HESLOP-HARRISON, J.S., 1990. Discrimination between closely related Triticeae species using genomic DNA as a probe. *Theoretical and Applied Genetics, 79:* 721–728.

ÁRNASON, Ú. & WIDEGREN, B., 1989. Composition and chromosomal localization of cetacean highly repetitive DNA with special reference to the blue whale, *Balaenoptera musculus. Chromosoma, 98:* 323–329.

BENNETT, M.D. & SMITH, J.B., 1976. Nuclear DNA amounts in angiosperms. *Philosophical Transactions of the Royal Society of London, B (Biology), 274:* 227–274.

BENNETT, S.T. & THOMAS, S.M., 1991. Karyological analysis and genome size in *Milium* (Gramineae) with special reference to polyploidy and chromosomal evolution. *Genome, 34:* 868–878.

BENNETT, S.T., KENTON, A.Y. & BENNETT, M.D., 1992. Genomic *in situ* hybridization reveals the allopolyploid nature of *Milium montianum* (Gramineae). *Chromosoma,* 101: 420–424.

BHATT, B. & McGEE, J.O'D., 1990. Chromosomal assignment of genes. In J.M. Polak & J.O'D. McGee (Eds), In situ *Hybridization. Principles and Practice:* 149–164. New York: Oxford University Press.

BICKHAM, J.W., 1984. Patterns and modes of chromosomal evolution in reptiles. In A.K. Sharma & A. Sharma (Eds), *Chromosomes in Evolution of Eukaryotic Groups:* 13–40. Florida: CRC Press.

BRANDHAM, P.E., 1983. Evolution in a stable chromosome system. In P.E. Brandham & M.D. Bennett (Eds), *Kew Chromosome Conference II:* 251–260. London: Allen & Unwin.

BRANDHAM, P.E. & KIRTON, P.R., 1987. The chromosomes of species and cultivars of *Narcissus* L. (Amaryllidaceae). *Kew Bulletin, 42:* 65–102.

BRANDRIFF, B.F., GORDON, L.A., SEGRAVES, R. & PINKEL, D., 1991. The male-derived genome after sperm-egg fusion: spatial distribution of chromosomal DNA and paternal–maternal genomic association. *Chromosoma,* 100: 262–266.

BRIGHTON, C.A., MATHEW, B. & RUDALL, P., 1983. A detailed study of *Crocus speciosus* and its ally *C. pulchellus* (Iridaceae). *Plant Systematics and Evolution, 142:* 187-206.

BRINK, R.A., COOPER, D.C. & AUSCHERMAN, L.E., 1944. A hybrid between *Hordeum jubatum* and *Secale cereale* reared from an artificially cultivated embryo. *Journal of Heredity, 35:* 67–75.

BRINKLEY, B.R., VALDIVIA, M.M., TOUSSON, A. & BRENNER, S.L., 1984. Compound kinetochores of the Indian muntjac: evolution by linear fusion of kinetochores. *Chromosoma, 91:* 1–11.

CHEN, T.L. & MANUELIDIS, L., 1989. SINES and LINES cluster in distinct DNA fragments of G-band size. *Chromosoma, 98:* 309–316.

CHRISTIDIS, L., 1990. Chromosomal repatterning and systematics in the passeriformes (songbirds). In K. Fredga, B.A. Kihlman & M.D. Bennett (Eds), *Chromosomes Today, volume 10:* 279–294. London: Unwin Hyman.

CLELAND, R.E., 1972. *Oenothera: Cytogenetics and Evolution.* New York: Academic Press.

DARLINGTON, C.D., 1958. *Evolution of Genetic Systems,* 2nd edition. Alva: Oliver & Boyd.

DAVIES, A., JENKINS, G. & REES, H., 1990. The fate of multivalents during meiotic prophase in the hybrid *Gibasis consobrina* × *G. karwinskyana* Rafin. (Commelinaceae). *Genetica, 82:* 103–110.

DE GROUCHY, J., TURLEAU, C., ROUBIN, M. & COLIN, F.C., 1972. Chromosomal evolution of man and the primates (*Pan troglodytes, Gorilla gorilla, Pongo pygmaeus*). Chromosome identification – technique and applications in biology and medicine. *Nobel Symposium 23:* 124–131. New York, London: Academic Press.

DEUMLING, B. & GREILHUBER, J., 1982. Characterization of heterochromatin in different species of the *Scilla sibirica* group. *Proceedings of the National Academy of Sciences, U.S.A., 78:* 338–442.

DEWEY, D.R., 1984. The genomic system of classification as a guide to intergenomic hybridization within the perennial Triticeae. In J.P. Gustafson (Ed.), *Gene Manipulation in Plant Improvement:* 209–279. New York: Plenum.

DOVER, G.A. & FLAVELL, R.B., 1984. Molecular co-evolution: DNA divergence and the maintenance of function. *Cell, 38:* 622–623.

EVANS, G.M., 1988. Genetic control of chromosome pairing in polyploids. In P.E. Brandham (Ed.), *Kew Chromosome Conference III:* 253–260. London: HMSO.

EVANS, H.J., BUCKLAND, R.A. & SUMNER, A.T., 1973. Chromosome homology and heterochromatin in goat, sheep and ox studied by banding techniques. *Chromosoma, 42:* 383–402.

FINCH, R.A., SMITH, J.B. & BENNETT, M.D., 1981. *Hordeum* and *Secale* genomes lie apart in a hybrid. *Journal of Cell Science, 52:* 391–403.

FLAVELL, R.B., 1989. Variation in structure and expression of ribosomal DNA loci in wheat. *Genome, 31:* 963–968.

FLAVELL, R.B., O'DELL, M. & HUTCHINSON, J., 1981. Nucleotide sequence organization in plant chromosomes and evidence for sequence translocation during evolution. *Cold Spring Harbor Symposia in Quantitative Biology, 45:* 501–508.

GERSTEL, D.U., 1960. Segregation of new allopolyploids of *Nicotiana.* I. Comparison of 6x (*N. tabacum* × *tomentosiformis*) and 6x (*N. tabacum* × *otophora*). *Genetics, 45:* 1723–1734.

GERSTEL, D.U., 1963. Segregation in new allopolyploids of *Nicotiana.* II. Discordant ratios from individual loci in 6x (*N. tabacum* × *N. sylvestris*) *Genetics, 48:* 677–689.

GLEBA, Y.Y., PAROKONNY, A., KOTOV, V., NEGRUTIU, I. & MOMOT, V., 1987. Spatial separation of parental genomes in hybrids of somatic plant cells. *Proceedings of the National Academy of Sciences, U.S.A., 84:* 3709–3713.

GOODSPEED, T.H., 1954. *The genus* Nicotiana. *Origins, Relationships and Evolution of its Species in the Light of Distribution, Morphology and Cytogenetics.* Waltham, Massachusetts: Chronica Botanica Company.

GREILHUBER, J., 1977. Why plant chromosomes do not show G-bands. *Theoretical and Applied Genetics, 50:* 121–124.

GREILHUBER, J. & SPETA, F., 1976. C-banded karyotypes in the *Scilla hohenackeri* group, *S. persica,* and *Puschkinia* (Liliaceae). *Plant Systematics and Evolution, 126:* 149–188.

HAIDUK, M.W. & BAKER, R.J., 1982. Cladistical analysis of G-banded chromosomes of nectar feeding bats (Glossophaginae: Phyllostomidae). *Systematic Zoology, 31:* 252–265.

HATCH, F.T., BODNER, A.J., MAZRIMAS, J.A. & MOORE, D.H. II., 1976. Satellite DNA and cytogenetic evolution. *Chromosoma, 58:* 155–168.

HESLOP-HARRISON, J.S., 1991. The molecular cytogenetics of plants. *Journal of Cell Science, 100:* 15–21.

HOLMQUIST, G.P., 1989. Evolution of chromosome bands: molecular ecology of noncoding DNA. *Journal of Molecular Evolution, 28:* 469–486.

IMAI, H.T. & TAYLOR, R.W., 1989. Chromosomal polymorphisms involving telomere fusion, centromeric inactivation and centromere shift in the ant, *Myrmecia (pilosula) n* = 1. *Chromosoma, 98:* 456–460.

IMAI, H.T., CROZIER, R.H. & TAYLOR, R.W., 1977. Karyotype evolution in Australian ants. *Chromosoma, 59:* 341–393.

JENKINS, G., 1985. Synaptonemal complex formation in hybrids of *Lolium temulentum* × *Lolium perenne* (L.) 1. High chiasma frequency diploid. *Chromosoma, 92:* 81–88.

JOHN, B., 1990. Meiosis. In P.W. Barlow, D. Bray, P.B. Green & J.M.W. Slack (Eds), *Developmental and Cell Biology Series, no. 22.* Cambridge: Cambridge University Press.

JOHN, B. & MIKLOS, G.L.G., 1988. *The Eukaryotic Genome in Development and Evolution.* London: Allen & Unwin.

JOHNSON, M.A.T., KENTON, A.Y., BENNETT, M.D. & BRANDHAM, P.E., 1989. *Voanioala gerardii* has the highest known chromosome number in the monocotyledons. *Genome, 32:* 328–333.

JOHNSON, M.A.T., GARBARI, F. & MATHEW, B., 1991. In D. Phitar & W. Greuter (Eds), *Botanika Chronika, 10:* 827–839. Patras, Greece: Publications of the Botanical Institute, University of Patras.

JONES, K., PAPES, D. & HUNT, D.R., 1975. Contributions to the cytotaxonomy of the Commelinaceae. II. Further observations on *Gibasis geniculata* and its allies. *Botanical Journal of the Linnean Society, 71:* 145–166.

JONES, K., KENTON, A. & HUNT, D.R., 1981. Contributions to the cytotaxonomy of the Commelinaceae. Chromosome evolution in *Tradescantia* section *Cymbispatha. Botanical Journal of the Linnean Society, 83:* 157–188.

JUNG, K.Y., WARTER, S. & RUMPLER, Y., 1991. Confirmation, via *in situ* hybridization, of the occurrence of Robertsonian translocations during lemur evolution by localization of GLUDP1 DNA sequences on lemur chromosomes. *Cytogenetics and Cell Genetics, 57:* 59–62.

KARASAWA, K. & TANAKA, R., 1980. C-banding study on centric fission in chromosome of *Paphiopedilum*. *Cytologia, 45:* 97–102.

KENTON, A., 1978. Giemsa C-banding in *Gibasis* (Commelinaceae). *Chromosoma, 65:* 309–324.

KENTON, A., 1991. Heterochromatin accumulation, disposition and diversity in *Gibasis karwinskyana* (Commelinaceae). *Chromosoma, 100:* 467–478.

KENTON, A. & DRAKEFORD, A., 1990. Genome size and karyotype evolution in *Tradescantia* section *Cymbispatha* (Commelinaceae). *Genome, 33:* 604–610.

KENTON, A. & JONES, K., 1985. Autosyndetic pairing in *Gibasis* revealed by C-banding. *Chromosoma, 91:* 176–184.

KENTON, A., DAVIES, A. & JONES, K., 1987. Identification of Renner complexes and duplications in permant hybrids of *Gibasis pulchella*. *Chromosoma, 95:* 424–434.

KIDD, A.D., FRANCIS, D. & BENNETT, M.D., 1992. Replicon size, rate of DNA replication and the cell cycle in a primary hexaploid triticale and its parents. *Genome. 35:* 126–132.

KIMBER, G., ALONSO, L.C. & SALLEE, P.J., 1981. The analysis of meiosis in hybrids. 1. Aneuploid hybrids. *Canadian Journal of Genetics and Cytology, 23:* 209–219.

KOSTOFF, D., 1933. A contribution to the sterility and irregularities in the meiotic process caused by virus diseases. *Genetica, 15:* 103–114.

KOUKALOVÁ, B., REICH, J. & BEZDEK, M., 1990. A BamH1 family of tobacco highly repeated DNA: a study about its species specificity. *Biologia Plantarum (Praha), 32:* 445–449.

KUHROVÁ, V., BEZDĚK, M., VYSKOT, B., KOUKALOVÁ, B. & FAJKUS, J., 1991. Isolation and characterization of two middle repetitive DNA sequences of nuclear tobacco genome. *Theoretical and Applied Genetics, 81:* 740–744.

LEITCH, A.R., SCHWARZACHER, T., MOSGÖLLER, W., BENNETT, M.D. & HESLOP-HARRISON, J.S., 1991. Parental genomes are separated throughout the cell cycle in a plant hybrid. *Chromosoma, 101:* 206–213.

LEWIS, H., 1962. Catastrophic selection as a factor in speciation. *Evolution, 3:* 257–271.

LINDE-LAURSEN, I., 1982. Linkage map of the long arm of barley chromosome 3 using C-bands and marker genes. *Heredity, 49:* 27–35.

McCLINTOCK, B., 1951. Chromosome organization and gene expression. *Cold Spring Harbor Symposia in Quantitative Biology, 16:* 13–47.

MAYR, B., KALAT, M. & RÀB, P., 1990. Sequential counterstain-enhanced fluorescence chromosome banding in the fish *Anguilla anguilla*. *Caryologia, 43:* 277–281.

MEYNE, J., BAKER, R.J., HOBART, H.H., HSU, T.C., RYDER, O.A., WARD, O.G., WILEY, J.E., WURSTER-HILL, D.H., YATES, T.L. & MOYZIS, R.K., 1990. Distribution of non-telomeric sites of the $(TTAGGG)_n$ telomeric sequence in vertebrate chromosomes. *Chromosoma, 99:* 3–10.

MOLSEED, E., 1970. The genus *Tigridia* (Iridaceae) of Mexico and Central America. *University of California Publications in Botany, 54:* 1–127.

NARAYAN, R.K.J. & DURRANT, A., 1983. DNA distribution in the chromosomes of *Lathyrus*. *Genetica, 61:* 47–53.

OHNO, S., MURAMOTO, J., STENIUS, C., CHRISTIAN, L. & KETTRELL, W.A., 1969. Microchromosomes in holocephalian, chondrostean and holostean fishes. *Chromosoma, 26:* 35–40.

PARDUE, M.L. & GALL, J.G., 1970. Chromosomal localization of mouse satellite DNA. *Science, 168:* 1356–1358.

PAROKONNY, A.S., KENTON, A.Y., MEREDITH, L., OWENS, S.J. & BENNETT, M.D., 1992. Genomic divergence of allopatric sibling species investigated by molecular cytogenetics of their F_1 hybrids. *The Plant Journal,* 2: 695–704.

PATHAK, S., HSU, T.C. & ARRIGHI, F.E., 1973. Chromosomes of *Peromyscus* (Rodentia: Cricetidae). IV. The role of heterochromatin in karyotype evolution. *Cytogenetics & Cell Genetics, 12:* 315–326.

PATTON, J.L. & SHERWOOD, S.W., 1982. Genome evolution in pocket gophers (genus *Thomomys*). 1. Heterochromatin variation and speciation potential. *Chromosoma, 85:* 149–162.

REDI, C.A., GARAGNA, S., MAZZINI, G. & WINKING, H., 1986. Pericentromeric heterochromatin and A–T contents during Robertsonian fusion in the house mouse. *Chromosoma, 94:* 31–35.

ROTHFELS, K., SEXSMITH, E. & HEIMBURGER, M.O., 1966. Chromosome size and DNA content in *Anemone* and related genera. *Chromosoma, 20:* 54–74.

SCHNEDL, W., 1972. Giemsa banding techniques. In T. Caspersson & L. Zech, Chromosome identification – technique and applications in biology and medicine. *Nobel Symposium 23:* 34–37. New York, London: Academic Press.

SCHUBERT, I., 1990. Restriction endonuclease (Re-) banding of plant chromosomes. *Caryologia, 43:* 117–130.

SCHWARZACHER, T., LEITCH, A.R., BENNETT, M.D. & HESLOP-HARRISON, J.S., 1989. *In situ* localization of parental genomes in a wide hybrid. *Annals of Botany, 64:* 315–324.

SCHWEIZER, D., 1976. Reverse fluorescent chromosome banding with chromomycin and DAPI. *Chromosoma, 58:* 307–324.

SCHWEIZER, D., STREHL, D. & HAGEMANN, S., 1990. Plant repetitive DNA elements and chromosome structure. In K. Fredga, B.A. Kihlman & M.D. Bennett (Eds), *Chromosomes Today, volume 10:* 33–43. London: Unwin Hyman.

SMITH-WHITE, S., 1968. *Brachycome lineariloba:* a species for experimental cytogenetics. *Chromosoma, 23:* 359–364.

STEBBINS, G.L., 1971. *Chromosomal Evolution in Higher Plants.* London: Edward Arnold.

TRASK, B., 1991. Fluorescence *in situ* hybridization: applications in cytogenetics and gene mapping. *Trends in Genetics, 7:* 149–154.

UHL, C.H., 1978. Chromosomes of Mexican *Sedum.* II. Section *Pachysedum. Rhodora, 80:* 491–512.

VAN DEKKEN, H., PINKEL, D., MULLIKIN, J., TRASK, B., VAN DEN ENGH, G. & GRAY, J., 1989. Three-dimensional analysis of the organization of human chromosome domains in human and human-hamster interphase nuclei. *Journal of Cell Science, 94:* 299–306.

VERMA, S.C. & REES, H., 1974. Nuclear DNA and the evolution of allotetraploid Brassicae. *Heredity, 33:* 61–68.

WANG, H.C. & KAO, K.N., 1988. G-banding in plant chromosomes. *Genome, 30:* 48–51.

WAYE, J.S. & WILLARD, H.F., 1989. Concerted evolution of alpha-satellite DNA: evidence for species specificity and a general lack of sequence conservation among alphoid sequences of higher primates. *Chromosoma, 98:* 273–279.

WHITE, J. & REES, H., 1985. The chromosome cytology of a somatic hybrid petunia. *Heredity, 55:* 53–59.

WHITE, M.J.D., 1973. *Animal Cytology and Evolution.* Cambridge: Cambridge University Press.

WHITE, N.S., BENNETT, S.T., KENTON, A.Y., CALLIMASSIA, M.A. & FRICKER, M.D., 1991. Characterising plant chromosomes and their 3-D organisation using CLSM. *Scanning, 13 (Suppl. 1):* 1/128–1/129.

Part III

Species and Speciation

CHAPTER

13

Animal species and sexual selection

HUGH E.H. PATERSON

Introduction 210

Natural selection and adaptation 210

Genetical species 212

Sexual selection 214

The problem of language 216

Case histories 218

Discussion 224

Acknowledgements 226

References 226

Keywords: Genetical species – natural selection – sexual selection – SMRS.

Abstract

In considering the possible role of sexual selection in speciation, it is necessary to reconsider the nature of natural selection and its relationship to sexual selection. Darwin and later evolutionists have not always been able to distinguish natural selection from sexual selection, and many authors have used a terminology which is misleading. Authors frequently attribute to sexual selection phenomena to do with the specific mate recognition system (SMRS), for example. Since sexual selection is defined by Darwin as an intraspecific process, it is clear that the SMRS must first act before sexual selection is possible. Thus, sexual selection is independent of the SMRS. This means that speciation is essentially independent of sexual selection. I argue that, in many cases, much of the debate over sexual selection can be attributed to the difficulties in distinguishing sexual selection from other phenomena, including stabilizing selection. It is argued that it is scarcely useful to attempt to separate these human-contrived categories, and it is more useful to consider how characters influence the effector processes of evolution, namely differential survival and differential reproductive success.

Evolutionary Patterns and Processes
ISBN 0-12-440895-8

> It is the desire for explanations which are at once systematic and controllable by factual evidence that generates science; and it is the organization and classification of knowledge on the basis of explanatory principles that is the distinctive goal of the sciences.
>
> (E. Nagel, 1961)

INTRODUCTION

In attempting to understand the roots of the diversity of the natural world I have been obliged to understand the origin and perpetuation of the units of diversity, the species. With this motivating interest it is necessary to approach species in genetical terms while remaining sensitive to approaches to species by workers with other motivating interests, such as taxonomists and ecologists. Since 1973 I have criticized the prevailing biological species concept, demonstrating that it conflates two distinct views of species that I have called the isolation concept (Dobzhansky, 1951; Mayr, 1963) and the recognition concept (Paterson, 1978, 1980, 1985). Furthermore, I have strongly criticized the opinion that natural selection is directly involved in speciation (Paterson, 1978, 1982, 1985), and have supported the alternative view that species are incidental consequences of sex. Species are not 'adaptive devices' as claimed by Dobzhansky (e.g. 1976) and Mayr (e.g. 1949). I have also attempted to disassociate my views from discordant side issues of species theory, such as group selectionist views (e.g. 'genetic integrity of species'), the viewing of species in terms of sterility (Paterson, 1988), and 'speciation' by polyploidy (Paterson, 1981, 1988). I have also disassociated the recognition concept from theories involving 'mate choice' (Paterson, 1982). In this paper I shall explain further the reasons for my avoidance of sexual selection theory in developing the recognition concept of species.

In theoretical evolutionary biology we set up models that can explain the biological diversity of which we are aware. We erect models of the processes that may be involved in producing the observed pattern. The critical constraining point to notice is that the models should be able to explain the pattern of the living world. The models should take due cognizance of our knowledge of chemistry, physics, geology, climatology, probability, genetics, cytology, physiology and biochemistry, besides the sciences more directly implicated. I shall attempt to follow the advice given long ago by Michael Ghiselin (1974) when he drew attention to the necessity of working from fundamentals in order to limit errors of conception in theoretical biology: "Yet unless one goes back to fundamentals, misconceptions are all but inevitable."

In preparing my argument I need to look first at a number of basic ideas such as natural selection, adaptation, and the ideas underlying sexual selection. I shall also discuss some problems of scientific exposition which I believe are pertinent to discussing the role of sexual selection.

NATURAL SELECTION AND ADAPTATION

> In the present study, we begin by deriving selection theory from its most basic elements. Then, and only then, it will be possible to see how the diverse modes of reproduction are appropriate to particular situations.
>
> (Ghiselin, 1974:40)

I find that Tuomi, Vuorisalo & Laihonen (1988) have earlier traversed some of the ground that I have covered in this section and reached conclusions not unlike mine in some respects. Their treatment should be compared with mine for perspective.

The rise of population genetics has led to a widespread perception of natural selection in terms of differential reproduction. Darwin himself at first (1859:5) mentioned only differential survival, but soon included differential reproduction as well (1859:62). Later, Darwin (1874:939) wrote: "natural selection depends on the success of both sexes, at all ages, in relation to the general conditions of life". This statement, if taken seriously, adds another criterion to Darwin's conception of natural selection, because it appears to imply that anything to do with reproduction, or one sex only, is excluded from natural selection. Probably it should be regarded as a lapse, since Darwin contradicts the remark elsewhere in the same book. We should take into account the fact that the remark occurs in the general summary of the *The Descent of Man*, and that it seems likely that at this point he was attempting to draw as sharp a line as possible between natural and sexual selection. Evidence will be cited below which reveals that Darwin had a very real problem in distinguishing the two processes in particular cases. It is clearly vital to know whether natural and sexual selection can be effectively distinguished; for, if they cannot, sexual selection should be dispensed with.

Another problem with natural selection is its name. The term natural selection is far from ideal, as has often been pointed out before. In 1866 Wallace wrote to Darwin urging him to abandon the term natural selection because of its deceptive properties (Darwin & Seward, 1903). Darwin in 1860 had already been aware of the shortcomings of his term and had confessed in a letter to W.H. Harvey (Darwin & Seward, 1903) that he had come to prefer the term 'natural preservation', which had been used in the subtitle of *On the Origin of Species*. Nevertheless, in replying to Wallace, he made it plain that he was reluctant to abandon the term natural selection because he had thought (Darwin, 1859:61), and still thought, "that it was a great advantage to bring into connexion natural and artificial selection". He did not foresee that in so doing he would impose a conceptual handicap on all succeeding generations of evolutionists, for his metaphor was not an apt one. A breeder, when selecting, deterministically chooses individual organisms bearing the characteristics that he has in mind, and permits only these individuals to reproduce to provide the next generation. What happens under natural selection is quite different from what happens under artificial selection, being essentially stochastic. In nature, all organisms which have survived to reproductive age will contribute to the next generation provided they are fertile, and can attract and recognize a mate. Furthermore, humans breed animals and plants by artificial selection to serve human needs, whereas 'natural selection' can lead to changes which benefit the individuals expressing them. In his 'Big Book', Darwin notes a number of differences between artificial and natural selection (Stauffer, 1975:223). It is obviously both confusing and undesirable to group together two distinct processes (differential survival and differential reproduction) under a name, which, in turn, suggests a relationship to a third, but different, process (artificial selection). As I shall show below, the use of 'selection' is similarly a handicap when speaking of sexual selection. In fact, it is generally advantageous to bear in mind the underlying processes, differential survival to maturity and differential reproduction, and to avoid the terms involving 'selection'.

Central to Darwin's conception of selection is the idea of competition. Indeed, to many evolutionists and ecologists, even today, competition is the motor that drives evolution. Darwin invoked competition when he first introduced natural selection (1859:5), and continued to do so thereafter. It is often forgotten that differential survival and differential reproduction are effective processes even in the absence of competition. It is pertinent to read what Darwin (1859:76) wrote in first articulating the ideas behind the competitive exclusion principle, and note his confession of ignorance of causes whilst still using the most extreme imagery of direct conflict:

> We can dimly see why the competition should be most severe between allied forms, which fill nearly the same place in the economy of nature; but probably in no one case could we precisely say why one species has been victorious over another in the great battle for life.

Though he was here seeking to emphasize interspecific competition, Darwin soon realized that his vision implied that the most intense competition of all is intraspecific and intrasexual. Although 'competition' was central to his thoughts on natural selection, the term is a slippery one (e.g. see Ghiselin, 1974:50–51), and symptomatic of a particular mind-set as will be discussed below.

In 1966 George Williams wrote an important book in which he set out to examine critically the process of adaptation through natural selection. After making clear that adaptation is an "onerous concept" to be invoked only with much caution, he provided guiding principles which would, if followed, greatly improve evolutionary discussion. This is not the place for a detailed exposition of Williams's now well-known suggestions, and so I shall simply quote from a paragraph from his introductory chapter (Williams, 1966:8–9) because of its pertinence to natural selection. It will provide an introduction to the terminology that I shall use in this paper.

> Adaptation is often recognized in purely fortuitous effects, and natural selection is invoked to resolve problems which do not exist. If natural selection is shown to be inadequate for the production of a given adaptation, it is a matter of basic importance to decide whether the adaptation is real. Any biological mechanism produces at least one effect that can properly be called its goal: vision for the eye, reproduction and dispersal for the apple. There may also be other *effects* such as the apple's contribution to man's economy . . . . The designation of something as the *means* or *mechanism* for a certain *goal* or *function* or *purpose* will imply that the machinery involved was fashioned by selection for the goal attributed to it. When I do not believe that such a relationship exists I will avoid such terms and use words appropriate to fortuitous relationships such as *cause* and *effect*.

GENETICAL SPECIES

Thoughtful observers in general have long been impressed with the positive assortative mating of familiar sexual organisms. When song thrushes pair and breed with song thrushes, and mistle thrushes pair and breed with mistle thrushes the exchange of genes is restricted to occurring within a 'gene pool'. The group of organisms constituting such an inclusive 'field of gene combination' has long been called a species by population geneticists. In this case the two genetical bird species mentioned happen to coincide, respectively, with the taxonomic species *Turdus philomelos* and *Turdus viscivorus*. The consideration of such commonplace observations led to the recognition concept of species. Clearly mating is restricted to biparental (sexual) organisms, which means that sex is central to understanding positive assortative mating, and, hence, species. Understanding the functions of the characters of the fertilization system is also critical to understanding species.

Sex involves three complex processes: meiosis, fertilization and syngamy (here restricted to the process of fusion of the half-nuclei of the fused gametes). The key to understanding genetical species can be limited further to the process of fertilization. Male and female gametes cannot on their own find each other and then fuse. Invariably, the gametes are brought together in complex and diverse ways ('fertilization system'), which are appropriate to achieving fertilization in the normal environment of the various organisms.

In all fertilization systems one finds that a system of signalling is involved. In unicellular organisms this is a simple chemical signal–response system involving a chemical signal from one partner and a specific receptor in the other. A positive response to a signal is what is here referred to as 'recognition'. This form of recognition is general in signalling between sperm and ovum, or pollen and stigma, or trichogyne and conidium, etc. In motile organisms more elaborate signalling is found, which brings potential mating partners together from a distance, and makes possible the orientation of the sexual organs for the transfer of gametes. I group all these diverse processes involving signal/response interactions together under the name specific mate recognition systems (SMRS). Accordingly, species, under the recognition concept are viewed as follows: A genetical species is that most inclusive group of biparental organisms that share a common fertilization system.

The fertilization system includes the SMRS, which comprises the signals/responses between mating partners or their cells (e.g. sperm and ovum), but not the signalling that occurs between flowers and their pollinators (these are part of the fertilization system). The aptness of the SMRS components to the organisms' usual environment is evident from comparative studies, and from detailed studies of organisms in their normal environment. So concerned have many authors been with sexual selection that this more fundamental aspect of sex has often been eclipsed. The extensive pollination studies in plants have been poorly matched in zoology. The role of environment in shaping fertilization systems is very evident from pollination studies in forests, alpine regions (Hedberg, 1969), deserts, countries at high latitudes, windswept terrestrial areas (Eisikowitch, 1973), and aquatic habitats where angiosperms such as *Zostera* spp. and *Vallisneria* spp. (Proctor & Yeo, 1973) occur. Compared with plants, studies which focus on how fertilization systems relate to the environment of species, are far fewer among animals (Morton, 1975; Marten & Marler, 1977; Marten, Quine & Marler, 1977), where studies on reproductive behaviour have been dominated by an excessive concern with sexual selection. It is informative to contrast the perspectives of animal and plant studies even in the work of a particular author. Darwin, the author of sexual selection, when dealing with orchids was able to abandon his competition-driven perspective, which is such a feature of his work with animals. This is quite evident from the long sentence with which he opened his book, *The Various Contrivances by Which Orchids are Fertilised by Insects* (1877):

> The object of the following work is to show that the contrivances by which Orchids are fertilised, are as varied and almost as perfect as any of the most beautiful adaptations in the animal kingdom; and, secondly, to show that these contrivances have for their main object the fertilisation of the flowers with pollen brought by insects from a distinct plant.

It is telling that Darwin evidently saw no need for a book about 'The Various Contrivances by Which Animals are Fertilised', and yet these are just as impressive as the ones in plants that so intrigued him. Again, the views of Williams (1966) provide the key to fertilization studies, enabling us to distinguish, with some confidence, functional characters such as fertilization system characters, from effects such as 'isolating mechanisms', hybrid sterility, and 'cohesion' (Templeton, 1989).

The following are some consequences of the recognition concept that should be emphasized in the context of this paper:

1. Genetical species are realities of nature.
2. Genetical species are not direct products of differential survival and differential reproduction, but incidental consequences of adaptive evolution during which the fertilization system, among other environment-related features, adapts to the features of the normal environment of a population.
3. Reproductive isolation between species is an incidental effect in Williams's (1966) sense.
4. The cohesion of a species is an incidental effect in Williams's sense.
5. Sterility is an incidental effect in Williams's sense.

These conclusions are fundamentally important, with broad significance for both evolutionary theory and theoretical ecology. Their implications require deep consideration because they are very different from those arising from the isolation concept (Paterson, 1985). Understanding speciation is only possible when the genetical nature of species is understood. Consequently, each concept of species imposes different constraints on models of speciation. An outline of how speciation might occur under the recognition concept can be found in Paterson (1982, 1985) and other papers.

SEXUAL SELECTION

> It is important to remember that language itself is a moral medium, almost all uses of language convey value .... Life is soaked in the moral, literature is soaked in the moral .... Value is only artificially and with difficulty expelled from language for scientific purposes.
>
> (Iris Murdoch in Magee, 1982:248)

We shall see that at times Darwin had difficulty in distinguishing sexual selection from natural selection. It is thus important to understand why he felt the need for a second process to account for evolution. In probing this question it will be necessary to notice Darwin's general views and his sensitivity to the prevailing intellectual pressures from educated Victorian society. These pressures often influenced his choice of metaphors and terms, and so these matters will also receive some brief attention.

The idea of sexual selection can be found in Darwin's 'sketch' of 1842, the 'essay' of 1844, and in the paper read before the Linnean Society in 1858 (de Beer, 1958: 48, 120, 263). In *On the Origin* sexual selection was outlined in some detail (Darwin, 1859:87–90), but it was dealt with comprehensively in *The Descent of Man* (1874: 319ff.). Darwin (1874:324) explained his reasons for introducing the term sexual selection:

> Since in such cases the males have acquired their present structure, not from being better fitted to survive in the struggle for existence, but from having gained an advantage over other males, and from having transmitted this advantage to their male offspring alone, sexual selection must here have come into action. It was the importance of this distinction which led me to designate this form of selection as Sexual Selection.

Later, in the general summary of the same book he expands this to:

> Sexual selection depends on the success of certain individuals over others of the same sex, in relation to the propagation of the species; whilst natural selection depends on the success of both sexes, at all ages, in relation to the general conditions of life. The sexual struggle is of two kinds; in the one it is between the individuals of the same sex, generally the males, in order to drive away or kill their rivals, the females remaining passive; whilst in the other, the struggle is likewise between the individuals of the same sex, in order to excite or charm those of the opposite sex, generally the females, which no longer remain passive, but select the most agreeable partners.
>
> (Darwin 1874:939)

As already suggested above, it seems that, in this summarizing passage at the end of the book, Darwin was attempting to contrast natural selection and sexual selection in the strongest possible way. The definition of natural selection is in contrast with what he wrote earlier (Darwin, 1874:321), and is certainly not acceptable as it stands. This quotation clearly reveals that the two forms of sexual selection that Darwin had in mind were:

1. Competition for mates among males with the females remaining passive.
2. Competition for mates among males involving active choice of mates by females.

Once again, a single term, 'sexual selection', is used for two different processes, a procedure that invites confusion. Notice that the matter is clarified if we speak in terms of differential reproduction, and, perhaps, differential survival. In *The Descent of Man*, Darwin (1874:349) pointed out that "the acquirement through sexual selection of conspicuous colours, appears to have been sometimes checked from the danger thus incurred".

In considering particular cases from nature Darwin (1874:323) found it difficult distinguishing between natural and sexual selection, and provided examples of how the framing of the argument could shape his final decision. He wrote:

> When the male has found the female, he sometimes absolutely requires prehensile organs to hold her; thus Dr. Wallace informs me that the males of certain moths cannot unite with the females if their tarsi or feet are broken. The males of many oceanic crustaceans, when adult, have their legs and antennae modified in an extraordinary manner for the prehension of the female; hence we may suspect that it is because these animals are washed about by the waves of the open sea, that they require these organs in order to propagate their kind, and if so, their development has been the result of ordinary or natural selection.

On the next page Darwin presented a second version that modifies his account of the shaping of the prehensile organs in favour of sexual, instead of natural selection:

> So again, if the chief service rendered to the male by his prehensile organs is to prevent the escape of the female before the arrival of other males, or when assaulted by them, these organs will have been perfected through sexual selection, that is by the advantage acquired by certain individuals over their rivals.

While writing these words Darwin felt sufficiently insecure to add: "But in most cases of this kind it is impossible to distinguish between the effects of natural and sexual selection." I believe that the real problem confronting Darwin was one of preconception arising from his tendency to use metaphors deriving from his social world. If he had been thinking in terms of differential survival and differential reproduction I doubt that he would have distinguished between natural and sexual selection. One can classify factors affecting differential survival and differential reproduction in a number of ways, but he chose only to distinguish sexual selection from the rest (natural selection). The number of cases which he and later workers (e.g. Huxley, 1938; Ghiselin, 1974) have found it difficult to classify provide a hint that the two classes, natural and sexual selection, are human contrivances: human-made, not natural categories. A feature of such cases is that they always give rise to demarcation disputes: witness other biological examples like the limits of taxa (say, 'species' and 'subspecies'), 'parasites', 'symbionts', 'weeds', 'semispecies' and 'biotypes'. Darwin (1874), in introducing his chapter 8, showed that he was aware of the general problem when he discussed Hunter's terms, primary and secondary sexual characters. In all such cases procrustean treatments are inevitable.

Why, then, did Darwin feel the overwhelming need for two processes to direct the course of evolution? There is reason to believe that he was influenced primarily by the breeding practices of animal breeders, just as he had been with natural selection. In the sketch of 1842 we first find the following abbreviated thought in the paragraph devoted to what was to be called sexual selection: "Compared to man using a male alone of good breed" (de Beer, 1958). In the presentation to the Linnean Society in 1858 (de Beer, 1958), this was expanded to become the closing sentence:

> The result of this struggle amongst the males may be compared in some respects to that produced by those agriculturists, who pay less attention to the careful selection of all their young animals, and more to the occasional use of a choice male.

No doubt he also attended to what occurred in his own species and his own society, and this is supported by his observations throughout *The Descent of Man* (1874:42ff., etc), and his emphasis on the aesthetic aspects of female choice. No doubt Darwin, like most of us, was subject to a "predisposition to certain modes of conception".

The problem of language

In the further analysis of Darwin's sexual selection, it is necessary to examine the language he pressed into service in order to communicate his particular viewpoint. Of course it is a commonplace to note the influence of economic ideas on his thinking. Nevertheless, 'competition' is central to his mode of thought and its role must be taken into account. His ideas on competition determined many of the words and terms he used. A number of 'key words' (Lambert & Hughes, 1987) in evolutionary biology are major contributory causes of misunderstanding, and are, accordingly, best avoided. 'Competition' and 'choice' are examples. Dobzhansky (1950) wrote: "Competition and struggle are emotionally loaded words, which are best avoided in discussions of causes of evolution", and so are many others. Darwin has loaded his exposition of natural selection with images of competition and conflict, yet we know that differential survival and reproduction can occur in the absence of competitors, as has been pointed out above. We talk of 'choice' of individuals with 'good genes', forgetting that 'good' is a relative term which takes on meaning only in the presence of what is 'bad'. Ignoring those with 'bad genes' is often called 'choice' of the 'good', though this option is taken by default. In standard population genetical terms, we could well speak here of 'quality control' through 'stabilizing selection'.

In our attempts at scientific exposition, we are all constrained by our dependence on a language. Metaphor and analogy add colour and power to our writings, and can be helpful by using the familiar to illuminate the unfamiliar (Beer, 1983). However, both should be used with great caution if we are to avoid obfuscation. After careful definition I (e.g. Paterson, 1982) use 'recognition' as a term, and yet I object to 'choice' as used in the writings on sexual selection. I justify this by pointing out that 'recognition' is less anthropomorphic. An antibody *recognizes* a certain antigen; the 'search' facility on my computer word-processing program *recognizes* any word that I stipulate. 'Recognize' is even used in inorganic chemistry (Davey *et al.*, 1991). The word 'choice', as applied to mates or habitats, seems less apt and more misleading because it invokes images of higher cognitive processes when, in fact, the process being described can be accounted for in terms of a positive response to a signal as I believe occurs with 'recognition'.

Readers may feel impatient with such arguments, but I justify them by pointing to the patent inability of scientists to communicate broadly. Scientific communication normally occurs between specialists within narrow fields and with similar perspectives: sociobiologists, cladists, pheneticists, structuralists, population geneticists, molecular biologists, ornithologists, lepidopterists, anthropologists, etc. Often each field has its own journal or journals, but communication often fails, even with their fellow specialists. Because of the constraints imposed on communication, such specialization makes the broad understanding of evolution very difficult. The synthesis of ideas is inhibited. Misleading terms should not be persisted with simply because they are the norm.

In discussing reproductive behaviour it is critical to elucidate the function of characters, and to recognize their secondary functions, and to distinguish their incidental effects. Attending to this unravels the problems of reproductive behaviour. The fertilization system is a group of characters which generally brings about effective fertilization under natural conditions. The system includes the SMRS which comprises those characters which are involved with signalling between mating partners or their cells. Associated with the SMRS is a distinct phenomenon: individual mate recognition – the recognition by an individual of its particular mate. To use the term 'mate recognition system' (e.g. Butlin & Ritchie, 1989), which fails to distinguish between these two distinct systems, is to degrade a valuable term by blurring its meaning.

Response to a signal may be the simple act of continuing the exchange (= 'recognition') (Paterson, 1982), and may not be evident to an observer. Similarly, decamping by a female constitutes 'termination', and may not involve 'recognition', or may even occur because of the absence of 'recognition'.

Authors speak loosely of 'courtship', and may even synonymize it with the SMRS of a species. This, too, is not acceptable if effective communication of ideas is to occur. In contrast to the SMRS, courtship does not include the signals and responses between an ovum and a sperm, or a stigma and pollen grain. Besides including the fertilization system, 'reproductive behaviour' may include signals that are only effective between conspecific males, or between parents and offspring, or behaviour by males which induces ovarian maturation in the female as in doves and pigeons.

Another term that is often used in this context is 'species recognition'. This is entirely unacceptable because of its imprecision and because using it introduces conceptual confusion. 'Species recognition' can only be aspired to by one species: *Homo sapiens*. What is 'recognized', using 'recognized' in the sense defined above, is a mating partner of the same kind, hence 'specific mate recognition system'.

Hybridization should receive consideration since it occurs in nature. It should be noticed that, in a sense, a hybridization incident is not the 'breakdown' commonly portrayed by advocates of the isolation concept, because a mate of the opposite sex has, in fact, been effectively recognized, albeit not a 'specific mate'. That a signal–response reaction chain, which is efficient under the conditions prevailing in the environment of a particular species, should sometimes result in bringing together gametes of two related species should not be surprising since related species often share a number of signals in their specific mate recognition systems. A few commonplace examples of groups in which signals are common to more than one species are the Pelicaniformes (van Tets, 1965) and the genera *Anas* (Johnsgard, 1965) and *Drosophila* (Manning, 1959).

A character of the SMRS may have the secondary function of a male identificatory signal as in the case of the 'moustache stripe' of the flicker, *Colaptes auratus* (Noble, 1936), and the red body colour of the stickleback, *Gasterosteus aculeatus* (Wootton, 1976). In such cases it is context that determines function.

The effective attribution of a function to a character requires an understanding of its part in the life of the organism, and this initially entails a working hypothesis. It is here that preconception can be very misleading. In the following section some well-known studies of reproductive behaviour are re-examined to illustrate this problem, which is central to any assessment of sexual selection.

CASE HISTORIES

In this section I propose re-examining a number of case histories, well known studies that act as exemplars in the study of sexual selection. My aim is to demonstrate that plausible alternative explanations exist in most cases, though they have been either ignored or not given equal consideration before being rejected.

In order to understand how any particular character evolved, it is essential to identify correctly the function that the character serves. After all, we will not understand the *raison d'être* of a character which has evolved to function in attracting a mate if we believe that its role is to prevent hybridization. The problems of distinguishing the functions of particular characters, and distinguishing functions from effects, lie at the root of many evolutionary debates, and, not least, the debate on sexual selection. This is because it is at this point that preconception often intrudes. These examples will serve to emphasize further the difficulty of distinguishing sexual selection from natural selection, and to strengthen the case for abandoning attempts at distinguishing the two human-contrived categories.

The reproductive behaviour of the three-spined stickleback, *Gasterosteus aculeatus*, has been much studied (Tinbergen, 1951, 1953; Wootton, 1976, etc.) These intensive investigations provide the detail which improves the quality of guesses as to the functions of the different components of the behaviour.

It should be remembered that no pair formation occurs in this species, so that males are polygynous. The male builds a nest, often on a muddy substrate, in clear water with sparse vegetation. The nest is within a territory held by the male. In intersexual signalling, the colouration of the male appears to serve the function of attracting females in reproductive condition from some distance. This signal is enhanced in effectiveness by characteristic movements. Females form groups which move about undeterred by male territories. The male recognizes an approaching female in reproductive condition by her bodily profile (Tinbergen, 1951; Rowland, 1989b). The ensuing signals lead the female to the nest entrance.

The signal–response reaction chain to this point has brought the sexes together and induced the female to enter the nest. Next the male provides a signal, 'vibrating', which stimulates the female to deposit her ova. This done, she leaves and the male enters the nest to fertilize the eggs. Later he may resume patrolling above his territory, and repeat the whole process with one or more other females before he commences egg fanning behaviour.

The males react to the red colour of other males. In fact a male will respond to anything red that enters his field of vision. Other mature males near the territory of a male elicit agonistic behaviour from him. Descriptions of interactions between males imply one male recognizing another, but interactions between males and females are generally referred to in terms of 'choice' for no good reason. Descriptions provide no explicit evidence of 'choice' of 'competing' males by gravid females. Females form small schools that criss-cross male territories so it could be assumed that choice among males would be possible. However, this is far from demonstrating that any choice is made.

Reproductive success is scored in terms of the numbers of offspring that reach reproductive age scored over the male's lifetime. Providing convincing evidence of reproductive advantage to males, or females through 'female choice', is extremely difficult.

The components of reproductive behaviour of *G. aculeatus* can readily be understood in terms of being evolved to achieve fertilization under the conditions prevailing in the usual habitat of the species. I am not aware of any evidence which demands the invoking of either form of sexual selection. I justify this statement on the grounds that it is extremely rare, for good reasons, to find any life-long data on reproductive success of elephant seals, ruff, pheasants, peacocks or grouse, and on the additional grounds that so seldom are alternative explanations ever explored at all, or, if they are, the attempt is often no more than a gesture. Already there are many situations in the literature which run counter to the current dogma on female choice, and many empirical studies which are not in accord with predictions (see Passmore & Telford, 1983; Hamilton, 1984; Breitwisch & Whitesides, 1987; Höglund & Säterberg, 1989; Kirkpatrick, 1989; Björklund, 1990; Schroeder, 1991, etc.).

It is difficult to tease out significant alternative interpretations, and difficult to test the alternatives discerned. As usual, one gathers the impression in many studies that there is a considerable degree of preconception involved.

A careful reading of a documented case of female 'choice' (Milinski & Bakker, 1990) in sticklebacks illustrates this point. The 'condition factor' used in this study is a fisheries measure, 'heaviness', as stated. Parasitism affects condition, and hence, supposedly, the male's ability to incubate the eggs, as well as the intensity of colour. Females are supposed to detect males of superior condition on the intensity of their red colouration.

However, male *Gasterosteus* seem able to distinguish size differences between themselves and rival males in artificial contests for territories (Rowland, 1989a), so one wonders why an oblique method of judging relative condition between males in terms of colour should have evolved in females. Such laboratory results are one thing, but in the absence of data from the field, one is justified in entertaining a certain scepticism. This is particularly so in a species where intensity of colour is influenced by a number of other factors.

As an alternative, are the available facts explicable under the recognition concept? This could be examined in terms of signal quality (response = recognition), but, as usual, the possibility was not addressed. What concerns me is the lack of attention accorded to a simple and comprehensive alternative explanation for the reproductive behaviour of *G. aculeatus*, that is yet fully in keeping with the experimental studies of Tinbergen (1951, 1953) and others. Perhaps relevant here is the fact that I used the work on *G. aculeatus* and on *Pygosteus pungitius* (ten-spined stickleback) (Morris, 1958) as exemplars when first thinking through the recognition concept. Tinbergen's accounts of the system may be somewhat idealized (Wootton, 1976) – for example, the chain may not be as regular as stated – but this does not have significant bearing on the recognition concept. Studies with models have shown that males in which the red colour was replaced with other colours still attract some females. This might be construed in terms of the females exerting 'choice', but these data are not inimical to the simpler view that the red colour is a signal eliciting a particular response, and it was thus interpreted by Tinbergen (1951). I am generally of the opinion that the simplest explanation should be persevered with until it is shown by compelling evidence to be inadequate.

Due to the experimental complexities underlying the ideas on the genetic consequences of 'female choice', I tend to take this conservative position on the many contemporary studies. While we can feel confident that a system must exist which results in positive assortative mating of the sort that leads to the formation of a field for gene recombination, we can feel no such certainty about the ideas of 'female choice' which call for something more than simple stabilizing selection.

A celebrated paper by Bateman (1948) can be used to illustrate some of the problems inherent in the concept of female choice. Bateman's discussion is quite well-balanced compared with some contemporary discussions of sexual selection, and pays attention to many of the problems pointed out earlier by Huxley (1938). These include the occurrence of strong sexual dimorphism in monogamous species such as species of sunbirds (Nectariniidae) and parrots, and the fact that elaborate displays may occur after pairing has occurred, when competition is no longer an issue.

Pairing and copulation occur simultaneously in *Drosophila melanogaster*. "Courtship behaviour determines the number of mates and, therefore, enters into intra-sexual selection" (Bateman, 1948:349). A number of the examples of intrasexual selection within *Drosophila* cited by Bateman (1948:table 1) are, however, examples of other phenomena: positive assortative mating among species and stabilizing selection. It can only confuse the field of sexual selection further if such examples are to be admitted as examples of sexual selection. To the best of my knowledge Darwin always considered sexual selection as an intraspecific phenomenon. For example, Darwin (1874:322) wrote: "This depends on the advantage which certain individuals have over others of the same sex and species solely in respect of reproduction."

One of the cited examples in Bateman (1948:table 1) (*D. pseudoobscura* and *D. miranda*) can be used to exemplify the use of 'female choice'. Let us consider what is usually involved in such an experiment. In one experimental arena are placed equal numbers of virgin females of both species together with males of *D. pseudoobscura*. In a second arena the same numbers of virgin females of both species are combined with the same number of *D. miranda* males. The experimental arrangement is generally left until about half the females have been inseminated. At this point the males in both arenas are removed and the females examined to see how many of each species in each arena have been fertilized. This examination tells us how many conspecific and how many interspecific matings have occurred in each arena. Were these matings the consequence of female 'choice'? Or were they the consequence of 'mate recognition'? These alternatives must be faced and assessed, and not ignored.

In science we often put words to new uses for which they were not originally intended. Such words often carry with them their earlier meanings and evoke inappropriate nuances in their new roles. Even when redefined (Halliday, 1983:4; see below), a word like 'choice' (or 'selection') inevitably invokes an intellectual judgement scarcely appropriate for organisms other than humans. To avoid any hint of such higher mental processes I have described (above and 1982) as 'recognition' the positive response of a physiologically mature female, say, to a particular signal of a mature male. Sometimes a particular signal is to be found in more than one species. If the female of one species responds to this signal broadcast by a male that is not conspecific, it is still a case of 'recognition'. Thus, in the experiment involving *D. pseudoobscura* and *D. miranda*, any matings that do occur between the species are still acts of 'recognition': the SMRS of both species share sufficient components for females to 'recognize' males whether they are conspecific or not. In nature, in the context of a species' particular habitat and distributional range, the SMRS generally leads to the regular bringing together of males and females of the same population. This non-cognitive process generally leads to effective fertilization under natural circumstances. As I have frequently emphasized, it is likely that the environmental characteristics of a species are specific, whether we consider them broad ('generalist') or narrow ('specialist'), and that they probably reflect the conditions under which the species arose (Paterson, 1978, 1980, 1985, 1986). Again, to my mind, the phenomenon described by Bateman is most appropriately viewed in terms of the 'recognition' of signals rather than in terms of 'choice'.

Let us look at another case, based on the work of Rendel, mentioned by Bateman in his table 1: the case involving wild-type *Drosophila subobscura* and the mutant yellow (y) of the same species. In tests with these two stocks, yellow males are relatively ineffective in inseminating both kinds of female, whereas wild-type males inseminate equal numbers of both yellow and wild-type females. Is it appropriate to say that both yellow and wild-type females 'choose' wild-type males, or is it more precise to say that both types of female 'recognize' wild-type males? Work since Bateman's paper has shown that the SMRS signals of yellow males in *Drosophila melanogaster* are aberrant (e.g. Bastock, 1956; Burnet & Wilson, 1980). It is, therefore reasonable to assume that the aberrant signals of the yellow males are not recognized by either the yellow or wild-type females, whereas both recognize the wild-type males. Bateman (1948:350), after providing experimental evidence to show that yellow females mated very readily with wild-type males, and that wild-type females mated infrequently with yellow males, commented as follows: "The males courted equally vigorously in both matings. Thus, wild type females found yellow males objectionable, but wild type males failed to discriminate between yellow and wild type females."

Such anthropocentric language as "wildtype females found yellow males objectionable" seems unjustifiable. If one bears in mind the fact that the SMRS is a signal–response reaction chain, it is surely more comprehensible to speak of 'recognition' as I have defined it, or 'non-recognition' when speaking of response or non-response to signalling by a male? This is tantamount to speaking of the female responding or not responding to a signal, which is surely preferable to Bateman's phraseology or speaking of 'choice'. In effect, the non-recognition of aberrant males, in this case yellow males, constitutes 'stabilizing selection', not 'sexual selection'.

Finally, Bateman's work can be used to warn against the use of oblique evidence to support assumed examples of sexual selection. Bateman used differences in variance in mating success between males and females as an indication of non-random mating due to female choice. However, non-random mating is also an expectation of the recognition concept, and Sutherland (1985) has shown that Bateman's data can be explained under random mating.

The open grasslands of eastern Africa are the natural habitat of the long-tailed widowbird (*Euplectes progne*). It was the subject of an important study by Malte Andersson (1982) which has been very influential for its bearing on sexual selection. Majerus (1986), for example, described this study as "Perhaps the most dramatic illustration that females may indeed have a preference for increased development of a secondary sexual character." I shall look briefly at this species in terms of recognition concept theory.

The male in the non-breeding season is mostly drab brown like the female, except for its coloured epaulets. At this period males and females occur in flocks. In the breeding season the males change radically to take on a velvety black plumage which contrasts sharply with the colourful epaulets, which are retained. The most striking feature is the long tail; 6–8 of the 12 tail feathers are around 50 cm long (Andersson, 1982). In the slow display flight over his territory, the male spreads his tail, effectively broadening it. Andersson experimented with a number of males, artificially lengthening or shortening the tail, and scored the effect his manipulations had on the reproductive success of males as scored by the number of females attracted to nest in each territory. Although this scoring system has shortcomings because it does not actually measure life-long reproductive success, it can, in the absense of anything better, be accepted as a rough measure of relative reproductive performance.

Male sticklebacks display over their territories to attract females which criss-cross the territories of the males in groups. Similarly, *E. progne* males perform a slow display flight over their territories. In doing so they broaden their long black tails, making themselves visible over considerable distances across the open grasslands. This appears to be a very appropriate distance attractant in open country, where wind interferes with auditory signals (Morton, 1975). Shortening the tail experimentally is likely to reduce the effectiveness of the display flight as a distance attractant, while lengthening it is likely to improve it.

It will be evident that Andersson's findings are fully in accord with the view that the broad function of the fertilization system is the achieving of fertilization under the conditions prevailing in an organism's normal environment. However, Andersson interpreted his results in terms of sexual selection involving 'female choice'. He did briefly consider the interpretation in terms of environmental aptness of the SMRS in the following words:

> Highly adorned males can be favoured by active mate choice, where females compare males before accepting one, but also by easier detection (Parker, 1982). This latter advantage may have contributed to the evolution of the long tail and the flight display in the long-tailed widow.
>
> The lateral surface of the displaying male is enlarged 2–3 times by the tail, making him correspondingly easier to discover from a distance in the open habitat. However, neighbouring males often display simultaneously, and females sometimes visit several males in rapid succession with ample opportunity for comparisons. The long tail is therefore probably maintained at least partly through active female discrimination among males.

A little later he summed up the paper as follows, without any further reference to the alternative hypothesis:

> The results presented here support Darwin's (1874) hypothesis that certain male ornaments are favoured by female mate choice, and probably evolved through it.

This conclusion does not seem justifiable on the basis of his remarks just quoted. 'Visiting', might mean 'swiftly flying over', and one wonders about the justification for the comment "with ample opportunity for comparisons": what is known about the time required for a process that we are not certain even occurs? It is my opinion that the data provided by Andersson is at least as fully supportive of the recognition concept view as of the 'female choice' position. In his article, Majerus did not even mention the alternative possibility briefly considered by Andersson.

Some years ago, Halliday (1983:10) conceded:

> The concept of mate choice for good genes is fraught with problems and must be regarded as an open question. Perhaps the only instance where it is tenable, on the basis of existing studies, are those where there is evidence that females choose older males, whose ability to survive may have a heritable basis.

Halliday was probably thinking of male Anura as the possible exception to his uncompromising remark. With female choice in mind, biologists committed to sexual selection have long sought mechanisms by means of which the fitness of their offspring could be enhanced. As Halliday stated, perhaps the most credible suggestion involves the selection of older, and, hence, proven males. In Anura, where growth continues with age, and where larger males emit lower frequency calls than smaller males, a connection between a signal detectable by the female and fitness would seem feasible.

This idea has long been popular (Trivers, 1972; Howard, 1978, 1979; Wilbur, Rubenstein & Fairchild, 1978; Ryan, 1980), but has not been easy to test because of the scarcity of data relating age to size in male frogs. However, Passmore & Telford (1983) were able to assess the utility of this approach using the frog species *Hyperolius marmoratus* in which young males in their first breeding season are distinguishable from older males. These authors were able to demonstrate that female *H. marmoratus* did not mate selectively among males on the basis of age, despite a correlation between size and sound-spectral structure. It was demonstrated that sound intensity was also positively correlated with size in males, a significant factor in interpreting female mating behaviour. Furthermore, Höglund & Säterberg (1989), working with *Bufo bufo*, found no correlation between size and age in male toads.

DISCUSSION

Comprehending the genetical nature of species is an essential prerequisite to understanding how a new species can arise. For this reason I have spent much time thinking about the genetical properties of species, their stability and their ability to change (Paterson, 1985). My concern with speciation has, of course, led me to consider the nature of the underlying mechanism of evolutionary change, 'natural selection', and its bearing on speciation. Because I have argued that genetical species are best understood as fields of gene combination that have their limits set by the fertilization systems of organisms, it seems obvious to some authors (e.g. Verrell, 1988; Eldredge, 1989) that there should be an important role for sexual selection in the theory of the recognition concept.

Darwin, as I have shown, clearly believed that the two processes, which he brought together under the heading of 'sexual selection', involve members of the same species. In other words, before sexual selection can occur, the males and females have already become associated by the functioning of the earlier stages of the specific mate recognition system. If female choice, in the strict sense, really exists, the female is generally 'choosing' from among an array of males that share a common fertilization system (i.e. conspecific males). For this reason, I have, hitherto (e.g., Paterson, 1982), simply left sexual selection on one side as of secondary interest in understanding the nature of species (Paterson, 1985).

In this paper, I have argued, like others before me, that the term 'natural selection' is based on an inappropriate analogy with artificial selection, which is distinctly misleading. Under 'natural selection', Darwin admitted two actually distinct processes: differential survival and differential reproduction under the prevailing environmental conditions. I believe that we should be wise to avoid the ambiguous term of 'natural selection', just as we now avoid Spencer's term, 'survival of the fittest', and concentrate on the two processes that lead to evolution.

In population genetics the two processes are generally blurred and not distinguished. Simplification is achieved though at the price of more detailed understanding (cf. Tuomi, Vuorisalo & Laihonen, 1988). Only recently have attempts been initiated to explore the consequences of distinguishing between the processes.

Darwin distinguished, or sought to distinguish, sexual selection from natural selection. Yet, both act through the processes of differential survival and differential reproduction. Darwin, in the general summary of his book, *The Descent of Man*, attempted to distinguish sharply between natural and sexual selection, though, as I have shown, he had earlier in the book been pessimistic about being able to do so. More recently, Michael Ghiselin (1974) has trodden the same path. When seeking to establish the differences between natural and sexual selection he wrote:

> In artificial selection, the selective agent (that which does the selecting) is man; in natural selection it is the forces active in the 'environment'; in sexual selection it is a mate or a rival. A failure to understand the differentia underlying Darwin's terminology is one of the main reasons why many have treated sexual selection as if it were just a special case of natural selection.
>
> (Ghiselin, 1974:130)

But, a little later in the same book, he, too, struck, difficulties. When discussing dwarf males in organisms such as in some barnacles, angler fish and rotifers, he confessed:

> Exactly how much reduction of the male [in dwarf males] should be attributed to sexual and how much to natural, selection, in such cases is problematic. Indeed, the distinction tends to become a semantic one in this context, for the various modes of selection, although rightly kept conceptually distinct, merge into one another by insensible degrees.
>
> (Ghiselin, 1974:194)

Thus, the differentia underlying Darwin's terminology are not as straightforward as Ghiselin at first implied. In fact, Ghiselin's statement (1974:130), illustrates other problems which need emphasizing. His remark that in sexual selection the selecting agent is a mate or a rival requires critical assessment. As I have pointed out earlier in this paper, the two forms of sexual selection are conceptually quite different, and represent two distinct phenomena tenuously connected by a nexus with reproduction. Therefore, Ghiselin's attribution of the selecting agent in sexual selection to either a "mate or a rival" is deceptive, implying as it does that 'inter-male competition' and 'female choice of mates' are in some way comparable phenomena. Probing what lies behind Ghiselin's "forces active in the 'environment'" would reveal other problems because such a phrase implies a very heterogeneous group of factors.

It is worth considering in broad terms the factors that influence differential survival and differential reproduction. Genetically determined variation occurs that affects the following activities:

1. Accommodation to the environment
2. Food finding and food processing
3. Avoidance of predators and aggressors
4. Reproductive activities

In basic terms, then, characteristics under all these headings affect the two processes of evolution. It is somewhat inconsistent that we should single out characters to do with reproductive activities for special emphasis under the name 'sexual selection'. It is also misleading that two distinct processes should be grouped under this name. The imprecision of the concept of sexual selection is such that some authors assign all factors that affect differential reproduction to sexual selection. This is something Darwin specifically argued against, and something which most evolutionists would find unacceptable. If this practice were generally tolerated, sexual selection would then constitute an even greater rag bag than it is at present.

Not only is it difficult for us to decide what phenomena should fall under 'sexual selection', and what under 'natural selection', but phenomena are attributed to 'sexual selection' that should be grouped under the SMRS, or included simply through a play on words, as Darwin (1874) illustrated with the two scenarios from *The Descent of Man*. I have attempted to illustrate these practices with examples from well-known papers on sexual selection. Evolutionary biology would be much improved and better focused if biologists were to give due consideration to alternative explanations, and design tests to distinguish between them.

To summarize: in seeking to understand evolutionary change and stability, including the origin of species, we should concentrate on determining the factors bearing on survival and reproductive success, avoiding the deficient Darwinian terminology of natural and sexual selection. Conscientiously practised, this would lead to the removal of yet another layer of confusion from evolutionary theory.

Since 'sexual selection' is by definition an intraspecific process, it can, in any case, only operate after the SMRS has brought the sexes together. This demonstrates the essential independence of the two processes.

ACKNOWLEDGEMENTS

It is a pleasure to acknowledge the discussions with John Endler, David Lambert, Gimme Walter, Erkki Haukioja, and Chris Pavey, some of whom also read and commented helpfully on the paper. Shirley Paterson and a referee have been very helpful with comments. I have found Haukioja (1982) and a paper by Richard Sylvan to appear in *Revista di Biologia* most helpful.

REFERENCES

ANDERSSON, M., 1982. Female choice selects for extreme tail length in a widowbird. *Nature, 299:* 818–820.

BASTOCK, M., 1956. A gene mutation which changes a behavior pattern. *Evolution, 10:* 421–439.

BATEMAN, A.J., 1948. Intra-sexual selection in *Drosophila. Heredity, 2:* 349–368.

BEER, G., 1983. *Darwin's Plots – Evolutionary Narrative in Darwin, George Eliot and Nineteenth Century Fiction.* London: Routledge & Kegan Paul.

BJÖRKLUND, M., 1990. Mate choice is not important for female reproductive success in the common rosefinch (*Carpodacus erythrinus*). *Auk, 107:* 35–44.

BREITWISCH, R. & WHITESIDES, G.H., 1987. Directionality of singing and non-singing behavior of mated and unmated northern mockingbirds. *Animal Behaviour, 35:* 331.

BURNET, R., & WILSON, R., 1980. Pattern mosaicism for behaviour controlled by the *yellow* locus in *Drosophila melanogaster. Genetical Research, 36:* 235–247.

BUTLIN, R.K. & RITCHIE, M.G., 1989. Genetic coupling in mate recognition systems: what is the evidence? *Biological Journal of the Linnean Society, 37:* 237–246.

DARWIN, C., 1859. *On the Origin of Species by Means of Natural Selection.* London: John Murray.

DARWIN, C., 1874. *The Descent of Man and Selection in Relation to Sex*, 2nd edition. London: John Murray.

DARWIN, C., 1877. *The Various Contrivances by Which Orchids are Fertilised by Insects*, 2nd edition. London: John Murray.

DARWIN, F. & SEWARD, A.C. (Eds), 1903. *More Letters of Charles Darwin, 1.* New York: D. Appleton & Co.

DAVEY, R.J., BLACK, S.N., BROMLEY, L.A., COTTIER, D., DOBBS, B. & ROUT, J.E., 1991. Molecular design based on recognition at inorganic surfaces. *Nature, 353:* 549.

DE BEER, G. (Ed.), 1958. *Evolution by Natural Selection.* Cambridge: Cambridge University Press.

DOBZHANSKY, T., 1950. Mendelian populations and their evolution. *American Naturalist, 84:* 401–418.

DOBZHANSKY, T., 1951. *Genetics and the Origin of Species.* New York: Columbia University Press.

DOBZHANSKY, T., 1976. Organismic and molecular aspects of species formation. In F.J. Ayala (Ed.), *Molecular Evolution:* 95–105. Sunderland, Massachusetts: Sinauer Associates.

EISIKOWITCH, D., 1973. Mode of pollination as a consequence of ecological factors. In V.H. Heywood (Ed.), *Taxonomy and Ecology:* 283–288. London: Academic Press.

ELDREDGE, N., 1989. *Macroevolutionary Dynamics: Species, Niches, and Adaptive Peaks.* McGraw-Hill, New York.

GHISELIN, M.T., 1974. *The Economy of Nature and the Evolution of Sex.* Berkeley: University of California Press.

HAMILTON, M.E., 1984. Revising evolutionary narratives: a consideration of alternative assumptions about sexual selection and competition for mates. *American Anthropologist, 86:* 651–662.

HALLIDAY, T.R., 1983. The study of mate choice. In P. Bateson (Ed.), *Mate Choice:* 3–32. Cambridge: Cambridge University Press.

HAUKIOJA, E., 1982. Are individuals really subordinated to genes? A theory of living entities. *Journal of Theoretical Biology, 99:* 357–375.

HEDBERG, O., 1969. Evolution and speciation in a tropical high mountain flora. *Journal of the Linnean Society (Biology), 1:* 135–148.

HÖGLUND, J. & SÄTERBERG, L., 1989. Sexual selection in common toads: correlates with age and body size. *Journal of Evolutionary Biology, 2:* 367–372.

HOWARD, R.D., 1978. The evolution of mating strategies in bullfrogs, *Rana catesbeiana. Evolution, 32:* 850–871.

HOWARD, R.D., 1979. Estimating reproductive success in natural populations. *American Naturalist, 114:* 221–231.

HUXLEY, J.S., 1938. The present stand of the theory of sexual selection. In G.R. de Beer (Ed.) *Evolution:* 11–42. Oxford: Oxford University Press.

JOHNSGARD, P.A., 1965. *Handbook of Waterfowl Behavior*. Ithaca: Cornell University Press.

KIRKPATRICK, M., 1989. Is bigger always better? *Nature, 337:* 116–117.

LAMBERT, D.M. & HUGHES, A.J., 1987. Keywords: The importance of language in structural biology. *Rivista di Biologia, 80:* 188–191.

MAGEE, B., 1982. *Men of Ideas*. Oxford: Oxford University Press.

MAJERUS, M.E.N., 1986. The genetics and evolution of female choice. *Trends in Ecology and Evolution, 1:* 1–7.

MANNING, A., 1959. The sexual behaviour of two sibling *Drosophila* species. *Behaviour, 15:* 123–145.

MARTEN, K. & MARLER, P., 1977. Sound transmission and its significance for animal vocalization I. Temperate habitats. *Behavioral Ecology and Sociobiology, 2:* 271–290.

MARTEN, K., QUINE, D. & MARLER, P., 1977. Sound transmission and its significance for animal vocalization II. Tropical forest habitats. *Behavioral Ecology and Sociobiology, 2:* 291–302.

MAYR, E., 1949. Speciation and systematics. In G.L. Jepsen, E. Mayr, & G.G. Simpson (Eds), *Genetics, Paleontology, and Evolution:* 281–298. Princeton: Princeton University Press.

MAYR, E., 1963. *Animal Species and Evolution*. Cambridge, Massachusetts: Harvard University Press.

MILINSKI, M. & BAKKER, C.M., 1990. Female sticklebacks use male colouration in mate choice and hence avoid parasitized males. *Nature, 344:* 330–333.

MORRIS, D., 1958. The reproductive behaviour of the ten-spined stickleback (*Pygosteus pungitius* L.) *Behaviour Suppl. 6:* 1–154.

MORTON, E.S., 1975. Ecological sources of selection on avian sounds. *American Naturalist, 109:* 17–34.

NAGEL, E., 1961. *The Structure of Science: Problems in the Logic of Scientific Explanation*. London: Routledge & Kegan Paul.

NOBLE, G.K., 1936. Courtship and sexual selection of the flicker (*Colaptes auratus leuteus*). *Auk, 53:* 269–282.

PASSMORE, N.I. & TELFORD, S.R., 1983. Random mating by size and age of males in the painted reed frog, *Hyperolius marmoratus. South African Journal of Science, 79:* 353–355.

PATERSON, H.E.H., 1978. More evidence against speciation by reinforcement. *South African Journal of Science, 74:* 369–371.

PATERSON, H.E.H., 1980. A comment on 'mate recognition systems'. *Evolution, 34:* 330–331.

PATERSON, H.E.H., 1981. The continuing search for the unknown and the unknowable: a critique of contemporary ideas on speciation. *South African Journal of Science, 77:* 113–119.

PATERSON, H.E.H., 1982. Perspectives on speciation by reinforcement. *South African Journal of Science, 78:* 53–57.

PATERSON, H.E.H., 1985. The recognition concept of species. In E.S. Vrba (Ed.), *Species and Speciation:* 21–29. Pretoria: Transvaal Museum.

PATERSON, H.E.H., 1986. Environment and species. *South African Journal of Science, 82:* 62–65.

PATERSON, H.E.H., 1988. On defining species in terms of sterility: problems and alternatives. *Pacific Science, 42:* 65–71.

PROCTOR, M. & YEO, P., 1973. *The Pollination of Flowers*. London: Collins.

ROWLAND, W.J., 1989a. The effect of body size, aggression and nuptial coloration on competition for territories in male threespine sticklebacks, *Gasterosteus aculeatus. Animal Behaviour, 37:* 282–289.

ROWLAND, W.J., 1989b. The ethological basis of mate choice in male threespine sticklebacks, *Gasterosteus aculeatus. Animal Behaviour, 38:* 112–120.

RYAN, M.J., 1980. Female mate choice in a Neotropical frog. *Science, 209:* 523–525.

SCHROEDER, M.A., 1991. Movement and lek visitation by female greater prairie-chickens in relation to predictions of Bradbury's female preferance hypothesis of lek evolution. *Auk, 108:* 896–903.

STAUFFER, R.C., 1975. *Charles Darwin's Natural Selection Being the Second Part of His Big Species Book Written From 1856 to 1858.* Cambridge: Cambridge University Press.

SUTHERLAND, W.J., 1985. Chance can produce a sex difference in variance in mating success and explain Bateman's data. *Animal Behaviour, 33:* 1349–1352.

TEMPLETON, A.R., 1989. The meaning of species and speciation: a genetic perspective. In D. Otte & J.A. Endler (Eds), *Speciation and its Consequences:* 3–27. Sunderland, Massachusetts: Sinauer Associates.

TINBERGEN, N., 1951. *The Study of Instinct.* Oxford: Oxford University Press.

TINBERGEN, N., 1953. *Social Behaviour in Animals.* London: Methuen.

TRIVERS, R.L., 1972. Parental investment and sexual selection. In B. Campbell (Ed.), *Sexual Selection and the Descent of Man 1871–1971.* London: Heinemann.

TUOMI, J., VUORISALO, T. & LAIHONEN, P., 1988. Components of selection: an expanded theory of natural selection. In G. de Jong (Ed.), *Population Genetics and Evolution:* 109–118. Berlin: Springer-Verlag.

VAN TETS, G.F., 1965. *A Comparative Study of Some Social Communication Patterns in the Pelecaniformes:* 1–88. Lawrence, Kansas: American Ornithologists Union.

VERRELL, P.A., 1988. Stabilizing selection, sexual selection and speciation. *Systematic Zoology, 37:* 209–215.

WILBUR, H.M., RUBENSTEIN, D.I. & FAIRCHILD, L., 1978. Sexual selection in toads: the roles of female choice and male body size. *Evolution, 32:* 264–270.

WILLIAMS, G.C., 1966. *Adaptation and Natural Selection.* Princeton: Princeton University Press.

WOOTTON, R.J., 1976. *The Biology of Sticklebacks.* New York: Academic Press.

CHAPTER

14

A reaffirmation of Santa Rosalia, or why are there so many kinds of *small* animals?*

GUY L. BUSH

Introduction 230

Body size, life history traits and sympatry 231

Body size and mate recognition 232

Body size and behaviour 232

Body size and developmental complexity 232

Body size, ecological isolation and speciation 236

The species problem and non-allopatric speciation 242

Acknowledgements 245

References 245

Keywords: Body size – mate recognition – mating behaviour – ecological specialization – habitat selection – competition – diversity – developmental complexity – non-allopatric speciation – species concepts.

Abstract

Hutchinson and others have proposed that the origin and apportionment of animal species within a trophic community is strongly influenced by body size. I explore the relationship between body size and a number of parameters such as resource specialization, competition, behaviour, mate recognition and developmental complexity, to determine how the interactions might affect the origin and diversity of species. Large and small animals differ greatly in the way they perceive and utilize resources. Sister species of large animals usually share similar ecological needs and

*This paper is dedicated to the memory of G. Evelyn Hutchinson

Evolutionary Patterns and Processes
ISBN 0-12-440895-8

because of competitive exclusion seldom occur sympatrically. Ethological rather than ecological isolation is the most important feature of mate recognition systems of large metazoans. Their high level of developmental complexity places constraints on the kind and extent of phenotypic change tolerated during speciation, which occurs primarily by geographic isolation. These biological features contrast sharply with very small animals which are usually resource specialists, with sister species often coexisting in close sympatry with a minimum of competition. Ecological cues serve as major components of their mate recognition systems. Their developmental systems, which are far more diverse than those of large animals, are more tolerant of phenotypic change. Speciation in small animals like insects, mites and nematodes, frequently occurs without geographic isolation.

> The higher animals are not larger than the lower because they are more complicated. They are more complicated because they are larger.
>
> (J.B.S. Haldane, 1985 [1927])

INTRODUCTION

Seldom has a paper caught the imagination and stimulated so much attention and interest among evolutionary biologists over the years as G. Evelyn Hutchinson's Presidential Address to the American Society of Naturalists in 1959 entitled *'Homage to Santa Rosalia, or why are there so many kinds of animals?'*. To answer this question, Hutchinson cogently argued that not only is a complex trophic organization of a community more stable than a simple one, but also that the pattern of trophic complexity and community structure as well as the origin and apportionment of species are strongly influenced by body size. He pointed out that as body size increases, so obviously does the need for more space and resources. Because space and resources are inevitably limited, not only does the number of large species sharply decline, but also related species, which are likely to compete for those resources, seldom can be found sharing the same space. Furthermore, as animals become larger important features of their behaviour and life history change. As they pass from primarily innate to highly modifiable behaviour, ecological factors become less important in their mate recognition system, their generation times increase, reproductive rates decrease, and the likelihood of extinction increases. More recently, Bonner (1988) has emphasized that an increase in the size of metazoans has also fostered an increase in the complexity of their developmental systems.

Hutchinson and others stressed that because body size has such a profound effect on so many key biological attributes of metazoans, evolutionary processes in small animals can be expected to differ markedly from large ones in many important ways (Hutchinson, 1959; Hutchinson & MacArthur, 1959). This is particularly true with respect to the way large and small animals speciate, an evolutionary event which determines the level of species diversity within a taxon and which may be associated with varying degrees of morphological change (Eldredge & Gould, 1972). With some notable exceptions (Bush *et al.*, 1977), body size strongly influences the mode and rate of speciation as well as what kind of phenotypic changes are tolerated during the process.

BODY SIZE, LIFE HISTORY TRAITS AND SYMPATRY

One of the most obvious outcomes of body size is its effect on the spatial relationships among related species. As a rule, large animals clearly pre-empt a greater share of resources and perceive the environment in quite a different way than small ones (Strong, Lawton & Southwood, 1984; Brown & Maurer, 1986, 1987; McKinney, 1990). What is regarded as a rather homogeneous environment to a large animal may appear to be heterogeneous to a small one (Levins, 1968). Not only do larger individuals eat foods smaller ones cannot, while the reverse is not true (exploitative asymmetry), but they can also actively impede the feeding of smaller taxa (interference asymmetry) (McKinney, 1990). Another important attribute of large animals, particularly those that are closely related, is that few if any will be competitively sympatric. Their resource needs are usually sufficiently similar that they competitively exclude one another from a habitat (Lack, 1971; MacArthur, 1972; Haffer, 1986). Most habitats have reached saturation levels for large species. Co-existence of related species of large animals, when it does occur, as exemplified in such well-documented cases as seals, birds (Lack, 1947, 1971) and rodents (Grant, 1972), generally involves species which exploit distinctly different habitats or resources (Rosenzweig, 1981, 1985; Begon, Harper & Townsend, 1990), an axiom which can be demonstrated experimentally (Grant, 1972; Neet & Hausser, 1990).

But how about all those tiny creatures like insects, mites and other small invertebrates? Insects alone are represented by at least 5 million species, with some estimates based on recent sampling in tropical rain forests reaching 30 million (May, 1990). These metazoans are the consummate specialists, exploiting each other as well as all other terrestrial species of plants and animals in extraordinarily varied and inventive ways. Their small body size, short generation times and ability to readily contact and colonize new resources represent a combination of life-history traits that promote high speciation and low extinction rates. The result is increased species diversity (Marzluff & Dial, 1991). As plants evolved more complex structures in the Palaeozoic and early Mesozoic, insects, mites and nematodes devised new ways of exploiting these structures. With the rise of the Angiosperms to dominance in the late Cretaceous, the number of insect species proliferated explosively (Strong *et al.*, 1984).

Small animals perceive, use and partition resources on a much finer spatial scale than larger ones (Lawton, 1989). A bison's perception and use of a prairie is quite different from that of a monophagous chrysomelid beetle feeding on only one part of a single host plant species. Because of their high degree of specialization, insects and mites have been able to finely subdivide the relatively broad habitats of large animals into an extremely diverse array of 'microhabitats', among which even very closely related species may co-exist in comfortable sympatry with a minimum or apparent absence of competition. Communities of phytophagous insects are thus unlikely to be saturated (Strong *et al.*, 1984), and the same may well hold true for most parasites and parasitoids as well. Finally, insects and mites, by the very nature of their close association with plants, have promoted increased plant diversity in a recursive process that generates new resources for exploitation by phytophagous insects as well as by those that parasitize and prey upon them (Strong *et al.*, 1984). The take home lesson is that there is always room for one more species. If speciation can occur, resources are available for exploitation with minimal competition.

BODY SIZE AND MATE RECOGNITION

The degree to which related species co-exist in close sympatry and the way they perceive their environment also has a strong influence on the way they locate and select mates. Sister species of large animals seldom occupy the same territory because they share similar habitat requirements. Specific ecological cues, therefore, do not often play a direct role in mate choice although broad environmental features may be used to narrow the search for mates. Mate recognition relies primarily on ethological courtship rituals and mating signals to ensure proper pair-formation.

This contrasts sharply with many small animals such as insects and mites. Their fine subdivision and exploitation of microhabitats or 'niche space' often provide an important means for locating and recognizing a suitable mate. Many sister species find themselves in close sympatry while dispersing and resolve the problem of locating the right mate by making microhabitat fidelity at mating time a prominent, pivotal feature of their mate recognition system. This is especially true of many parasites, parasitoids and phytophagous insects which together constitute approximately 70% of all insect species (Price, 1980) and thus the majority of the animal kingdom. These microhabitat specialists tend to restrict mating only to their animal or plant hosts (Bush, 1975a; Zwölfer & Bush, 1984) which serve as rendezvous sites for pair-formation. This may be true even when long-range pheromones are used to attract mates (Menken, Herrebout & Wiebes, 1991).

BODY SIZE AND BEHAVIOUR

The fact that small animals like insects rely to a much greater extent than large ones on innate, pre-programmed behaviour to locate a suitable habitat or to find a mate has long been recognized (Wilson, 1975). Their relatively small body size and simple body plan places a premium on efficient and economical use of limited neural resources for behaviour. Habitat selection in an insect, mite or nematode may be elicited by relatively few chemical cues emitted by a host plant or animal, and mating behaviour generally is highly stereotyped and relatively inflexible. In contrast, environmental relationships and reproductive behaviour are far more flexible and adaptable in a fish, reptile, bird or mammal. These differences in flexibility must also reflect the way habitat selection and mating behaviour are genetically controlled. The response of a small insect to a chemical cue involved in host choice on which mating occurs might be altered by a simple mutation (Lilly & Carlson, 1990). Such a mutation might modify a gene that encodes for or regulates the expression of a chemosensory protein or alter the pattern of central processing of received stimuli. Habitat or mate choice in bison, on the other hand, requires a far more complex suite of stimuli, and must rely on a battery of polygenetically controlled traits whose overall expression is less likely to be affected significantly by a single mutation.

BODY SIZE AND DEVELOPMENTAL COMPLEXITY

Development is the least well-understood yet potentially the most important factor in orchestrating the speciation process. Again, size appears to be involved in the evolution of the organization, operation and complexity of the developmental system. In most animal lineages, the general trend for adult body size to increase during the course of evolution (Cope's rule of phyletic size increase) (Stanley, 1973) has required the evolution of new organ systems and cell types to deal with this size increase. In the process of getting bigger, developmental systems inevitably also became more complex (Bonner, 1988). Because all speciation events require the phenotypic alteration of one or more physiological, behavioural or morphological traits by mutations that affect some aspect of the development process, the mode and level of developmental complexity can impose constraints on what kind of genetic and phenotypic changes may be acceptable during speciation. Unless we understand how development shapes the phenotype, our theories of evolution will necessarily remain incomplete and the interpretation of the speciation process suspect. This is the common theme of four recent small, but intriguing books authored by Buss (1987), Thomson (1988), Bonner (1988), and Arthur (1988), as well as articles by Raff (e.g. Raff & Wray, 1989) on the role of development in phenotypic evolution. Although we are still far from a comprehensive theory of development, enough has emerged from recent molecular genetic studies to draw tentative inferences about how certain aspects of the developmental process might relate to evolution and, in some cases, impinge on the speciation process.

Two aspects of development are particularly pertinent to the problem of how animals evolve and speciate. The first is the hierarchical way developmental systems are organized in multicellular organisms. The second relates to how these hierarchical systems of development are controlled and modified.

Buss (1987), Thomson (1988) and Raff & Wray (1989) have provided important clues as to how mutations might affect the phenotype in different developmental systems. Although their focus is primarily morphological, their reasoning pertains equally well to changes at the physiological and behavioural levels. The extent to which a mutation will alter the phenotype depends not only on where in the developmental hierarchy it occurs, but also whether or not temporal sequences in this process are causally linked. Although developmental events may occur in sequence, their order may be inverted if they are not mechanistically dependent upon one another. This sequential pattern of development provides the opportunity for introducing radical changes in early development and mosaic evolution, and the independent evolution of body parts. Furthermore, genes expressed early may determine the final form and function of a large number of cells in one or more cell lineages. Mutations in these early acting genes can have a far greater phenotypic effect than mutations in genes responsible for fine-grained cytodifferentiation that are mobilized during the final stages of morphogenesis.

The evolutionary step from unicellular to multicellular organization required the establishment of controls over the heritability of cell lineages (Buss, 1987). As metazoan body size increased, it became essential to organize the control of expression of these cell lineages into an increasingly complex hierarchical cascade of binary events capable of coordinating each ontological step. Such restraints on the evolution of cell lineages were essential to reduce suborganismal competition among cell lineages and thus maintain functional harmony. As each basic body plan evolved during early metazoan evolution, their developmental systems became more and more resistant to change and heterochrony became the primary means of introducing evolutionary change (Buss, 1987). Also, once basic body plans were established, genetic variants that affected the course of ontogeny could only be passed on if they arose in the stem cell lineages producing the germ line (Buss, 1987). Thus, the stage at which the germ line is set aside during ontogeny determines when and in which cell lineage a mutation must arise if it is to be passed on to the next generation. This feature of some metazoan developmental systems provides a means to ensure orderly development and to limit the production of variants (Buss, 1987).

The timing of germ line determination, the level of developmental complexity, and the degree of linkage and dissociability between developmental processes also strongly influences the kind of phenotypic changes likely to be adaptively acceptable during the course of evolution and speciation. Where a distinct germ line is lacking, as in somatic embryogenesis, a cell lineage retains the potential of producing gametes or somatic cells throughout ontogeny. This form of development occurs in several metazoan phyla such as the Porifera, Cnidaria, the acoel Platyhelminthes and the Bryozoa as well as all plants and fungi (Buss, 1987: 20). Somatic mutations may arise throughout ontogeny and can be transmitted to the next generation if they are beneficial both to the cell lineage and to the individual. Somatic embryogenesis thus provides a potent mechanism in relatively simple organisms for generating and testing novel epigenetic interactions (Buss, 1987). The range of adaptively acceptable mutations and phenotypic plasticity can be greater in such organisms than in taxa with more complex developmental systems.

In clonal bryozoans, for example, new mutations may arise in any member of a colony and establish a new phenotype if the variant favours both its own replication and that of the organism harbouring it. Individuals bearing the new mutation may out-compete and displace the parental phenotype. Alternatively, competition may be avoided if the new phenotype promotes a shift to a new habitat or resource. Such habitat shifts may give rise to a new race or even species under suitable conditions. This is particularly so for those taxa with low vagility and a tendency for inbreeding, as is the case with cheilostome bryozoans (Jackson & Coates, 1986). This feature, coupled with the ability to produce novel phenotypes because of somatic embryogenesis, may contribute to the apparent lack of cryptic species in these organisms (Jackson & Cheetham, 1990), and the punctuated nature of their appearance and persistence (Cheetham, 1986).

The opposite extreme of germ line determination occurs in organisms with preformist (or mosaic) development. In phyla such as Nematoda, Onychophora and some Mollusca, Arthropoda, Chordata and others the primordial germ cells are set aside in very early ontogeny (Buss, 1987). Because mutations that can be passed on to the next generation can only arise in very few cells, preformist development is a very effective way of restricting the production of variants. The destiny of most cells is determined during the early stages of pattern formation in relatively small, simple animals, such as a nematode, which is composed of relatively few organ systems and is assembled predominantly from independent parts. Embryonic cell lineages in such organisms are virtually invariant with the pattern of cell divisions and cell fates similar in every individual. Each cell has been pre-programmed to differentiate in a specific way, usually, but not exclusively, prior to the cleavage of the fertilized egg by maternal genes, a process called maternal predestination. Cell differentiation in such small animals does not depend to any great extent on the behaviour or even the existence of neighbouring cells, but relies on genetic instructions and events emanating from within each cell. Thus, variants that arise during early ontogeny are unlikely to be expressed phenotypically in subsequent generations unless they alter genes responsible for maternal control. As a consequence, such taxa are morphologically very conservative (Buss, 1987).

From an adaptive standpoint small animals with strictly preformistic development have only limited flexibility in phenotypic fine-tuning (Thomson, 1988). Because of the modular nature of the developmental system, mutations can affect a specific phenotypic trait without substantially altering others. In such a system a small number of master or switch genes initiate, control or execute a developmental programme and the differentiation of individual cell lineages (Kimble & Schedl, 1988). Mutations in such genes may have a profound effect on the phenotype. Heterochronic mutations have been characterized in the nematodes, *Caenorhabditis elegans*, whose cell lineage is known completely (Sulston *et al*., 1983), and in *Panagrellus redivivus* which resemble the kinds of phenotypic differences observed between nematode species and genera (Sternburg & Horvitz, 1981; Ambros & Horvitz, 1984). Speciation in such organisms may, therefore, be accompanied by relatively large phenotypic changes, although such phenotypic variants may also exist as polymorphisms not necessarily coupled directly with speciation.

As animals became larger and more complicated, preformistic control soon reached a point where further complexity could not be entirely encoded by maternal mRNA in the oocyte and fertilized egg. This limitation was overcome in some annelids, molluscs, ascidians, echinoderms, some arthropods and some chordates by epigenetic (or regulative) development (Buss, 1987: 75). Germ line sequestration in such animals is usually coupled with secondary somatic differentiation. Somatic cells in one state of differentiation can give rise to other somatic cells in a very different state of differentiation and function. Differentiation in these organisms is partially or completely dependent on the interaction between neighbouring cells, and a clearly differentiated germ line appears only after the primordia of major organ systems of the adult are in place. Phenotypic variants may arise during the course of ontogeny, but must occur in the cell lineage leading to the germ line before the adult stage if they are to be transmitted to the next generation. Animals with epigenetic development have a greater opportunity for phenotype fine-tuning, as well as the production of major phenotypic variants, than those with a strictly preformistic system, as mutations may have a variety of effects on the phenotype depending on what point in the developmental cascade they exert their influence. Phenotypic changes associated with speciation in these organisms may range from slight to large.

Most animals are seldom fully mosaic or regulative, but are assembled by a mixture of the two processes. Which system predominates depends on the organism's size and physical complexity, although even epigenetic development is ultimately under intrinsic genetic control. Once the early stage of pattern formation is complete and the zygotic genome is completely activated, the fate of almost all cell lineages is fixed (Thomson, 1988).

In mammals, predestination is largely irrelevant and early development proceeds in a regulative fashion, a system which permits extensive fine adjustments to the phenotype. Because of the extreme complexity of the developmental process, mutations occurring early in the developmental cascade are unlikely to be compatible and are only very rarely incorporated into the developmental process. Phenotypic variants that influence the terminal process of cell differentiation are most likely to be the kind that are involved in adaptation and speciation. Early embryonic development in *Drosophila*, although subject to somewhat more maternal pre-programming, particularly of the germ line, progresses by a mixture of maternal and zygotic genes with zygotic control over the later stages of differentiation. Unlike mammals, morphogenesis in *Drosophila* is modular. The process of assembling the modules appears to be controlled by a series of molecular switches which apparently govern the hierarchy of genes controlling the morphogenesis of individual modules. This mode of development opens the door for major phenotypic alteration during evolution (Watson *et al.*, 1987). Changes in modular number and function have certainly occurred during the course of insect evolution.

It is now clear that mutations of the developmental cascade represent a continuum in the effects they may have on phenotype. They may range from those whose impacts are so severe that only very rarely will one be capable of surviving and reproducing as heterozygotes to others that are very slightly beneficial or even neutral with respect to fitness. The extreme representatives of this continuum have been labelled macro- and micromutations in the past, but, as Thomson (1988) has stressed, because they represent only the extremes, they misrepresent the way mutations actually affect the developmental system. The use of these terms tends to obscure the important role of developmental mutations in evolution and speciation. Most neo-Darwinians such as Mayr (1963, 1988) and Carson (1975), as well as Lewontin (1974) and other population geneticists reject macromutations in favour of micromutations as the sole means of adaptive evolution and speciation, ignoring the fact that they are dealing with a continuum. They hold the view that all evolution occurs by the accumulation of many genetic changes with small effects because, they argue, mutations producing large phenotypic changes will be eliminated by natural selection. But current evidence fails to support this view (Buss, 1987; Bonner, 1988; Thompson, 1988; Wallace, 1988). In the light of our understanding of developmental genetics, such a narrow position no longer seems tenable.

Modification of early ontogenetic events, which result in substantial phenotypic modification, have clearly occurred in several taxa such as flatworms, nematodes and insects (Buss 1987). Even in vertebrates a slight alteration in early development may account for the sudden appearance of a novel *Bauplan* such as the turtle carapace (Burke, 1989). Acceptable mutations affect not only the terminal process of cell differentiation, the level scrutinized by most evolutionary geneticists, but also other important binary switches, earlier in the developmental hierachy. Such mutations, rare as they may be, have obviously played evolutionarily important roles during the course of metazoan (Buss, 1987) and plant evolution (Gottlieb, 1984). It is also quite likely that heterochronic mutations with a wide range of phenotypic effects have accompanied or been instrumental in promoting the process of speciation by providing the bases for exploiting new habitats and resources. It would, therefore, be extremely fruitful to focus on the biological conditions necessary for the incorporation of such mutations into the genome.

BODY SIZE, ECOLOGICAL ISOLATION AND SPECIATION

The lack of strong ecological isolation and the relative unimportance of ecological factors in mate recognition between related species of most large animals, coupled with a complex developmental system, have important effects on how most large animals speciate. It is now generally accepted that mate recognition behaviours, which facilitate reproduction among individuals within a population (Paterson, 1985, and Chapter 13; Templeton, 1989), evolve during periods of isolation sufficiently free of gene flow to allow populations to diverge. In the absence of strong ecological isolation, the principal way gene flow between populations of most large animals can be reduced is by geographic separation, although non-allopatric (parapatric and sympatric) exceptions to this rule associated with habitat shifts have been proposed (Wilson *et al.*, 1975; Bush *et al.*, 1977; Bush, 1981; Grant & Grant, 1989; Lynch, 1989; Meyer, 1990). But an absence of ecological factors in mate recognition is certainly not the norm for many small, microhabitat specialists in which inherent biological traits alone can often provide sufficient ecological and temporal isolation for the formation of ecological races, and eventually the evolution of species, without the benefit of geographic isolation (Bush, 1975a; Bush & Diehl, 1982; Diehl & Bush, 1989; Rice, 1987; Tauber & Tauber, 1989).

It has long been my contention that ecological isolation without the benefit of geographic barriers has engendered a far greater percentage of insect and mite species than most evolutionary biologists appreciate or care to acknowledge (Bush, 1969; Bush & Howard, 1986). This lack of appreciation for the importance of ecological isolation in speciation stems in part from the failure of theorists to recognize it as an effective, genetically based, intrinsic barrier to gene flow between closely related microhabitat specialists. Efforts to correct this view have been made recently by incorporating the biological attributes of ecological specialists into population genetic models and experiments of race formation and speciation (Rice, 1987; Rice & Salt, 1990; Diehl & Bush, 1989; Wilson, 1989).

One factor contributing to this situation is that appropriate biological details of very few small animals have been well-documented. Microhabitat and resource specialists, particularly small ones like insects and mites, are difficult to study and therefore often ignored in preference to subjects such as *Drosophila* which are more amenable to the laboratory culture bottle. Furthermore, little effort has been made to identify traits that are responsible for mate recognition or involved with speciation. It should come as no surprise to those working on evolutionary processes in natural populations that one's perspective on how speciation occurs in the animal world is likely to be very strongly coloured by the kind of animal studied. Unfortunately, this has meant that the 'modern synthesis' was formulated by individuals whose expertise was *Drosophila* and vertebrates (Buss, 1987). The remaining >99.9% of the animal kingdom has been given short shrift.

Possibly because of their size, ease of study and phylogenetic proximity to humans, our attention seems naturally to be drawn to big, conspicuous vertebrates. On average, they establish independently evolving lineages quite differently from very small animals. This is not to say that large and small animals always speciate in different ways. Many grasshoppers and artiodactyl herbivores grazing on prairie grasses may well share many similar requisites for speciation. But I know of no mammal whose life history and attendant biological attributes and needs is even remotely as fastidious or unique as those of a highly specialized fig wasp (Ramírez, 1970), hummingbird flower mite (Colwell, 1986), apple maggot fruit fly (Bush, 1969), or an *Enchenopa* treehopper (Wood & Guttman, 1982) to name only a few of what must be many millions of examples. Mayr, Carson and other advocates of the modern synthesis have strongly promoted the view over the last five decades that geographic isolation is a necessary prerequisite for speciation in most if not all sexually reproducing animals, a view that has been the centrepiece of the neo-Darwinian paradigm. Results beginning to emerge from studies noted above cast doubt on this widely held conventional wisdom (Barton, Jones & Mallet, 1988).

The importance of ecological isolation between species has long been acknowledged, but its rôle as a factor in speciation has been strongly contested. Although Darwin (1859) recognized the role of geographic isolation in speciation, he also concluded it was neither essential nor necessary. Much to the chagrin of latter-day neo-Darwinians, Darwin was convinced that ecological specialization, particularly in locations where competition for resources occurs, was sufficient to prevent interbreeding and permit new species to arise in the absence of geographical barriers, although his concepts appear to conform more to a parapatric than a sympatric mode of divergence (see Sulloway, 1979 for details). Mayr (1988) and others have gently but firmly rebuked Darwin for his position and have systematically rejected models of non-allopatric speciation. They have emphasized repeatedly that under no circumstances can intrinsic genetic changes affecting habitat or resource choice reduce gene flow sufficiently to permit non-allopatric speciation to occur. I think nature tells us otherwise.

Shortly after the publication of *On the Origin of Species*, Walsh (1864), following Darwin's lead, developed the first true model of sympatric speciation via ecological isolation. He noted that very closely related but distinct forms of phytophagous insects often occurred on plants growing in close proximity, yet did not interbreed. From this observation and a detailed knowledge of insect natural history, Walsh postulated that what he called "phytophagic varieties", but what we would now refer to as ecological or host races, feeding on different plants might undergo heritable changes that would predispose them to remain on specific species of plants for feeding and mating. These host races, he reasoned, would eventually evolve into species adapted to different hosts. He stressed that "laws of inheritance" and what we now call conditioning were likely involved in the transmission of a tendency of a host race to select a particular plant for feeding, oviposition and mating. This model of sympatric speciation by ecological isolation through habitat or host shifting is essentially the same as those independently 'discovered' or resurrected on several occasions beginning early in the 20th century (Brues, 1924; Smith, 1941; Thorpe, 1945; Emerson, 1949; Dethier, 1954; Haldane, 1959; Maynard Smith, 1966; Bush, 1969; Wood & Guttman, 1982; Rice, 1984). It should be noted that most of these examples involve insects, and are based on changes in a few habitat preference genes with major effects. Although such genes are recognized, just where and how they act in the developmental cascade is not known. They are certainly not the kind of maintenance genes currently used in most population genetic studies of natural populations. Such genes are rarely, if ever, involved in the origin of phenotypic novelty or speciation.

Although non-allopatric speciation is often viewed as potentially rapid and easy to accomplish, other factors may reduce the likelihood that it will occur. The quality and abundance of the microhabitat can play an important role in how many species it can support. Not all microhabitats are equally suitable for colonization, and rare microhabitats or ones that are unstable over time are less likely to be utilized by a specialized herbivore or support as much species diversity as more common ones. The number and kind of competitors, predators, parasites and disease could also affect the success of a colonization event (Price, 1980). Thus, a habitat shift is most likely to occur when new, relatively unexploited habitats become available through the modification of the environment by natural causes and, more recently, by man. This could entail the introduction of one or more new plant or animal species, a catastrophic geophysical modification of the biome or the extinction of species by disease, competition, or a number of other factors followed by recolonization (Bush, 1975b).

Three biological conditions are essential for non-allopatric race formation and speciation to occur in small, sexually reproducing metazoans. First, mate choice must be dependent directly and primarily upon habitat or resource selection. That is, males and females use their microhabitat such as a host plant or animal as a rendezvous for courtship and mating. Second, habitat selection and fidelity must be under genetic control. The third condition follows necessarily from the other two: speciation must involve a genetically based shift to a new habitat or resource. Temporal isolation of resources, low vagility, inbreeding and other factors may accelerate the process of divergence (Bush, 1975a). One or the other or sometimes all three of these criteria are overlooked, ignored, or labelled as irrelevant in discussions of non-allopatric speciation. Mayr (1947, 1963), for instance, while recognizing the importance of ecological differences between species, for many years consistently rejected the possibility that changes in genetically based microhabitat or host fidelity could establish and maintain sufficiently strong isolation for speciation to occur. Although he now concedes that the most probable cases of sympatric speciation involve host-specific plant and animal parasites, he continues to emphasize that the origin of sympatric sister species exploiting different hosts are best explained by peripatric rather than sympatric speciation (Mayr & Ashlock, 1991).

Felsenstein (1981), in a paper widely cited as demonstrating the unlikelihood of non-allopatric speciation, based his genetic simulation models of non-allopatric speciation on assortative mating and ignored habitat selection as a factor in mate choice. In his models individuals disperse at random between two habitats, then mate assortatively according to their genotype within whichever habitat they occupy. Selection then acts upon alternative alleles at two fitness loci. It is not clear what kind of organism Felsenstein's model is meant to simulate or how habitat fidelity, which much evolve to avoid competition, might arise. Under conditions of his model speciation occurs only when assortative mating is essentially error free and tight linkage exists between assortative mating and fitness loci. Linkage is required to cope with the antagonistic effects between selection and recombination. If this were the only way non-allopatric divergence could occur, Felsenstein and others would indeed be justified in concluding that non-allopatric speciation is unlikely to occur.

There are alternative ways of ensuring assortative mating that do not rely on classical assortative mating genes. If dispersal to alternative habitats or resources is genetically determined and mating occurs at random within each habitat followed by selection for fitness, the selection–recombination antagonism is eliminated. Substantial progress towards speciation is then likely to occur (Bush & Diehl, 1982; Diehl & Bush, 1989). This is true even under realistic levels of selection on fitness, and moderate penetrance of the habitat-preference locus. Such a system simultaneously establishes different mate recognition systems and provides the conditions for adaptation and specialization to different habitats in the two ecological races while promoting the reduction of competition and gene flow during the course of divergence (Bush, 1969, 1975b). Assortative mating based on host use is common in many groups of insects and mites.

As emphasized, 70% of all insects can be considered parasites (Price, 1980). Many of these parasites mate only on or near their animal or plant hosts. Furthermore, recent phylogenetic studies of insect–plant interactions indicates that clades of phytophagous insects are consistently more diverse than their primitively non-phytophagous sister groups, supporting the postulate that herbivory has fostered diversity in a relatively unsaturated adaptive zone (Mitter, Farrell & Futuyma, 1991). The analysis of phylogenetic relationships between 14 insect families and genera and their host plants carried out by Mitter and co-workers indicated that most (10 out of 14) showed little or no concordance while only one suggested a pattern of cospeciation. Thus, among phytophagous insect parasites, speciation is likely to be most frequently accompanied by a shift to a new host. Because some non-parasitic groups of animals also have strong microhabitat preferences, the potential number of animals speciating as a result of ecological isolation is very large. For example, Hamilton (1978) has pointed out that evolution and speciation are particularly fast and the production of novel phenotypes is promoted through inbreeding in highly specialized 'phytonecrophilic' insects living under the bark of decaying trees, plant litter and soil. If it could be quantified, I predict that microhabitat or host fidelity would represent the single most important factor restricting gene flow between sympatric ecological races and between recently evolved closely related sister species of insects, mites, nematodes and fungi, groups in which most species use their hosts or specific microhabitats as a rendezvous for locating mates and reproduction.

Much evidence exists to support Darwin's claim for a pivotal role of ecological isolation in some speciation events and for Walsh's proposal for the non-allopatric origin of ecological races and species. It is no coincidence that most of the claims come from individuals working on insects, mites or other relatively small invertebrates (Tauber & Tauber, 1989) as well as fungi (Brasier, 1987). There are now many excellent examples of taxa that meet the three criteria described above, and in which speciation is most logically and parsimoniously explained by non-allopatric processes via a habitat shift. A sampling of recently published examples include such diverse insects and other invertebrates as Diptera (*Rhagoletis*, Feder, Chilcote & Bush, 1988; Feder & Bush, 1989; McPheron, Smith & Berlocher, 1988; *Drosophila*, R'kha, Capy & David, 1991), Neuroptera (*Chrysopa* lacewings, Tauber & Tauber, 1977), Homoptera (*Enchenopa* treehoppers, Wood & Keese, 1990; *Tetraneura* and Cryptomyzuz aphids, Akimoto, 1990; Guldemond, 1990); Hymenoptera (socially parasitic ants, Buschinger, 1990; Bourke & Franks, 1991; parasitic wasps, Askew, 1971; Whitfield, 1990); Lepidoptera (*Spodoptera* moths, Pashley, 1989; *Heliconius* butterflies, Mallet, in press); mites (*Proctolaelaps* and *Rhinoseius* hummingbird flower mites, Colwell, 1986); Crustacea (parasitic copopods, de Meeus, Renaud & Gabrion, 1990) and Mollusca (pulmonate snails, Gittenberger, 1988).

I wish to emphasize that geographic isolation is sufficient but not necessary for speciation to occur in sexually reproducing animals. Given enough time even large populations physically isolated by a vicariant event will eventually accumulate enough genetic differences to render their genomes and mate recognition systems incompatible should they ever re-establish contact. Conditions favouring peripatric divergence (i.e. via founder effects) may accelerate the process of speciation (Mayr, 1963; Carson,

1975; Carson & Templeton, 1984). Although peripatric speciation may occur under certain conditions, the genetic evidence supporting such a mode of allopatric speciation in natural populations has been difficult to document (Barton & Charlesworth, 1984; Barton, 1989; Coyne, 1990). It is also clear that allopatric isolation is frequently associated with speciation events that do not involve an appreciable ecologial or habitat shift, particularly in vertebrates. It has long been recognized that differences in the mate recognition systems between geographically isolated sister species may arise as a byproduct of genetic changes that have accumulated in isolation. In *Drosophila*, a well-studied group, the effect of these changes may result in post-mating incompatibility as frequently as premating mate recognition between geographically isolated populations, while divergence in premating mate recognition predominates among sympatric species (Coyne & Orr, 1989). Although the widespread occurrence of allopatric speciation is not in doubt, the claim of the neo-Darwinians that geographic speciation is the most common if not the only mode of cladogenesis in animals is refuted by an increasing body of evidence. It is time that those espousing the universality of geographic speciation in metazoans be held to the same rigour of proof that has been demanded of proponents of non-allopatric speciation.

The following non-allopatric speciation scenario for small animals like insects and mites emerged from studies of host race formation and speciation in *Rhagoletis* and other invertebrate taxa as well as from genetic computer simulations. When microhabitat preference is under genetic control and mating is restricted to a preferred microhabitat, intrinsic genetic changes that alter microhabitat preference cause individuals bearing the altered genomes to move preferentially to a new microhabitat. Because positive assortative mating occurs among individuals with similar genetically based microhabitat preferences, ecological isolation is rapidly established between parent and daughter populations. These conditions initiate the speciation process by reducing gene flow sufficiently to enable further genetic divergence during the course of adaptation to the new habitat or host. Initially some gene flow will occur, but divergent selection on loci with habitat-sensitive alleles will tend to relegate such alleles to one habitat or the other. Selection will also favour new mutations that improve fitness and habitat fidelity and positive assortative mating among members of the new and old ecological races in their respective preferred microhabitats (i.e. divergent adult eclosion times). This adaptive process also indirectly strengthens existing differences in the mate recognition system, and in doing so progressively reduces gene flow over time between the diverging populations. Adaptation to a new microhabitat and the evolution of the mate recognition system thus co-evolve in non-allopatric speciation.

At least three predictions can be made about the genetic structure, phenotypes and distribution of sister species that have evolved without geographic isolation. (1) Because gene flow occurs between sympatric ecological races during the early stages of divergence, recently evolved parent and daughter species will share almost all of their alleles. No genetic bottleneck will occur as a result of 'founder effects' and there should be no evidence of a genetic revolution. Recently evolved sister species will differ, at least initially, only in those key habitat-related preference and fitness loci which may be quite few in number and, unless identified by ecological and behavioural studies, difficult to identify. (2) Morphologically, non-allopatrically evolved sister species are very similar and in some cases identical, although loci affecting traits subject to sexual selection may diverge rapidly as different mate recognition systems become established. (3) From a biogeographic standpoint the parent and daughter races and sister species will be either sympatric if the microhabitats exhibit a diffuse, patchy or mosaic distribution as is often the case for parasites, or parapatric if differentiation has occurred across an ecotone.

Unfortunately, it is difficult to distinguish between primary and secondary zones of overlap in parapatric situations (Endler, 1977), sometimes rendering them less useful for studies on speciation processes unless the nature of their origin can be established.

THE SPECIES PROBLEM AND NON-ALLOPATRIC SPECIATION

I have used the term species without defining exactly what I mean or how I have applied it to the process of allopatric and non-allopatric speciation. It is widely assumed by most evolutionary biologists that this concept is so fundamental that it must be established before the process of speciation can be investigated, let alone understood or discussed (e.g. Mayr, 1963; Templeton, 1989). This perspective strikes me as wrong minded, as quite the opposite is true. It is unlikely that species can be defined unless we first understand how populations actually diverge and establish separate breeding systems in nature. Much has been written on the species problem and to review these concepts is beyond the scope of this paper. Yet it is clear that how one defines species directly influences one's views of how they arise in nature (Chandler & Gromko, 1989). I will therefore touch on certain aspects of the species problem which I feel thwart our attempts to understand how speciation occurs in nature.

The 'biological species concept' (BSC), first proposed by Walsh (1863:220) and Wallace (1865:6) and later articulated clearly by Mayr (1942, 1963) and others is currently much in favour as a means of delineating species boundaries and testing hypotheses of speciation in sexually reproducing animals. Sokal & Crovello (1970), Levin (1979), Paterson (1985), Cracraft (1987), Templeton (1989) and others have presented a litany of well-founded objections and alternative definitions to the BSC. My concern with adopting the BSC is that, by definition, species status can only be conferred when populations are reproductively isolated from one another. Exactly at what point such isolation occurs has never been explicitly defined or determined, although Mayr has implied that it occurs when gene flow no longer occurs between them (Mayr, 1963).

This assumption presents a fundamental problem. Sympatric populations, even though they maintain differences in important aspects of their biology, cannot be considered species under the BSC as long as there is gene flow between them. By strict application of the BSC, only sympatric populations that do not exchange genes can be recognized as species. Unfortunately, nature often is not aware of this requirement and there is mounting evidence that many closely related taxa, long accepted as 'good species', may exchange genes at a low level (Howard, in press).

In fact, a state of complete genetic isolation is usually reached after, probably in most cases long after, two populations are biologically committed to separate evolutionary lineages. With the exception of 'instantaneous' sympatric speciation by polyploidy or hybridization, speciation is a continuous, although sometimes a relatively brief process. There must inevitably be a period of time when populations have the potential either to fuse or to continue to diverge. The length of this period and the outcome will differ greatly depending on the inherent genetic architecture of each population, their response to evolutionary forces, environmental contingencies, and the degree and pattern of physical and ecological isolation each experience over time. Body size, with its associated constellation of biological attributes, must strongly influence this process, as I have emphasized. Also, a period of time must elapse during which hybrids between individuals of the old and new genotypes are at least partially or even substantially viable and fecund, a time during which some gene exchange is possible. We do not understand how long this period lasts, how much phenotypic change is possible, under what conditions it will occur, or how gene flow is eventually eliminated in any sexually reproducing metazoan. Yet the outcome of this process is the essence of the debate between those that argue for or against the likelihood of non-allopatric speciation.

It is obviously impossible to pinpoint the precise time, place, or circumstance when two or more sister populations pass through this diffuse yet critical threshold of genome change and become irrevocably committed to different evolutionary paths. More importantly, each threshold and the conditions that bring it about (i.e. the level of gene flow, intensity of selection, role of drift, constraints of the developmental system, etc.) are inevitably unique to every speciation event and not dependent on some arbitrarily selected stage in the process of differentiation. To borrow a phrase from Gould (1989), if we could run the speciation 'tape' over and over again for the same pair of sister species, the outcome would never be the same. A systematist, palaeontologist or population geneticist may gain some comfort and satisfaction by invoking the BSC to distinguish taxa in the mistaken perception that it is biologically meaningful and thus somehow precise and unambiguous, when in fact it in no way provides a clear representation of the actual status of populations in nature.

Since it is impossible to characterize exactly under what conditions fusion or fission will or has occurred for the vast number of metazoans, advocates of the BSC insist that we must wait until the process of divergence is over before species status can be conferred. This places the emphasis on the end product rather than on the process and tends to obscure rather than clarify our understanding of what actually constitutes real species differences and the process by which they diverge. Because the BSC focuses on a level of divergence between populations that has nothing to do with the actual phase in which populations become committed to one lineage or another, it is not a particularly useful concept to apply in speciation studies. As emphasized by Chandler & Gromko (1989), species concepts are logically distinct from specific processes of speciation, a distinction most evolutionary biologists fail to recognize.

Our preoccupation with species definitions has focused our attention on the wrong end of the problem. Is it necessary or biologically relevant to define the term species as such other than for expedient or utilitarian reasons? Certainly none of the many currently available definitions satisfactorily covers all situations, and no single definition probably ever will. Because each speciation event in sexually reproducing organisms is, in most instances, a unique and continuous process, I believe the problem of trying to define species or establish exactly when a speciation event occurs is analogous to, and just as difficult as, the problem of defining conditions for the origin of life. This has been put in the right perspective by Küppers (1990).

> A complete definition (of life), that is one embracing both necessary and sufficient criteria, would only be possible if the transition from inanimate to animate were discontinuous. However, the definition would then inevitably contain at least *one* irreducible concept, which would express the ontological difference between living and nonliving systems. Since this concept would by definition be life-specific, every holistic definition of 'life' must be inherently tautologous. It is absolutely characteristic of physically oriented biology that it cannot give a complete definition of the phenomenon 'life.' If such a definition existed, this fact would contradict the premises of the reductionist program.

The same argument can be made against attempts to distinguish between populations, races and species. Speciation is a continuous process and, to make it even more difficult, never occurs twice under the same set of conditions – each event is unique.

While I accept the reductionist views on this matter, I also appreciate the structuralist's views of White, Michaux & Lambert (1990) that species are self-defining. There is no reason why, as Bonner has emphasized, one cannot be a reductionist and a holist at the same time (Bonner, 1988). Clearly, from a reductionist point of view, to understand the speciation process it is imperative to know what kind (structural, developmental, regulatory, etc.) and how many genes affect gene flow. This should be a major goal of any long-term research on speciation. But such facts are only pieces of a very complicated, and in every instance of speciation, singular puzzle. They are of little use unless interpreted within the context of how such genes interact at the developmental level to shape the phenotype and how the phenotype interacts with the biotic and physical world to create unique mate-recognition systems and sterility barriers to gene flow. Similarly, a purely holistic approach without the puzzle's pieces is equally unsatisfactory. To me the two levels of analysis are so obviously inseparable when it comes to understanding biological systems that I don't understand why the question of which approach is better has generated so much polemic.

Obviously I do not share the scepticism expressed by Felsenstein (1981) and others for the revelations by Santa Rosalia, the patroness saint of evolutionary studies, about the cause of species diversity. In their view, the conditions for non-allopatric speciation are far rarer than Santa Rosalia and her apologist, G. Evelyn Hutchinson, implied. Felsenstein (1981), for instance, argues that if species really did arise in the absence of geographic isolation, many more kinds of animal species should be roaming the earth's surface. But, there are, after all, a great many animal species, albeit mostly very small and rather inconspicuous ones, inhabiting the earth today. They deserve to be recognized and counted as major contributors to the economy of nature just as surely as birds and mammals. I would hazard a guess that the abundance and diversity of small animals like insects, mites, nematodes and other small invertebrates far exceed the numbers that could be generated by some process of geographic isolation. Unfortunately, all the controversy over numbers of species and their modes of speciation is after all just a byproduct of educated guess work, as we really know very little about the actual rate and process of speciation and extinction in the vast majority of animals (Bush, 1982). Partial details of speciation have been more or less inferred for roughly 1 in every 300 000 metazoans. Therefore, it seems a bit premature to proclaim that the vast majority of those reproducing sexually speciate by geographic isolation. I suspect that a great number, possibly even a majority, have evolved, some even rather rapidly, in the absence of any physical barriers. Our preoccupation with large animals has biased our view of the speciation process. Body size, as Hutchinson so prophetically stated, has a profound effect on the origin of species.

ACKNOWLEDGEMENTS

I am indebted to Stewart Berlocher, Donald Hall, William Hamilton, Daniel Howard, William Kopachik, Peter Price and James Smith, for thoroughly reviewing the manuscript and for their suggestions and constructive criticisms. Support while working on this paper was generously provided by a Visiting Fellowship from All Souls College, Oxford and from Michigan State University.

REFERENCES

AKIMOTO, S., 1990. Local adaptation and host race formation of a gall-forming aphid in relation to environmental heterogeneity. *Oecologia, 83:* 162–170.

AMBROS, V. & HORVITZ, H.R., 1984. Heterochronic mutants of the nematode *Caenorhabditis elegans. Science, 226:* 409–416.

ARTHUR, W., 1988. *A Theory of the Evolution of Development.* Bury St. Edmunds: John Wiley & Sons.

ASKEW, R.R., 1971. *Parasitic Insects.* New York: Elsevier.

BARTON, N.H., 1989. Founder effect speciation. In D. Otte & J.A. Endler (Eds), *Speciation and Its Consequences:* 229–256. Sunderland, Massachusetts: Sinauer Associates.

BARTON, N.H. & CHARLESWORTH, B., 1984. Genetic revolutions, founder effects, and speciation. *Annual Review of Ecology & Systematics, 15:* 133–164.

BARTON, N.H., JONES, J.S. & MALLET, J., 1988. No barriers to speciation. *Nature, 336:* 13–14.

BEGON, M., HARPER, J.L. & TOWNSEND, C.R., 1990. *Ecology: Individuals, Populations and Communities*, 2nd edition. Oxford: Blackwell Scientific.

BONNER, J.T., 1988. *The Evolution of Complexity by Means of Natural Selection.* Princeton: Princeton University Press.

BOURKE, A.F.G. & FRANKS, N.R., 1991. Alternative adaptations, sympatric speciation, and the evolution of parasitic, inquiline ants. *Biological Journal of the Linnean Society, 43:* 221–237.

BRASIER, C.M., 1987. The dynamics of fungal speciation. In A.D.M. Rayner, C.M. Brasier & D. Moore (Eds), *Evolutionary Biology of the Fungi*: 231–260. Cambridge: Cambridge University Press.

BROWN, J.H. & MAURER, B.A., 1986. Body size, ecological dominance and Cope's rule. *Nature, 324:* 248–250.

BROWN, J.H. & MAURER, B.A., 1987. Evolution of species assemblages: effects of energetic constraints and species dynamics on the diversification of the North American avifauna. *American Naturalist, 130:* 1–17.

BRUES, C.T., 1924. The specificity of food-plants in the evolution of phytophagous insects. *American Naturalist, 58:* 127–144.

BURKE, A.C., 1989. Development of the turtle carapase: Implications for the evolution of a novel bauplan. *Journal of Morphology, 199:* 363–378.

BUSCHINGER, A., 1990. Sympatric speciation and radiative evolution of socially parasitic ants – Heretic hypotheses and their factual background. *Zeitschrift für zoologiche Systematik und Evolutionsforschung, 28:* 241–260.

BUSH, G.L., 1969. Sympatric host race formation and speciation in frugivorous flies of the genus *Rhagoletis* (Diptera, Tephritidae). *Evolution, 23:* 237–251.

BUSH, G.L., 1975a. Modes of animal speciation. *Annual Review of Ecology and Systematics, 6:* 339–364.

BUSH, G.L., 1975b. Sympatric speciation in phytophagous parasitic insects. In P.W. Price (Ed.), *Evolutionary Strategies of Parasitic Insects and Mites:* 187–206. New York: Plenum Publishing.

BUSH, G.L., 1981. Stasipatric speciation and rapid evolution in animals. In W.R. Atchley & D.S. Woodruff (Eds), *Evolution and Speciation: Essays in Honor of M.J.D. White:* 201–218. Cambridge: Cambridge University Press.

BUSH, G.L., 1982. What do we really know about speciation? In R. Milkman (Ed.), *Perspectives on Evolution:* 119–128. Sunderland, Massachusetts: Sinauer Associates.

BUSH, G.L. & DIEHL, S.R., 1982. Host shifts, genetic models of sympatric speciation and the origin of parasitic insect species. In J.H. Visser & A.K. Minks (Eds), *Proceedings 5th International Symposium on Insect–Plant Relationships:* 297–305. Wageningen: Pudoc.

BUSH, G.L. & HOWARD, D.J., 1986. Allopatric and non-allopatric speciation; Assumptions and evidence. In S. Karlin & E. Nevo (Eds), *Evolutionary Processes and Theory:* 411–438. New York: Academic Press.

BUSH, G.L., CASE, S.M., WILSON, A.C. & PATTON, J.L., 1977. Rapid speciation and chromosomal evolution in mammals. *Proceedings of the National Academy of Sciences, U.S.A., 74:* 3942–3946.

BUSS, L.W., 1987. *The Evolution of Individuality*. Princeton: Princeton University Press.

CARSON, H.L., 1975. The genetics of speciation at the diploid level. *American Naturalist, 109:* 83–92.

CARSON, H.L., & TEMPLETON, A.R., 1984. Genetic revolutions in relation to speciation phenomena: The founding of new populations. *Annual Review of Ecology & Systematics, 15:* 97–131.

CHANDLER, C.R. & GROMKO, M.H., 1989. On the relationship between species concepts and speciation processes. *Systematic Zoology, 38:* 116–125.

CHEETHAM, A.H., 1986. Tempo of evolution in a Neogene bryozoan: rates of morphologic change within and across species boundaries. *Paleobiology, 12:* 190–202.

COLWELL, R.K., 1986. Population structure and sexual selection for host fidelity in the speciation of hummingbird flower mites. In S. Karlin & E. Nevo (Eds), *Evolutionary Processes and Theory:* 475–495. New York: Academic Press.

COYNE, J.A., 1990. Endless forms most beautiful. *Nature, 344:* 30.

COYNE, J.A. & ORR, H.A., 1989. Patterns of speciation in *Drosophila*. *Evolution, 43:* 362–381.

CRACRAFT, J., 1987. Species concepts and the ontology of evolution. *Biology and Philosophy, 2:* 63–80.

DARWIN, C., 1859. *On the Origin of Species* (A facsimile of the 1st edition). Cambridge, Massachusetts: Harvard University Press.

DE MEEUS, T., RENAUD, F. & GABRION, C., 1990. A model for studying isolation mechanisms in parasitic populations: The genus *Lepeophtherius* (Copepoda, Caligidae). *The Journal of Experimental Zoology, 254:* 207–214.

DETHIER, V.G., 1954. Evolution of feeding preferences in phytophagous insects. *Evolution, 8:* 33–54.

DIEHL, S.R. & BUSH, G.L., 1989. The role of habitat preference in adaptation and speciation. In D. Otte & J.A. Endler (Eds), *Speciation and its Consequences:* 345–365. Sunderland, Massachusetts: Sinauer Associates.

ELDREDGE, N. & GOULD, S.J., 1972. Punctuated equilibria: an alternative to phyletic gradualism. In T.J.M. Schopf (Ed.), *Models in Paleobiology:* 82–115. San Francisco: Freeman Cooper.

EMERSON, A.E., 1949. Ecology and Isolation. In W.C. Allee, A.E. Emerson, O. Park; T. Park & K.P. Schmidt (Eds), *Principles of Animal Ecology:* 605–630. Philadephia: Saunders.

ENDLER, J.A., 1977. *Geographic Variation, Speciation, and Clines*. Princeton: Princeton University Press.

FEDER, J.L. & BUSH, G.L., 1989. A field test of differential host-plant usage between two sibling species of *Rhagoletis pomonella* fruit flies (Diptera: Tephritidae) and its consequences for sympatric models of speciation. *Evolution, 43:* 1813–1819.

FEDER, J.L., CHILCOTE, C.A. & BUSH, G.L., 1988. Genetic differentiation between sympatric host races of the apple maggot fly *Rhagoletis pomonella*. *Nature, 336:* 61–64.

FELSENSTEIN, J., 1981. Skepticism towards Santa Rosalia, or why are there so few kinds of animals? *Evolution, 35:* 124–138.

GITTENBERGER, E., 1988. Sympatric speciation in snails; A largely neglected model. *Evolution, 42:* 826–828.

GOTTLIEB, L.D., 1984. Genetics and morphological evolution in plants. *American Naturalist, 123:* 681–709.

GOULD, S.J., 1989. *Wonderful Life.* London, Penguin Books.

GRANT, P.R., 1972. Interspecific competition among rodents. *Annual Review of Ecology and Systematics, 3:* 79–106.

GRANT, P.R. & GRANT, B.R., 1989. Sympatric speciation and Darwin's finches. In D. Otte & J.A. Endler (Eds), *Speciation and its Consequences:* 433–457. Sunderland, Massachusetts: Sinauer Associates.

GULDEMOND, J.A., 1990. Choice of host plant as a factor in reproductive isolation of the aphid genus *Cryptomyzus* (Homoptera, Aphidae). *Ecological Entomology, 15:* 43–51.

HAFFER, J., 1986. Superspecies and species limits in vertebrates. *Zeitschrift für zoologiche Systematik und Evolutionsforschung, 24:* 169–190.

HALDANE, J.B.S., 1959. Natural selection. In P.R. Bell (Ed.), *Darwin's Biological Work: Some Aspects Reconsidered:* 101–149. New York: Science Editions, Wiley & Sons.

HALDANE, J.B.S., 1985. On being the right size. In J. Maynard Smith (Ed.), *On Being the Right Size and Other Essays:* 1–8. Oxford: Oxford University Press. (Essay originally published in 1927.)

HAMILTON, W.D., 1978. Evolution and diversity under bark. In L.A. Mound & N. Waloff (Eds), *Diversity of Insect Faunas, 9:* 154–175. Oxford. For the Royal Entomological Society by Blackwell Scientific Publications.

HOWARD, D.J., in press. Reinforcement: the origin, dynamics, and fate of an evolutionary hypothesis. In R.J. Harrison (Ed.), *Hybrid Zones, and the Evolutionary Process.* Oxford: Oxford University Press.

HUTCHINSON, G.E., 1959. Homage to Santa Rosalia or why are there so many kinds of animals species? *American Naturalist, 93:* 145–159.

HUTCHINSON, G.E. & MacARTHUR, R., 1959. A theoretical ecological model of size distributions among species of animals. *American Naturalist, 93:* 117–126.

JACKSON, J.B.C. & CHEETHAM, A.H., 1990. Evolutionary significance of morphospecies: A test with cheilostome Bryozoa. *Science, 248:* 579–583.

JACKSON, J.B.C. & COATES, A.G., 1986. Life cycles and evolution of clonal (modular) animals. *Philosophical Transactions of the Royal Society of London, B, Biological Sciences, 313:* 7–22.

KIMBLE, J. & SCHEDL, T., 1988. Developmental genetics of *Caenorhabditis elegans.* In G.M. Malacinski (Ed.), *Developmental Genetics of Higher Organisms:* 171–192. New York: Macmillan.

KÜPPERS, B.-O., 1990. *Information and the Origin of Life.* Cambridge, Massachusetts: MIT Press.

LACK, D., 1947. *Darwin's Finches.* Cambridge: Cambridge University Press.

LACK, D., 1971. *Ecological Isolation in Birds.* Cambridge, Massachusetts: Harvard University Press.

LAWTON, J.H., 1989. What is the relationship between population density and body size in animals? *Oikos, 55:* 429–434.

LEVIN, D.A., 1979. The nature of plant species. *Science, 204:* 381–384.

LEVINS, R., 1968. *Evolution in Changing Environments.* Princeton: Princeton University Press.

LEWONTIN, R.C., 1974. *The Genetic Basis of Evolutionary Change.* New York: Columbia University Press.

LILLY, M. & CARLSON, J., 1990. *smellblind:* A gene required for *Drosophila* olfaction. *Genetics, 124:* 293–302.

LYNCH, J.D., 1989. The gauge of speciation: On the frequencies of modes of speciation. In D. Otte & J.A. Endler (Eds), *Speciation and its Consequences:* 527–553. Sunderland, Massachusetts: Sinauer Associates.

MacARTHUR, R.H., 1972. *Geographical Ecology.* New York: Harper & Row.

McKINNEY, M.L., 1990. Trends in body-size evolution. In K.J. McNamara (Ed.), *Evolutionary Trends:* 75–118. London: Belhaven Press (a division of Pinter Publishers).

McPHERON, B.A., SMITH, D.C. & BERLOCHER, S.H., 1988. Genetic differences between host races of *Rhagoletis pomonella. Nature, 336:* 64–66.

MALLET, J., in press. Colour pattern evolution in *Heliconius* butterflies: alternative parapatric models and evidence from hybrid zones. In R.G. Harrison (Ed), *Hybrid Zones and Evolutionary Process. Oxford:* Oxford University Press.

MARZLUFF, J.M. & DIAL, K.P., 1991. Life history correlates of taxonomic diversity. *Ecology, 72:* 428–439.

MAY, R.M., 1990. How many species? *Philosophical Transactions of the Royal Society of London, B, Biological Sciences, 330:* 293–304.

MAYNARD SMITH, J., 1966. Sympatric speciation. *American Naturalist, 100:* 637–650.

MAYR, E., 1942. *Systematics and the Origin of Species.* New York: Columbia University Press.

MAYR, E., 1947. Ecological factors in speciation. *Evolution, 1:* 163–288.

MAYR, E., 1963. *Animal Species and Evolution.* Cambridge, Massachusetts: Harvard University Press.

MAYR, E., 1988. *Toward a New Philosophy of Biology: Observations of an Evolutionist.* Cambridge, Massachusetts: Harvard University Press.

MAYR, E. & ASHLOCK, P.D., 1991. *Principles of Systematic Zoology*, 2nd edition. New York: McGraw-Hill.

MENKEN, S.B.J., HERREBOUT, W.M. & WIEBES, J.T., 1991. Small ermine moths (*Yponomeuta*): Their host relations and evolution. *Annual Review of Entomology, 37:* 41–66.

MEYER, A., 1990. Ecological and evolutionary consequences of the trophic polymorphism in *Cichlosoma citrinellum* (Pisces; Cichlidae). *Biological Journal of the Linnean Society, 39:* 1279–1300.

MITTER, C., FARRELL, B. & FUTUYMA, D.J., 1991. Phylogenetic studies of insect–plant interactions: Insights into the genesis of diversity. *Trends in Ecology and Evolution, 6:* 290–293.

NEET, C.R. & HAUSSER, J., 1990. Habitat selection in zones of parapatric contact between the common shrew *Sorex araneus* and Millet's shrew *S. coronatus. Journal of Animal Ecology, 59:* 235–250.

PASHLEY, D.P., 1989. Host-associated differentiation in armyworms (Lepidoptera: Noctuidae): An allozymic and mitochondrial DNA perspective. In H.D. Loxdale & J. den Hollander (Eds), *Electrophoretic Studies on Agricultural Pests. Systematics Association Special Volume, 39:* 103–114. Oxford: Clarendon Press.

PATERSON, H.E.H., 1985. The recognition concept of species. In E.S. Vrba (Ed.), *Species and Speciation:* 21–29. Pretoria: Transvaal Museum.

PRICE, P.W., 1980. *Evolutionary Biology of Parasites.* Princeton: Princeton University Press.

RAFF, R.A. & WRAY, G.A., 1989. Heterochrony: developmental mechanisms and evolutionary results. *Journal of Evolutionary Biology, 2:* 409–434.

RAMÍREZ, B.W., 1970. Host specificity of fig wasps (Agaonidae). *Evolution, 24:* 680–691.

RICE, W.R., 1984. Disruptive selection on habitat preference and the evolution of reproductive isolation: simulation studies. *Evolution, 38:* 1251–1260.

RICE, W.R., 1987. Speciation via habitat specialization: the evolution of reproductive isolation as a correlated character. *Evolutionary Ecology, 1:* 301–314.

RICE, W.R. & SALT, G.W., 1990. The evolution of reproductive isolation as a correlated character under sympatric conditions; experimental evidence. *Evolution, 44:* 1140–1152.

R'KHA, S., CAPY, P. & DAVID, J.R., 1991. Host plant specialization in the *Drosophila melanogaster* species complex: A physiological, behavioral, and genetical analysis. *Proceedings of the National Academy of Sciences, U.S.A., 88:* 1835–1839.

ROSENZWEIG, M.L., 1981. A theory of habitat selection. *Ecology, 62:* 327–335.

ROSENZWEIG, M.L., 1985. Some theoretical aspects of habitat selection. In M.L. Cody (Ed.), *Habitat Selection in Birds:* 517–540. Orlando: Academic Press.

SMITH, H.S., 1941. Racial segregation in insect populations and its significance in applied entomology. *Journal of Economic Entomology, 34:* 1–12.

SOKAL, R.R. & CROVELLO, T.J., 1970. The biological species concept: A critical evaluation. *American Naturalist, 104:* 127–153.

STANLEY, S.M., 1973. An explanation of Cope's Rule. *Evolution, 27:* 1–26.

STERNBURG, P.W. & HORVITZ, H.R., 1981. Gonadal cell lineages of the nematode *Panagrellus redivivus* and implications for evolution by the modification of cell lineage. *Developmental Biology, 88:* 147–166.

STRONG, D.R., LAWTON, J.H. & SOUTHWOOD, R., 1984. *Insects on Plants: Community Patterns and Mechanisms.* Oxford: Blackwell Scientific Publications.

SULLOWAY, F., 1979. Geographic isolation in Darwin's thinking: The vicissitudes of a crucial idea. *Studies in History of Biology, 3:* 23–65.

SULSTON, J.E., SCHIERENBERG, E., WHITE, J.G. & THOMPSON, J.N., 1983. The embryonic cell lineage of the nematode, *Caenorhabditis elegans. Developmental Biology, 100:* 64–119.

TAUBER, C.A. & TAUBER, M.J., 1977. Sympatric speciation based on allelic changes at three loci: Evidence from natural populations in two habitats. *Science, 197:* 1298–1299.

TAUBER, C.A. & TAUBER, M.J., 1989. Sympatric speciation in insects: perception and perspective. In D. Otte & J.A. Endler (Eds), *Speciation and Its Consequences:* 307–344. Sunderland, Massachusetts: Sinauer Associates.

TEMPLETON, A.R., 1989. The meaning of species and speciation: A genetic perspective. In D. Otte & J.A. Endler (Eds), *Speciation and Its Consequences:* 3–27. Sunderland, Massachusetts: Sinauer Associates.

THOMSON, K.S., 1988. *Morphogenesis and Evolution*. Oxford: Oxford University Press.

THORPE, W.H., 1945. The evolutionary significance of habitat selection. *Journal of Animal Ecology, 14:* 67–70.

WALLACE, A.R., 1865. On the phenomenon of variation and geographical distribution as illustrated by the Papilionidae of the Malayan region. *Transactions of the Linnean Society of London, 25:* 1–71.

WALSH, B.J., 1863. Notes by Benj. D. Walsh. *Proceedings of the Entomological Society of Philadelphia, 2:* 182–272.

WALSH, B.D., 1864. On phytophagic varieties and phytophagic species. *Proceedings of the Entomological Society of Philadelphia, 3:* 403–430.

WATSON, J.D., HOPKINS, N.H., ROBERTS, J.W., STEITZ, J.A. & WEINER, A.M., 1987. *Molecular Biology of the Gene*, 4th edition. Menlo Park: Benjamin/Cummings.

WHITE, C.S., MICHAUX, B. & LAMBERT, D.M., 1990. Species and neo-Darwinism. *Systematic Zoology, 39:* 399–413.

WHITFIELD, J.B., 1990. Symbiont-induced speciation and endoparasitoid insects. *Evolutionary Theory, 9:* 211–213.

WILSON, A.C., BUSH, G.L., CASE, S.M. & KING, M.C., 1975. Social structuring of mammalian populations and rate of chromosomal evolution. *Proceedings of the National Academy of Sciences, U.S.A., 72:* 5061–5065.

WILSON, D.S., 1989. The diversification of single gene pools by density- and frequency-dependent selection. In D. Otte & J.A. Endler (Eds), *Speciation and its Consequences:* 366–385. Sunderland, Massachusetts: Sinauer Associates.

WILSON, E.O., 1975. *Sociobiology: The New Synthesis.* Cambridge: Belknap Press of Harvard University Press.

WOOD, T.K. & GUTTMAN, S.I., 1982. Ecological and behavioral basis for reproductive isolation in the sympatric *Enchenopa binotata* complex (Homoptera: Membracidae). *Evolution, 36:* 233–242.

WOOD, T.K. & KEESE, M.C., 1990. Host-plant-induced assortative mating in *Enchenopa* treehoppers. *Evolution, 44:* 619–628.

ZWÖLFER, H. & BUSH, G.L., 1984. Sympatrische und parapatrische Artbildung. *Zeitschrift für zoologische Systematik und Evolutionsforschung, 22:* 211–233.

CHAPTER

15

Phylogenetic patterns of behavioural mate recognition systems in the *Physalaemus pustulosus* species group (Anura: Leptodactylidae): the role of ancestral and derived characters and sensory exploitation

MICHAEL J. RYAN & A. STANLEY RAND

Introduction	**252**
Anuran mate recognition	252
The *Physalaemus pustulosus* species group	253
Materials and methods	**254**
Results	**256**
Interspecific call preferences	256
Intraspecific call preferences	258
Discussion	**259**
Interspecific call preferences	259
Intraspecific call preferences	264
Summary	**265**
References	**265**

Keywords: Leptodactylidae – mate choice – mate recognition systems – *Physalaemus* – sensory exploitation – sexual selection – species recognition.

Abstract

We have investigated patterns of mate recognition in a group of closely related frogs in which, in most cases, divergence appears to have occurred in allopatry with no secondary contact. We infer the influence of derived and ancestral characters in communication between the sexes by determining female phonotactic responses to conspecific versus heterospecific calls, only heterospecific calls, and call characters

Evolutionary Patterns and Processes
ISBN 0-12-440895-8

that influence intraspecific mate choice. We test the phonotactic responses of females of two species to the calls of six species: four are in the same species group and two are closely related to the species group.

In all cases females preferred conspecific to heterospecific calls, suggesting that characters of the communication systems derived in allopatry are sufficient to ensure species mate recognition. Females are attracted to some heterospecific calls; thus shared, perhaps ancestral, characters of the systems can allow effective communication between heterospecifics. Finally, call characters that enhance mate attraction in two lineages also enhance call attraction in a lineage that lacks this call character. This intraspecific preference appears to be an ancestral character of the receiver that existed before the evolution of the call character that enhances attraction. This result is consistent with the hypothesis that males can evolve traits to exploit already existing preferences (sensory exploitation).

INTRODUCTION

Many animals exhibit assortative mating among species. This can result from conspecific males and females sharing a communication system in which males produce species-specific signals and females exhibit a preferential response to the signals of conspecifics relative to those of heterospecifics. Thus conspecific mate recognition is an interaction of both the specificity of the male signal and the female preference. Much debate on the evolution of mate recognition systems has concerned how this species specificity evolves during secondary contact (Littlejohn, 1981; Paterson, 1982; Mayr, 1988).

We emphasize several issues in our studies of the evolution of mate recognition systems that are often neglected in other studies. First, we consider the degree to which these systems diverge in allopatry rather than sympatry or parapatry, because we feel this is a common mode of speciation (Mayr, 1988). Second, we point out that although the divergence of mate recognition signals is of obvious importance, it is the interaction of the signal and the receiver that results in mate recognition. Third, although signals and receivers might exhibit species-specific and thus derived characters, each is a multivariate trait composed of a constellation of ancestral and derived characters, the importance of which might vary with the communication context. Fourth, although there is often close concordance between properties of the signal and receiver, the evolution of signals and receivers need not be genetically linked (Butlin, 1989), and in some cases the sequence of signal–receiver evolution could suggest that males evolve signals to exploit already existing biases in the receiver (Ryan, 1990a).

In this study we use female phonotaxis experiments to address these issues in a group of closely related frogs. Specifically, we determine: (1) if differences in the communication system are sufficient for species discrimination; (2) if similarities in these systems allow effective heterospecific communication; and, (3) if traits that characterize one lineage are effective for communication in another lineage.

Anuran mate recognition

Certain aspects of how conspecific mate recognition operates are well-understood in frogs (Fuzessery, 1988; Gerhardt, 1988; Rand, 1988; Walkowiak, 1988; Ryan, 1991). Males produce advertisement calls, one function of which is to attract females.

Although there can be substantial and biologically meaningful call variation within a species, it is usually small relative to the variation among species (Ryan, 1990b). In that sense (i.e. in partitioning intra- versus interspecific call variation, as opposed to attributing to the call an evolved function), the advertisement call is 'species specific'. Many studies of frogs have shown that when given a choice between calls of conspecifics and heterospecifics, females exhibit preferential phonotaxis to the conspecific call (Gerhardt, 1988). Furthermore, studies of anuran auditory physiology have revealed that neural properties of this system can exhibit a preferential bias towards the spectral and temporal properties of the conspecific advertisement call (Fuzessery, 1988; Walkowiak, 1988; Zakon & Wilczynski, 1988).

Females also exhibit mating preferences among conspecific males (Gerhardt, 1988; Ryan, 1991; Ryan & Keddy-Hector, 1992), and in some cases the neural properties of the auditory system can also explain intraspecific mating preferences. For example, in *Hyla versicolor* (Gerhardt & Doherty, 1988), *Physalaemus pustulosus* (Ryan *et al.*, 1990) and *Acris crepitans* (Ryan, Perrill & Wilczynski, 1992) females prefer calls with frequencies that are lower than the population average, and the female's peripheral auditory system is tuned to these lower frequencies.

The *Physalaemus pustulosus* species group

The genus *Physalaemus* (Family Leptodactylidae) as currently defined contains about 30 species (Frost, 1985). The *Physalaemus pustulosus* species group, as defined by Cannatella & Duellman (1984), consists of four species in two species pairs.

One species pair consists of *P. pustulosus* and *P. petersi*. *P. pustulosus* is distributed throughout much of Central America as well as Colombia, Venezuela and Trinidad (Fig. 1). *P. petersi* is distributed widely throughout Amazonia and is found in Colombia, Ecuador, Peru, Bolivia and Brazil. The ranges of the two species are allopatric; their closest proximity is in Colombia, where they occur within several hundred kilometres of one another (Fig. 1). Both species are sympatric with other members of the genus. *P. pustulosus* is sympatric with *P. enesefae* in the ilanos of Venezuela, and *P. petersi* is sympatric with *P. ephippifer* near the mouth of the Amazon River in the region of Belen, Brazil. Further exploration might show *P. petersi* to be sympatric or parapatric with other congenerics, this seems less likely for *P. pustulosus*.

The other two species of the group have more restricted distributions (Fig. 1). *P. coloradorum* is found in low to mid-elevation rain forests on the western slope of the Andes in Ecuador. *P. pustulatus* ranges from southern Ecuador to northern Peru in the drier regions along the Pacific coast. Preliminary evidence suggests that *P. pustulatus*, as currently described, might consist of two species; we foresee the population from Peru being described as taxonomically distinct (Cannatella *et al.*, unpublished data). Here, all experiments with calls of *P. pustulatus* refer to the Ecuadorian populations.

The ranges of *P. coloradorum* and *P. pustulatus (sensu lato)* are also allopatric, approaching within 200 km in western Ecuador. Being the only representatives of the genus west of the Andes, these species are isolated from any other congenerics, and probably have been since the late Tertiary (Cannatella & Duellman, 1984).

We use two species that are closely related to the *P. pustulosus* species group for outgroup comparisons. *P. enesefae*, as mentioned above, is sympatric with *P. pustulosus* in Venezuela. *P.* 'roraima' is an undescribed species from the northern-most state of

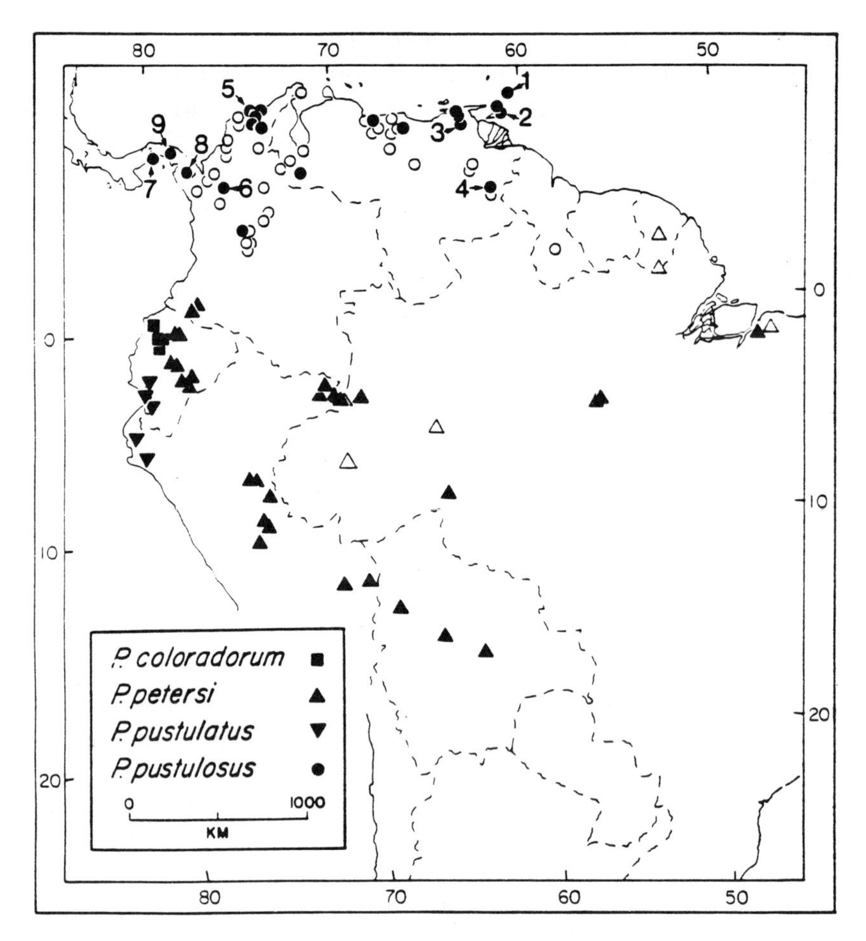

Figure 1 The distribution of members of the *Physalaemus pustulosus* species group. Localities of *P. pustulosus* in northern Central America are not included. (From Cannatella & Duellman, 1984.)

Roraima in Brazil. Its range is situated to the north of *P. petersi* and to the south of *P. pustulosus*. Although its taxonomic and phylogenetic status are not yet resolved, to us it appears most similar to *P. ephippifer*, and as such is probably closely related to but not a member of the *P. pustulosus* species group.

All species tested have advertisement calls that consist of a relatively low-frequency fundamental (< 1000 Hz) that sweeps downward to at least half its starting frequency over a duration of 80–300 ms (Ryan & Drewes, 1990; Fig. 2). All species differ in at least the initial and final frequency and duration of the call. *P. pustulosus* is able to add chucks to the whine. These chucks enhance the attraction of the call to females but are neither necessary nor sufficient for species recognition (Ryan, 1985). In some populations, *P. petersi* also adds chucks. Nothing is known of the function of the chuck in this species.

MATERIALS AND METHODS

Female *P. coloradorum* were studied in Tinalandia, near Santo Domingo de los Colorados, Ecuador. Female *P. pustulosus* were studied in Gamboa, Panama; results

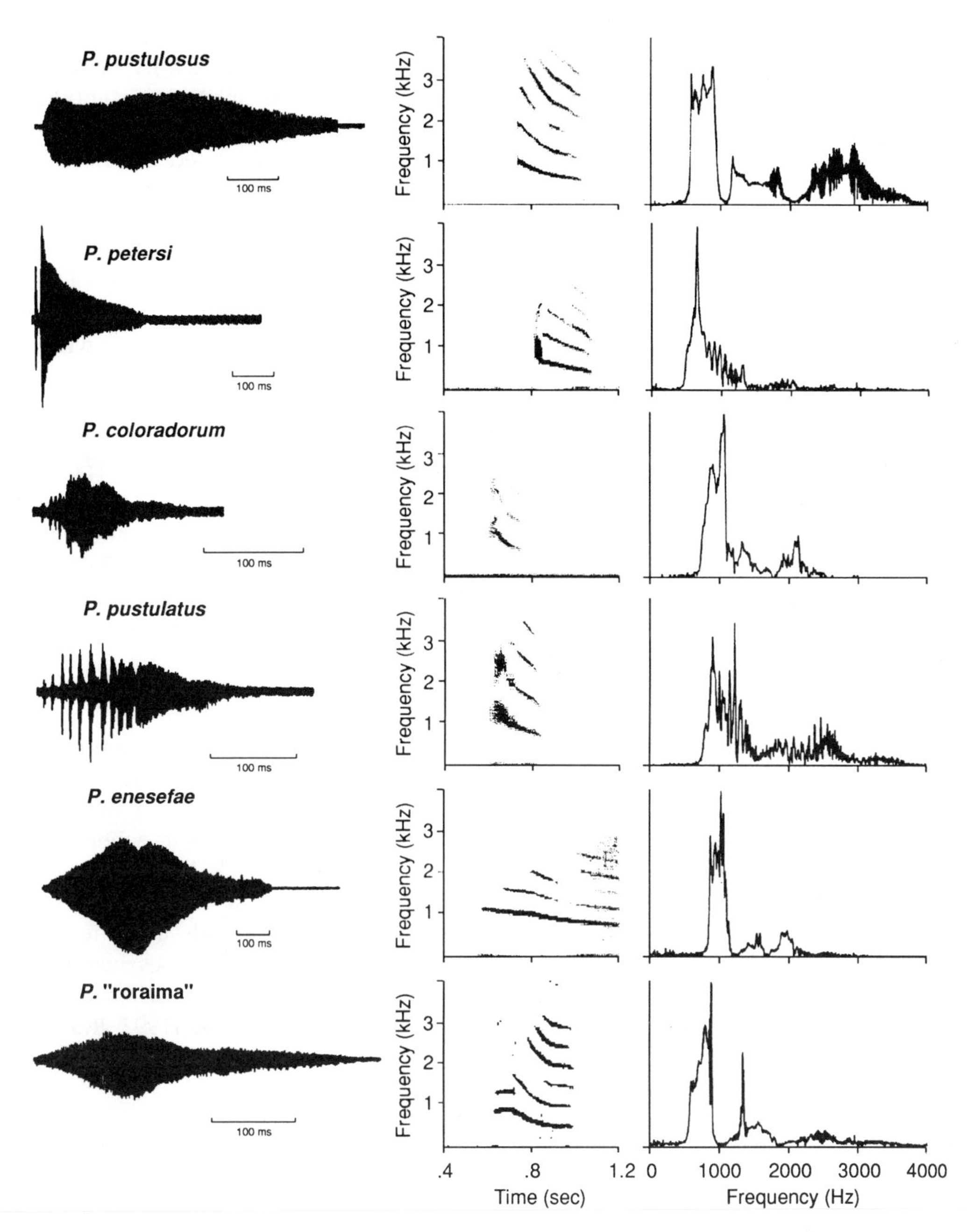

Figure 2 Oscillograms, sonograms, and fourier transforms of the advertisement calls of the species used in this study. (Figure 2a from Ryan & Drewes, 1990.)

of these studies have been published elsewhere (Ryan & Rand, in press), but are included here to facilitate comparisons. All calls used were recorded by us in the field or obtained from the U.S. National Museum of Natural History (*P.* 'roraima' and *P. enesefae*). The noise stimulus was randomly generated white noise with the duration and amplitude envelope of the 'whine' portion of the advertisement call of *P. pustulosus*.

Female phonotaxis experiments were conducted similarly in Ecuador and Panama (see also Ryan & Rand, 1990, in press; Rand, Ryan & Wilczynski, 1992). Experiments were conducted in 1990 and 1991. Calls were digitized on an Amiga computer (models 500 or 2000). Calls of each stimulus pair were played through one of the channels of the computer to an ADS L200C speaker. Calls were presented antiphonally, each call at a rate of 1 call per 2 s. Speakers were placed directly opposite one another (3 m apart in Panama, 1.4 m in Ecuador) in a small room the walls of which were covered with foam to reduce acoustic reverberation. Stimuli were broadcast at a peak intensity of 82 dB sound pressure level re 20 microPascals at the site of the female. A female was placed equidistant between the speakers under a funnel. After 2 min the female was released and a response was noted if she approached to within 10 cm of a speaker.

Statistical analysis was by exact binomial probability test. We used a one-tailed test when analysing the choices between conspecific and heterospecific calls, and between normal calls and calls to which characters that enhance attraction in other species had been added since there is an a priori prediction of the direction of the response.

Here, we do not treat the responses of females to heterospecific calls and noise statistically. This is because the null hypothesis is not 0.5 versus 0.5, and must be determined experimentally. Rand, Ryan & Wilczynski (1992) have done this for *P. pustulosus*, but it has not been done for *P. coloradorum*. Nevertheless, the patterns of response are quite clear without statistical treatment.

Interpreting female behaviour was more difficult when she was presented with a choice between a heterospecific call and noise. A significant preference for the heterospecific call shows that this is a stimulus sufficient for eliciting female phonotaxis. Random response to the heterospecific call and the noise would suggest that females are merely responding to sound, and that the heterospecific call is not a sufficient stimulus for mate attraction. However, the lack of phonotaxis could result either from females being unresponsive to either stimulus, for example due to lack of motivation, or because neither of the stimuli was biologically meaningful. If a female did not respond to either stimulus, this was counted as a 'no response' only if the female then showed a phonotactic response in a subsequent test. Such cases, we feel, demonstrate that lack of response is not due to lack of motivation, but rather to lack of appropriate acoustic stimulation (see also Rand, Ryan & Wilczynski, 1992; Ryan & Rand, in press.

RESULTS

Interspecific call preferences

As noted above, phonotaxis experiments with *P. pustulosus* have been published elsewhere (Ryan & Rand, in press) but are included here to facilitate comparisons. Females of both *P. pustulosus* and *P. coloradorum* showed statistically significant, almost unanimous preferences for the conspecific call over the calls of the other five heterospecifics (Figs 3 and 4). In ten separate experiments involving 88 phonotactic responses, only three times did a female respond to a heterospecific call.

In most cases, female *P. pustulosus* ignored both the heterospecific calls and the

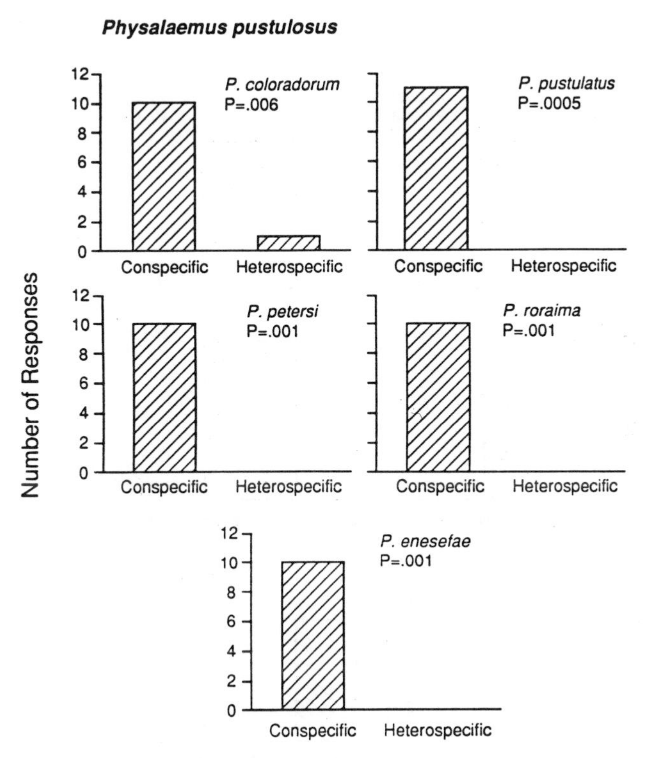

Figure 3 Phonotactic responses of female *Physalaemus pustulosus* to conspecific versus heterospecific calls.

noise stimulus. The calls of *P. pustulatus, P. petersi* and *P. enesefae* never elicited phonotaxis from female *P. pustulosus* in a total of 34 tests (Fig. 5). In all of those tests, a female responded to the speaker broadcasting noise only once. In the other 33 cases the females responded to stimuli in subsequent tests; thus their lack of response when confronted with heterospecific calls and noise was deemed to be due to the inappropriateness of the stimuli rather than lack of motivation due to physiological state.

In two cases female *P. pustulosus* showed statistically significant phonotactic responses to the heterospecific call (Fig. 5). In response to calls of *P. coloradorum* and *P.* 'roraima', females responded to the heterospecific call eight times, never to the noise, and only twice did females who approached neither stimulus subsequently exhibit phonotaxis in another test. Thus the calls of both *P. coloradorum* and *P.* 'roraima' are effective stimuli for eliciting phonotaxis from female *P. pustulosus.*

Female *P. coloradorum* differed from *P. pustulosus* in that they were more likely to approach a speaker broadcasting noise. Usually, in each experiment females approached the noise stimulus once or twice; in one experiment they approached the noise four times. They also showed more responses to the heterospecific calls (Fig. 6). However, unlike *P. pustulosus*, female *P. coloradorum* never showed more responses to the heterospecific call than 'no responses', as defined above.

It is intriguing that female *P. coloradorum* showed more responses to the *P. pustulosus* call than to any other heterospecific call. This bias is not statistically

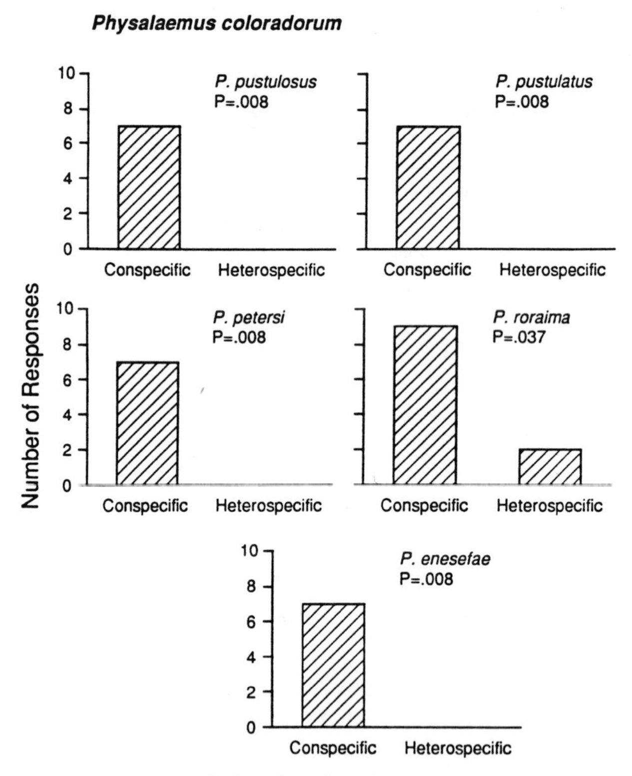

Figure 4 Phonotactic responses of female *Physalaemus coloradorum* to conspecific versus heterospecific calls.

significant, but the trend suggests a symmetry in efficacy of heterospecific signals between *P. pustulosus* and *P. coloradorum*.

Intraspecific call preferences

Male *P. pustulosus* add chucks to the whine component to increase the attractiveness of the call to females (Ryan, 1985). Chucks appear to be absent in other species of *Physalaemus* with the exception of some populations of *P. petersi*. We asked if female *P. coloradorum* would prefer chucks if they were to evolve. To do this we gave females a choice between a normal conspecific call and the same call to which three chucks from a *P. pustulosus* call were added (Fig. 7). Females showed a significant preference for the call with chucks. The preference for the chucks exhibited by the female *P. coloradorum* (9 v. 2) was not significantly different from the preference for chucks exhibited by the female *P. pustulosus* (19 v. 1) for their own calls with and without chucks (G=1.34, P=0.25; Ryan & Rand, 1990).

In response to vocalizations of other conspecific males *P. coloradorum* males produce calls in doublets or triplets with short inter-call intervals, usually less than 100 ms. Female *P. coloradorum* were given a choice between single and double calls. There was not a significant preference for the double calls (Fig. 8). However, the sample size is small and the trend is in the direction of preference for double calls. Male *P. pustulosus* do not produce calls in quick succession. Nevertheless, we

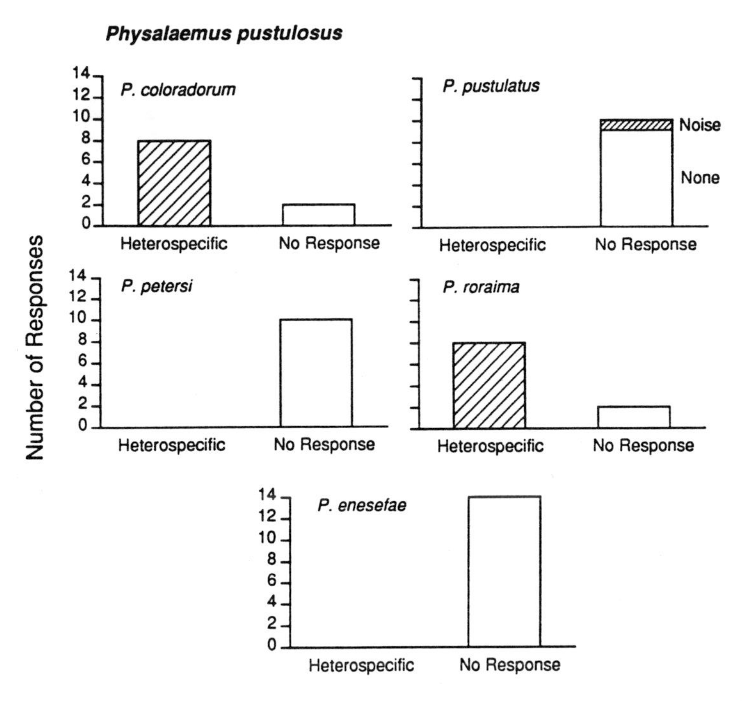

Figure 5 Phonotactic responses of female *Physalaemus pustulosus* to heterospecific calls versus white noise. A 'no response' is when females either respond to the noise or they show no phonotactic response in this test but show such a response in a subsequent test (see text for details).

asked if female *P. pustulosus* would prefer double calls to single calls if their males were to produce them. Females showed a significant preference for double calls (Fig. 8). Although this same preference exhibited by female *P. coloradorum* was not statistically significant, there was no significant difference between the females of each species in response to single and double calls ($G=1.98$, $P=0.16$).

DISCUSSION

Interspecific call preferences

The current geographic ranges of members of the *Physalaemus pustulosus* group and its two close relatives, *P.* 'roraima' and *P. enesefae*, are allopatric (with the exception of *P. pustulosus* and *P. enesefae*), and thus leads us to suggest that there has been no opportunity for character displacement or reinforcement between any pair of sister taxa. Nevertheless, females of *P. pustulosus* and *P. coloradorum* exhibit a strong preference for calls of conspecifics over those of heterospecifics. In this case the divergence of mate recognition systems during allopatry is sufficient to promote species-specific mate preferences in the absence of reinforcement fine-tuning such responses. These results support a widely held belief that the divergence of behaviours that cause reproductive isolation could result as an epiphenomenon of genetic

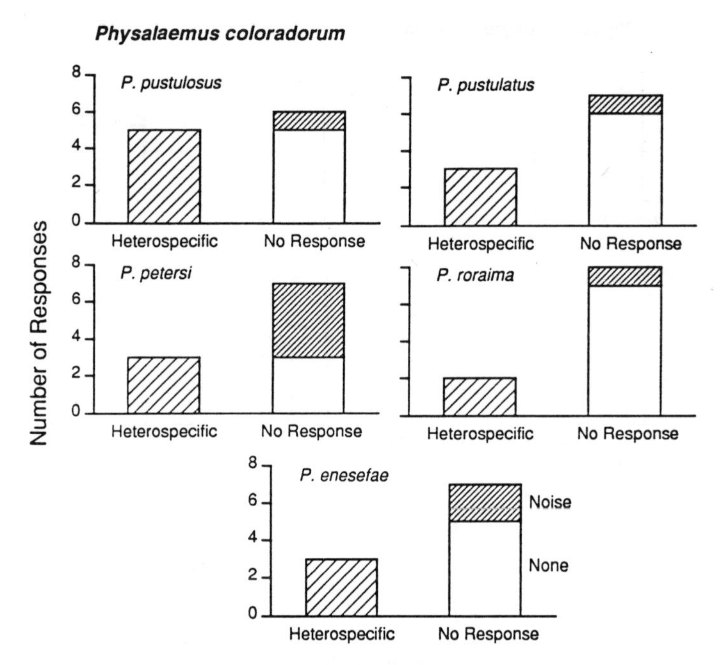

Figure 6 Phonotactic responses of female *Physalaemus coloradorum* to heterospecific calls versus white noise. A 'no response' is when females either respond to the noise or they show no phonotactic response in this test but show such a response in a subsequent test (see text for details).

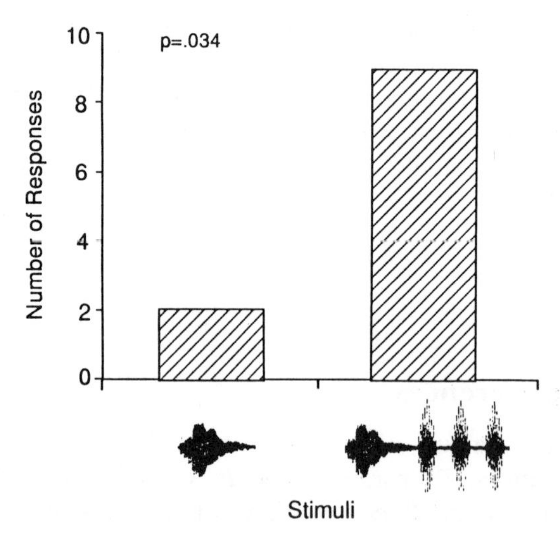

Figure 7 Phonotactic responses of female *Physalaemus coloradorum* to the normal, conspecific advertisement call and the same call to which three chucks of a *P. pustulosus* call have been added (calls represented in oscillograms).

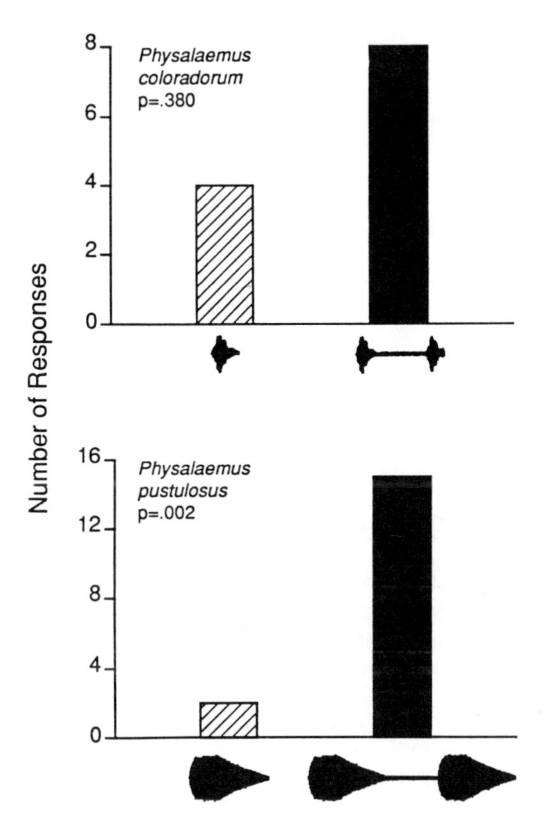

Figure 8 Phonotactic responses of female *Physalaemus coloradorum* and *P. pustulosus* to their own advertisement call presented either as a single or as a double call (calls represented in oscillograms).

divergence of allopatrically isolated gene pools (Darwin, 1859; Mayr, 1950, 1988; Paterson, 1985), as opposed to the theory that reinforcement of these mechanisms during secondary contact is crucial (Dobzhansky, 1937).

The species we tested can discriminate between conspecific and heterospecific calls, but another question asks if the heterospecific call is a viable communication signal. Ryan & Rand (in press) have discussed this issue in detail, and that discussion is summarized here.

Figure 9 illustrates various hypothetical relationships between a conspecific signal and the female preference function in univariate space, and emphasizes that discrimination between conspecific and heterospecific signals could result from several different relationships.

The preference functions a-d illustrated in Fig. 9 would result in female preference for the conspecific relative to the heterospecific signals illustrated, but how they would influence preferences among conspecifics and responses to heterospecific signals differ. Functions a–c (Fig. 9) would all result in no response to heterospecific signals; function a would not generate selection on conspecific signals, while functions b and c would generate stabilizing and directional selection, respectively. We do not know of any cases that would suggest function a (Fig. 9); this is probably because many studies concentrate either on species recognition or on sexual selection, but usually not both. For example, in our study the majority of experiments reveal

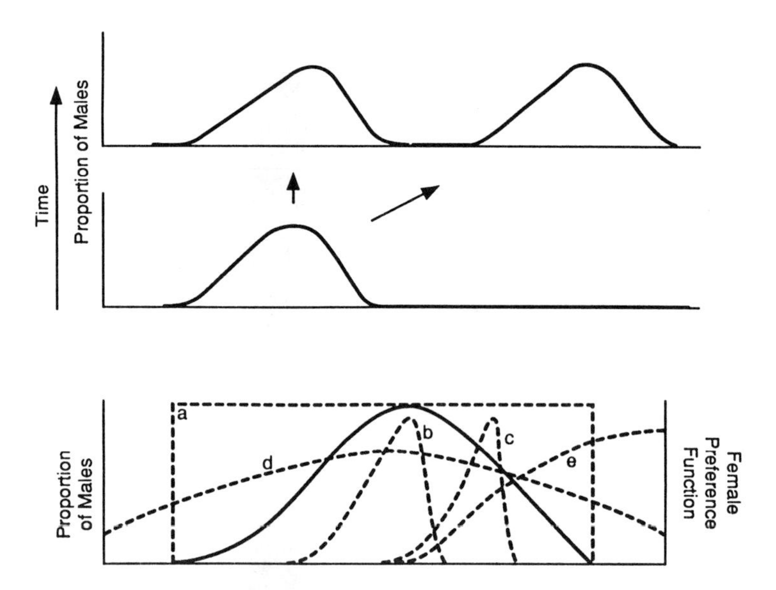

Figure 9 Top: Hypothetical distribution of species recognition signals in univariate space, illustrating the divergence of these signals in two populations during the process of speciation. Bottom: Solid line is the univariate distribution of male signals, and the dashed lines are various female preference functions relative to the male signal.

preference for conspecific signals and no response to heterospecifics. This would fulfil one of the predictions of the functions a–c (Fig. 9), but would not allow us to discriminate among those functions. There are several examples of the relationship illustrated by functions b and c (Fig. 9). For example, Gerhardt (1991) reviews call preferences in gray treefrogs, *Hyla versicolor*, and shows that preferences tend to generate stabilizing selection on some call properties and directional selection on others. Ryan & Keddy-Hector (1992) review numerous cases of directional female preferences for male courtship signals.

Function d (Fig. 9) predicts that females would prefer conspecific to heterospecific signals, but that a heterospecific signal would still elicit a response. In this study the responses of *P. pustulosus* (and possibly *P. coloradorum*) to heterospecifics are examples of such a phenomenon. Finally, function e (Fig. 9) predicts that females would respond to the conspecific signal but would prefer the heterospecific signal. Ryan & Wagner (1987) showed that female swordtails *Xiphophorus pygmaeus* preferred heterospecific males, *X. nigrensis*, to their own males. A similar example might be seen in this study. *P. coloradorum* females prefer their own calls to which a heterospecific signal has been added, to their own calls without such a signal. Also, female *P. pustulosus* prefer double calls to single calls, which is heterospecific rather than a conspecific manner of grouping calls. In both of these examples the difference in signals is presence/absence rather than a quantitative difference, as suggested in Fig. 9. Nevertheless, this is qualitatively the same phenomenon.

The relationship between signal and receiver properties has two important implications for understanding the evolution of communication systems. First, it tells us about the possible evolutionary patterns and prospects of signal–receiver evolution. For example: Are signal and receiver of the same species tightly matched, and different in all aspects from all other species, suggesting lock-step co-evolution? Are preferences fairly broad, suggesting the retention of ancestral features, or perhaps convergence, in addition to the derived features that must be responsible for conspecific preferences? Do these preferences exhibit sufficient latitude to suggest various signals not exhibited by the species would be integrated into the communication system if they were to evolve (also see below)?

These considerations also affect some more philosophical issues of species concept. Paterson (1985) suggests that a species is a set of individuals sharing a common mate recognition system, and suggests that mate recognition, i.e. a positive response to another individual, and not discrimination among individuals, is the diagnostic feature in his definition: "The response of one mating partner to a signal from the other is here regarded as an act of recognition. Recognition is thus a specific response by one partner to a specific signal of the other ... I strongly emphasize that I imply no act of judgement and no act of choice on the part of the responding partner." It is not clear how responses to heterospecifics can be integrated into Paterson's species definition; perhaps that definition would merely classify these taxa as conspecific.

If taxa have evolved (derived) characters that result in species-specific preferences, why is it that in some cases heterospecific signals are recognized? This suggests the communication systems of the two species either have converged in certain characters or, perhaps more parsimoniously, that they share ancestral characters that suffice in eliciting mate recognition. Our experiments show that there are cases in which heterospecific signals can attract females. *P. pustulosus* females were attracted to calls of both *P. coloradorum* and *P.* 'roraima', and *P. coloradorum* females showed a tendency to be attracted to calls of *P. pustulosus*.

It is difficult to speculate as to the specific call and preference parameters that cause females to respond to one heterospecific call but not another. Some calls are more similar in spectral properties (e.g. *P. pustulosus* and *P. coloradorum*) while others are more similar in temporal properties (e.g. *P. pustulosus* and *P. enesefae*). Also, we cannot assume either that all properties of the call are attended to by females, or that the degree of difference in one domain (e.g. spectral) would have the same influence on preferences as the same degree in another domain (e.g. temporal). To understand fully the mechanism and evolution of these mate recognition systems, which include both the signals and the receivers, we need to combine a detailed quantitative analysis of calls with a detailed determination of the females' preference functions across all axes of call variation for all species under consideration.

Our first set of experiments, which show that females prefer conspecific to heterospecific calls, is consistent with many studies of species-specific mate recognition systems. By itself, these experiments could suggest that preferences and traits are closely matched systems that have co-evolved in a fine-tuned manner. However, the fact that the call of one species can elicit a response from more than one species suggests this need not be the case. Instead, it seems that females of some species share some parameters of the recognition system. Since many of these species are closely related, it seems probable that these common features are shared through common descent, although convergence is also a possibility that cannot be rejected without rigorous outgroup comparisons. The efficacy of these shared, and perhaps ancestral, characters shows that distant evolutionary history of a lineage influences its communication system, and that mate recognition might not be performed totally in the domain of recently derived characters.

Intraspecific call preferences

In *P. pustulosus* females prefer calls to which chucks have been added. Although male *P. coloradorum* do not add chucks, these females prefer calls to which chucks are added artificially. Thus in *P. coloradorum* the response of the receiver can be elicited by signals that are not part of the communication system. If our interpretation of the evolutionary history of the chuck is correct, this also suggests that the chuck evolved after the divergence of the *P. pustulosus–P. petersi* and *P. coloradorum–P. pustulatus* lineages, and leads us to conclude that the preference for chucks in both *P. pustulosus* and *P. coloradorum* might be shared through a common ancestor and thus existed prior to the evolution of the chuck. This would support the notion of sensory exploitation, which states that males can evolve traits to exploit pre-existing female preferences (Ryan, 1990a; Ryan & Rand, 1990; Ryan *et al.*, 1990).

Responses to single and double calls reflect a similar pattern to preferences for chucks; female *P. pustulosus* show a strong preference for a trait exhibited by a closely related species but lacking in their own. In this case, it is not clear if the double call is a derived or an ancestral condition relative to the species group. Relative to the other species in its group, male *P. pustulosus* have much larger vocal sacs and appear to use more air during calling. Dudley & Rand (1991) showed that the time needed for the vocal sac to deflate was *c* 250 ms, while the intercall interval of the double calls of *P. coloradorum* was about 170 ms. It is possible that the use of a larger air reservoir for calling, and the concomitant increase in vocal sac size and vocal sac deflation time, precludes the ability to produce calls in quick succession. Nevertheless, our experiments show that female *P. pustulosus* exhibit a preference for a trait not produced by their males, a preference which could favour this trait if it were to evolve.

In preference for both calls with chucks and for double calls, there is no significant difference in the degree of the preference between species in which the trait is present and in which the trait is absent. This rejects the hypothesis that the preference has been subjected to further evolution due to the presence of the male trait. However, our sample sizes are too small to give much confidence in accepting the null hypothesis of no difference. The hypothesis of trait elaboration after sensory exploitation has been recently supported by studies of preference for song repertoire in oscine birds. Searcy (1992) has shown that female grackles prefer repertoires to single song types, despite the fact that male grackles only produce the latter. Searcy also showed that among oscines, the strength of the preference is positively correlated with repertoire size. This suggests that there is a pre-exisiting preference for repertoires but that the strength of this preference is further elaborated in species exhibiting repertoires.

Our results on intraspecific mate recognition also illuminate the importance of ancestral properties of the females' preference. This is especially pertinent to studies in sexual selection. A controversial question has been why females exhibit preferences for male secondary sexual characters in species in which mate choice has no immediate effect on female reproductive success. Two hypotheses that have received much attention are Fisher's theory of runaway sexual selection (Fisher, 1958) and various hypotheses based on 'good genes' (e.g. Zahavi, 1975; Hamilton & Zuk, 1982; Kodric-Brown & Brown, 1984). Both of these hypotheses suggest that the female trait evolves under indirect selection (Kirkpatrick & Ryan, 1991). In these models, female preferences are genetically correlated with male traits; the traits are under direct selection and the preferences co-evolve as correlated responses to the evolution of traits. Pre-existing preferences suggest that the male trait evolved after the preference. Thus models invoking the correlated co-evolution of traits and preferences can be rejected, and pre-existing preferences instead support the hypothesis of sensory exploitation – males evolve traits to exploit pre-existing preferences. Various forces inside and outside of the context of mate choice could be responsible for pre-existing preferences or for the further elaboration of the preference (Ryan, 1990a).

SUMMARY

Our results show that current mate recognition systems are sufficient for species-specific recognition among closely related species that are thought to have diverged in allopatry. This highlights the importance of derived characters in the mate recognition system. However, females of some species respond to heterospecific calls, suggesting the ancestral characters also can result in effective communication between the sexes.

We also examine traits that might be under sexual selection because they enhance the attractiveness of males relative to other conspecific males. *P. pustulosus* does this by adding chucks and *P. coloradorum* might do this by producing calls in doublets and triplets. Despite the fact that *P. pustulosus* does not produce double calls and that *P. coloradorum* does not produce chucks, the respective females prefer the trait that is lacking in their own males. This further emphasizes the importance of ancestral traits, in this case pre-existing female preferences.

Many studies have concentrated on how current, species-specific mate recognition systems ensure species integrity. Although our results do not refute this claim, we suggest that a view of the signals and receivers in mate recognition systems as constellations of derived and ancestral traits will yield a better understanding of communication between the sexes, whether among or within species.

REFERENCES

BOAKE, C.R.B., 1991. Coevolution of senders and receivers of sexual signals: genetic coupling and genetic correlations. *Trends in Ecology and Evolution, 6:* 225–227.

BUTLIN, R., 1989. Reinforcement of premating isolation. In D. Otte & J.A. Endler (Eds), *Speciation and Its Consequences:* 158–179. Sunderland, Massachusetts: Sinauer Associates.

CANNATELLA, D.C. & DUELLMAN, W.E., 1984. Leptodactylid frogs of the *Physalaemus pustulosus* group. *Copeia, 1984:* 902–921.

DARWIN, C., 1859. *The Origin of Species.* (reprint of original) New York: Random House.

DOBZHANSKY, T., 1937. *Genetics and the Origin of Species.* New York: Columbia University Press.

DUDLEY, R.D. & RAND, A.S., 1991. Sound production and vocal sac inflation in the túngara frog, *Physalaemus pustulosus* (Leptodactylidae). *Copeia, 1991:* 460–470.

FISHER, R.A., 1958. *The Genetical Theory of Natural Selection,* 2nd edition, New York: Dover Publications.

FROST, D.R. (Ed.). 1985. *Amphibian Species of the World.* Lawrence, Kansas: Allen Press.

FUZESSERY, Z.M., 1988. Frequency tuning in the anuran central auditory system. In B. Fritzsch, M. Ryan, W. Wilczynski, T. Hetherington & W. Walkowiak (Eds), *The Evolution of the Amphibian Auditory System:* 253–273. New York: John Wiley & Sons.

GERHARDT, H.C., 1988. Acoustic properties used in call recognition by frogs and toads. In B. Fritzsch, M. Ryan, W. Wilczynski, T. Hetherington & W. Walkowiak (Eds). *The Evolution of the Amphibian Auditory System:* 455–483. New York: John Wiley & Sons.

GERHARDT, H.C., 1991. Female mate choice in treefrogs: static and dynamic acoustic criteria. *Animal Behavior, 42:* 615–635.

GERHARDT, H.C., & DOHERTY, J.A., 1988. Acoustic communication in the gray treefrog. *Hyla versicolor*: evolutionary and neurobiological implications. *Journal of Comparative Physiology, 162:* 261–278.

HAMILTON, W.D. & ZUK, M., 1982. Heritable true fitness and bright birds: a role for parasites? *Science, 218:* 384–387.

KIRKPATRICK, M. & RYAN, M.J., 1991. The paradox of the lek and the evolution of mating preferences. *Nature, 350:* 33–38.

KODRIC-BROWN, A. & BROWN, J.H., 1984. Truth in advertising: the kinds of traits favored by sexual selection. *American Naturalist, 124:* 309–323.

LITTLEJOHN, M.J., 1981. Reproductive isolation: a critical review. In W.R. Atchley & D.S. Woodruff (Eds). *Essays in Honor of M.J.D. White,* Cambridge: Cambridge University Press.

MAYR, E., 1950. The role of the antennae in the mating behavior of female *Drosophila. Evolution 4:* 149–154.

MAYR, E., 1988. The why and how of species. *Biology and Philosophy, 3:* 431–441.

PATERSON, H.E.H., 1982. Perspectives on speciation by reinforcement. *South African Journal of Science, 78:* 53–57.

PATERSON, H.E.H., 1985. The recognition concept of species. In E. Vrba (Ed.), *Species and Speciation:* 21–29. Pretoria: Transvaal Museum.

RAND, A.S., 1988. An overview of anuran acoustic communication. In B. Fritzsch, M. Ryan, W. Wilczynski, T. Hetherington & W. Walkowiak (Eds), *The Evolution of the Amphibian Auditory System:* 415–431. New York: John Wiley & Sons.

RAND, A.S., RYAN, M.J. & WILCZYNSKI, W., 1992. Signal redundancy and receiver permissiveness in acoustic mate recogniton by the túngara frog, *Physalaemus pustulosus. American Zoologist, 32:* 81–90.

RYAN, M.J., 1985. *The Túngara Frog, A Study in Sexual Selection and Communication.* Chicago: University of Chicago Press.

RYAN, M.J., 1990a. Sensory systems, sexual selection, and sensory exploitation. *Oxford Surveys in Evolutionary Biology, 7:* 157–195.

RYAN, M.J., 1990b. Signals, species, and sexual selection. *American Scientist, 78:* 46–52.

RYAN, M.J., 1991. Sexual selection and communication in frogs. *Trends in Ecology and Evolution, 6:* 351–355.

RYAN, M.J. & DREWES, R.C., 1990. Vocal morphology of the *Physalaemus pustulosus* species group (Family Leptodactylidae): morphological response to sexual selection for complex calls. *Biological Journal of the Linnean Society, 40:* 37–52.

RYAN, M.J. & KEDDY-HECTOR, A., 1992. Directional patterns of female mate choice and the role of sensory biases. *American Naturalist*, 139: S4–S35.

RYAN, M.J. & RAND, A.S., 1990. The sensory basis of sexual selection for complex calls in the túngara frog, *Physalaemus pustulosus* (sexual selection for sensory exploitation). *Evolution, 44:* 305–314.

RYAN, M.J. & A.S. RAND, in press. Mate recognition, species recognition, and sexual selection: some clarifications. *Evolution.*

RYAN, M.J. & WAGNER, W., 1987. Asymmetries in mating preferences between species: female swordtails prefer heterospecific mates. *Science, 236:* 595–597.

RYAN, M.J., FOX, J.H., WILCZYNSKI, W., & RAND, A.S., 1990. Sexual selection for sensory exploitation in the frog *Physalaemus pustulosus. Nature, 343:* 66–67.

RYAN, M.J., PERRILL, S.A. & WILCZYNSKI, W., 1992. Auditory tuning and call frequency predict population-based mating preferences in the cricket frog, *Acris crepitans. American Naturalist, 139:* 1370–1383.

SEARCY, W.A., 1992. Song repertoire and mate choice in birds. *American Zoologist, 32:* 71–80.

WALKOWIAK, W., 1988. Central temporal coding. In B. Fritzsch, M. Ryan, W. Wilczynski, T. Hetherington & W. Walkowiak (Eds), *The Evolution of the Amphibian Auditory System:* 125–155. New York: John Wiley & Sons.

ZAHAVI, A., 1975. Mate selection: a selection for a handicap. *Journal of Theoretical Biology, 53:* 205-214.

ZAKON, H., & WILCZYNSKI, W., 1988. The physiology of the anuran eighth nerve. In B. Fritzsch, M. Ryan, W. Wilczynski, T. Hetherington & W. Walkowiak (Eds), *The Evolution of the Amphibian Auditory System:* 125–155. New York: John Wiley & Sons.

CHAPTER

16

Acoustic signals and speciation in cicadas (Insecta: Homoptera: Cicadidae)

THOMAS E. MOORE

Introduction	**270**
Extensive geographic range, little variation	**270**
Similar songs, plus morphological variation	**271**
Similar morphologies, strikingly different songs, complex ranges	**271**
Origin of North American cicadas	**282**
Conclusions	**282**
References	**283**

Keywords: 13-year cicadas – 17-year cicadas – cicada fossils – cicada songs – Cicadidae – *Cicada* – *Lyristes* – *Magicicada* – *Quesada* – *Tibicen*.

Abstract

Cicada species are separated by one or more of the following factors: geography, habitat, season, diurnal pattern of reproductive activity, and acoustic signals. A few species with unusually distinctive songs and broad habitat affinities have large geographic ranges, throughout which they show little acoustic or morphological variation. Most other species have far more restricted habitats and geographic ranges. *Tibicen lyricen* in eastern North America has evolved striking morphological varieties which all retain the same song.

The six species of periodical cicadas (*Magicicada*), with three distinctive but essentially invariant calling songs, have evolved extensive developmental and distributional complexity. Multispecies brood populations of both 13- and 17-year cicadas, isolated geographically and chronologically, have evidently persisted since the Pleistocene glaciation and have invaded glaciated regions. *Decim* cognate males' songs are heard by *Cassini* cognates even better than their own songs, which may

Evolutionary Patterns and Processes
ISBN 0-12-440895-8

enhance assembly as chorusing, mixed-species, selfish herds. Sometimes adult stragglers emerge within a few years of the main brood emergence date; the most common straggling times are 1 year early or late, or 4 years late.

Tertiary fossils, representing all of the major subgroups of cicadas in North America, are known from the western United States.

INTRODUCTION

Tymballing cicadas are not host specific, but in North America are clearly associated (1) with general ecological habitats (such as deserts, eastern deciduous forests, grasslands, etc.), (2) with seasons (late spring and early summer, or mid-summer and fall), and (3) with time of day of reproductive activity.

The evolution of cicada acoustic signals seems to be linked to (1) the presence of other cicada species (commonly, 2–5 species representing 1 or 2 genera are active in a local area at the same time), (2) how long that association has persisted, and (3) how distinctive the species' song was at the time of first encounter with other species. Thus, geographic distribution, temporal occurrence and acoustic behaviour are closely linked.

A few examples in three genera are dealt with here. One focus is on the correlation of distribution and evolution of calling songs of males. Callings songs are the primary congregating mechanisms of sexually receptive adult cicadas, functioning as premating isolating mechanisms and initial mate-recognition signals. These songs are their most common sounds, usually produced by lone males, but in a few species they are produced by groups of chorusing males. The other focus is on evolution of very long developmental and reproductive cycles along with complex multispecies geographic ranges, given prior selection for very distinctive acoustic signals.

EXTENSIVE GEOGRAPHIC RANGE, LITTLE VARIATION

Only one cicada species, *Quesada gigas* (Olivier), is known whose range includes both North and South America. This species is common during local summers from south central Texas, through Mexico and Central America, and from east to west from northern South America to south of Buenos Aires, Argentina. *Q. gigas* is typically active in late afternoon and at dusk, and is one of the few species that continues singing at night under bright moonlight or streetlights. Its song is long and distinctive from all other cicada sounds in those regions, beginning with a buzz, followed by a staccato phrase lasting up to several seconds, and ending with a whine lasting up to several seconds. Without interruption, the staccato and whine sections may be repeated several times in a row during one singing bout by a male. Throughout its large geographic range of several thousand kilometres of temperate and tropical conditions, there is only modest morphological variation and essentially no variation in song.

It is not likely that *Q. gigas* has been transported passively by man, for it is not closely associated with any crop plants. Only one other species, also South American, is known in the genus, and no closely related fossils are known. There is no evidence on the timing or direction (s) that this great dispersal of *Q. gigas* followed. Similarly extensive geographic ranges, but in temperate climes from the Mediterranean eastward across central Asia, also with little morphological or acoustical variation, are known for *Tibicen plebejus* (Scopoli) (=*Lyristes plebejus* (Scopoli)) and *Cicada orni Linnaeus* (Andrej Popov, personal communication; Claridge, Wilson & Singhrao, 1979).

SIMILAR SONGS, PLUS MORPHOLOGICAL VARIATION

Tibicen lyricen (De Geer) is a widespread summer species, active from July to the first frost in deciduous forests of eastern North America. It has three strikingly different morphs named as varieties (Fig. 1). Despite their distincitve appearances, these morphs show no clear geographic associations (Fig. 2), and there are no known barriers to introgression among the varieties. The generally larger variety with more heavily infuscated wings, *T. lyricen* var. *virescens*, has a more southerly and coastal distribution, but overlaps broadly with var. *lyricen*, in particular. Despite having evolved distinctive morphological varieties, as different as between many other species, the songs of males of all three varieties are virtually identical throughout eastern North America. They are composed of nondescript beginning and ending buzzes of lower intensity, and a middle distinctively buzzy vibrato that may last up to several seconds (Fig. 3, upper tracings). It is this middle 'vibrato' portion of songs in *Tibicen* that is distinguishing, and there are no other eastern species of *Tibicen* with closely similar songs.

The closest *Tibicen* with a similar song (Fig. 3, lower traces) occurs in the Bermuda Islands, associated with native trees there (*Juniperus bermudiana* L.). Adults are morphologically distinctive (Fig. 1, lower right), but the songs of *T. bermudianus* sound remarkably similar to those of *T. lyricen*. In both species the principle frequencies lie between 4 and 6 kHz. *Tibicen bermudianus* is the only cicada in the Bermuda Islands, so it may well have undergone little selection for differentiation of song through interaction with other species. The origin and age of the terrestrial flora and fauna of the oceanic subtropical Bermuda Islands, lying about 1000 km east of North America, is puzzling. Several endemic plant and animal species are known. Other native terrestrial organisms have their closest affinities with south-eastern U.S. coastal regions or the West Indies (Verrill, 1903; Britton, 1965:vii). The geological history of these islands is reasonably well-established, suggesting habitable land there for at least 200 000–125 000 years, and no prior land connections (Harmon *et al.*, 1983).

SIMILAR MORPHOLOGIES, STRIKINGLY DIFFERENT SONGS, COMPLEX RANGES

This final example involves some of the most distinctive and famous eastern North American animals: the six species of periodical cicadas (*Magicicada*). They are famous for: (1) their very similar morphology as adults and as juveniles; (2) the longest reproductive cycles of any animals (13–17 years); (3) their population density, sometimes exceeding 3 000 000 per acre; (4) their loud songs, produced by dense, commonly mixed-species choruses in early summer when few other insects are singing; (5) their apparent sudden appearance, short adult lives, and absence of defensive mechanisms, making them prime prey for nearly all general predators; (6)

Figure 1 Dorsal views of adult males of two *Tibicen* species with very similar songs, one species with striking colour morphs. Upper left, *T. lyricen* var. *engelhardti* (Davis); upper right, *T. lyricen* var. *lyricen* (De Geer); lower left, *T. lyricen* var. *virescens* Davis; lower right, *T. bermudianus* (Verrill).

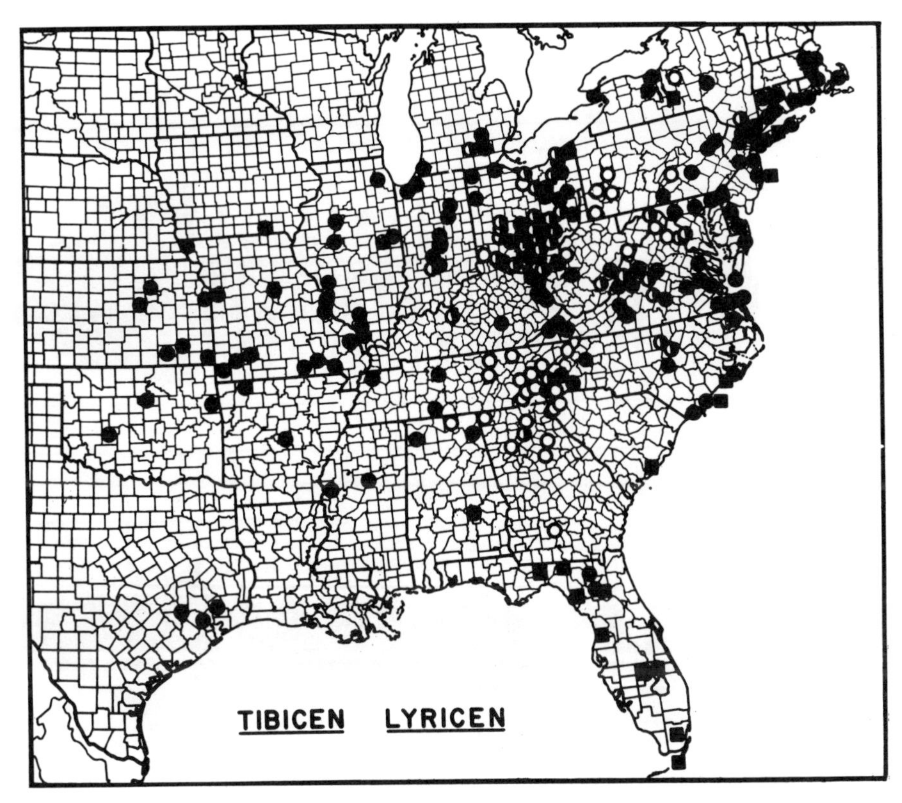

Figure 2 Geographic distribution of *Tibicen lyricen* (De Geer). Closed circles: var. *lyricen*; open circles: var. *engelhardti*; half-open circles, both var. *lyricen* and *engelhardti*; squares, var. *virescens*.

lack of distinctive habitat separations, daily redividing their habitats acoustically in unpredictable ways; (7) feeding regularly on xylem sap as juveniles and adults; and (8) their numerous and extensive mixed-species, chronologically and geographically disjunct populations called broods (Marlatt, 1923; Alexander & Moore, 1962; Williams & Smith, 1991).

The probable evolutionary relationships of the six species of periodical cicadas are shown in Fig. 4, based on morphological, developmental and behavioural characteristics. It is remarkable that each 17-year cicada has as its closest relative a 13-year cicada with essentially identical morphology and song. The songs and singing behaviour of the three 13-year species are shown in Fig. 5, and are essentially identical to the three 17-year species. The songs of the *Decim* cognates (*M. septendecim* and *M. tredecim*) centre between 1 and 3 kHz, those of both the *Cassini* (*M. cassini* and *M. tredecassini*) and the *Decula* (*M. septendecula* and *M. tredecula*) cognates centre between about 4 and 10 kHz. The songs of the *Decim* cognates are more musical, rising in pitch and in intensity before dropping in both characteristics at the end. The songs of the *Cassini* cognates sound like a series of rapid ticks, followed by a noisy buzz that rises in pitch and intensity before dropping in both at the end. The songs of the *Decula* cognates are about three times as long as a single song of the other two: a series of ticks separated by short buzzes, the buzzes becoming shorter and the ticks more closely spaced, ending with a series of ticks without buzzes between. These songs are about as different structurally as songs typically are

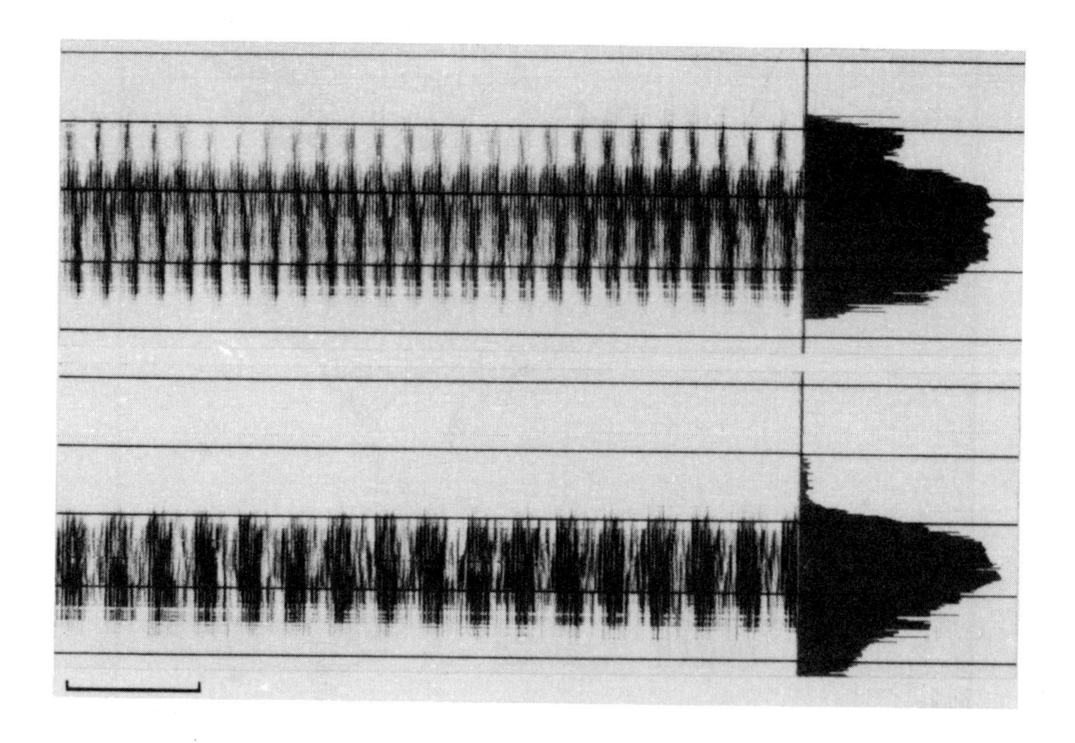

Figure 3 Audiospectrographs (left) and energy spectra (right) of vibrato middle portions of calling songs of *Tibicen lyricen* (upper) and *T. bermudianus* (lower). Time line for audiospectrographs, 0.1 s; amplitude scale relative; frequency scale marked at 2, 4, 6, 8 and 10 kHz.

between cicada genera, so they are well-distinguished by these signals, despite such similar adult appearance that the species were not distinguished until 1962 (Alexander & Moore, 1962). It is common for males of each species to sing one or a few songs, and then fly a short distance.

The three species of either 17-year or 13-year cicadas are thoroughly intermixed locally throughout most of their ranges, segregated together as mixed-species associates in chronologically and geographically isolated brood populations. It seems likely that some of these mixed-species populations have persisted since before Wisconsinan glaciation (Alexander & Moore, 1962; Cox & Carlton, 1988; Martin & Simon, 1990a; Cox, 1992). There are only 15 extant broods (Fig. 6), 12 of 17-year cicadas and 3 of 13-year cicadas, all designated by Roman numerals: I – XVII reserved for broods in any 17-year sequence, XVIII – XXX for those in any 13-year sequence. Brood VII, in upstate New York, includes only one species, *Magicicada septendecim*. All other broods include three species. One brood with the same single species, former brood XI in the Connecticut River valley, was still abundant, but reduced in range to one or at most a very few woodlots near Storrs, Connecticut, by 1954; by 1971 it had become extinct. Brood VII appears to be nearing a similar fate.

All three species of 17-year cicadas hear each other quite well, and hear a wide variety of other sounds, including birds, around them. In fact, *Magicicada cassini*

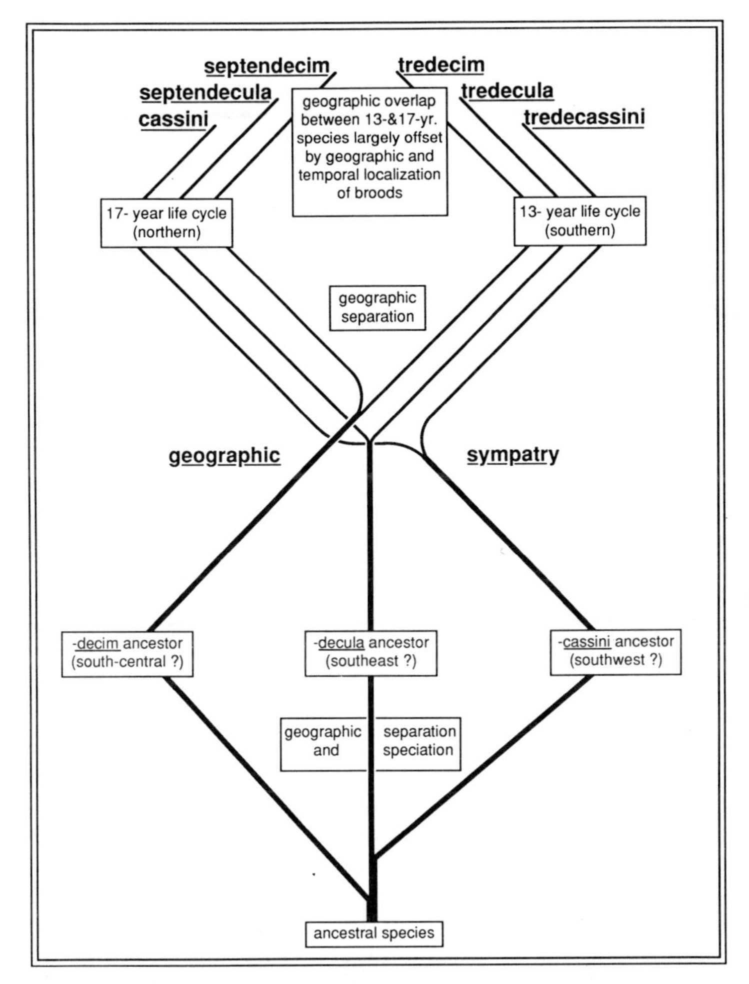

Figure 4 Diagram of probable phylogenetic relationships of 17-year and 13-year cicadas (*Magicicada*). (Modified from Alexander & Moore, 1962.)

hears *M. septendecim* (the species with which it most extensively occurs, whose peak of chorusing is reached earliest in the day, and whose song has the least intensity of the three) better than it hears its own songs (Huber *et al.*, 1990). This possibly enhances the formation of the typical complex mixed-species selfish herds (Hamilton, 1971) found in periodical cicadas, as the two later-chorusing species (*Cassini* and *Decula* cognates) daily move among the chorusing males and responding females once *Decim* males have begun to sing.

The only defences these species have against general predators are (1) surprise, emerging in huge numbers before their shorter-lived (and therefore inexperienced) potential predators have detected them, (2) accomplishing much of their reproductive activity quite early in an emergence, (3) emerging in such great numbers that they can satiate the predators once they are discovered, and (4) taking advantage of the buffering effect of the relative numbers of the other visually and behaviourally similar, sympatric, periodical cicada species by flocking together. Straggling individuals emerging in off years are at special risk, for their numbers are almost always so small

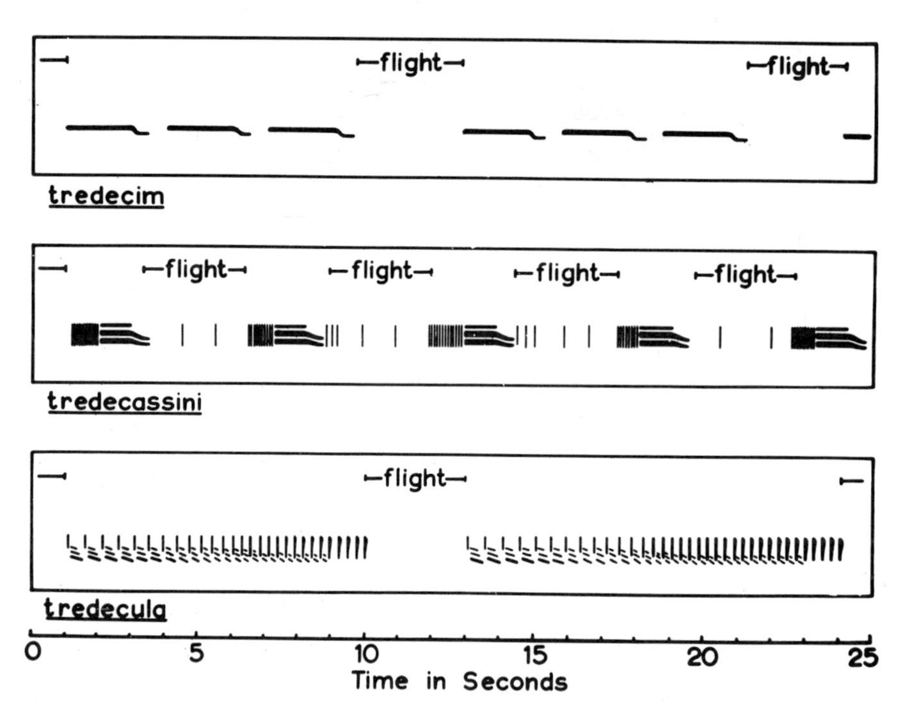

Figure 5 Diagrams of calling song audiospectrographs and typical chorusing behaviour in periodical cicadas (*Magicicada*). Song and behaviour patterns are essentially identical between cognate 13-year and 17-year cicadas: between *M. tredecim* (Walsh & Riley) and *M. septendecim* (L.), *M. tredecassini* (Alexander & Moore) and *M. cassini* (Fisher), and between *M. tredecula* (Alexander & Moore) and *M. septendecula* (Alexander & Moore). (From Alexander & Moore, 1962.)

that these populations are all eaten by predators within a few days, before they can lay sufficient eggs to sustain a subsequent emerging population of adults. This is particularly evident for the scattered stragglers in years after the main emergence.

The *Cassini* cognates usually emerge first in a brood (Williams & Smith, 1991; Moore, unpublished), and it is perhaps surprising that they are not more often the predominant species in an area, given the repeated opportunities to take advantage of predator surprise and the best singing and oviposition sites.

Whenever 17-year and 13-year broods occur in the same year, it is 13 × 17, or 221, years before that happens again; and since Wisconsinan glacial times, a maximum of about 50 co-occurrences could have taken place between any such pairs of broods. Few such broods have the possibility of overlapping geographically as well as chronologically, however (cf. Fig. 6). Alexander & Moore (unpublished) brought large numbers of all three 13-year species (brood XXIII) from south-eastern Missouri into the range of all three 17-year species (brood III) in north central Illinois during the simultaneous emergence of adults of the two broods in 1963. By recording individual matings of marked animals, it was established that freely roaming cognates of 13- and 17-year cicadas would interbreed (*M. cassini* with *M. tredecassini, M. septendecim* with *M. tredecim*), that males and females of these species sometimes mate more than once, that subsequent matings by females may follow oviposition bouts, and that matings are not always preceded by courtship signals. The facts that intermatings will take place freely, that those few populations that could overlap in time and space

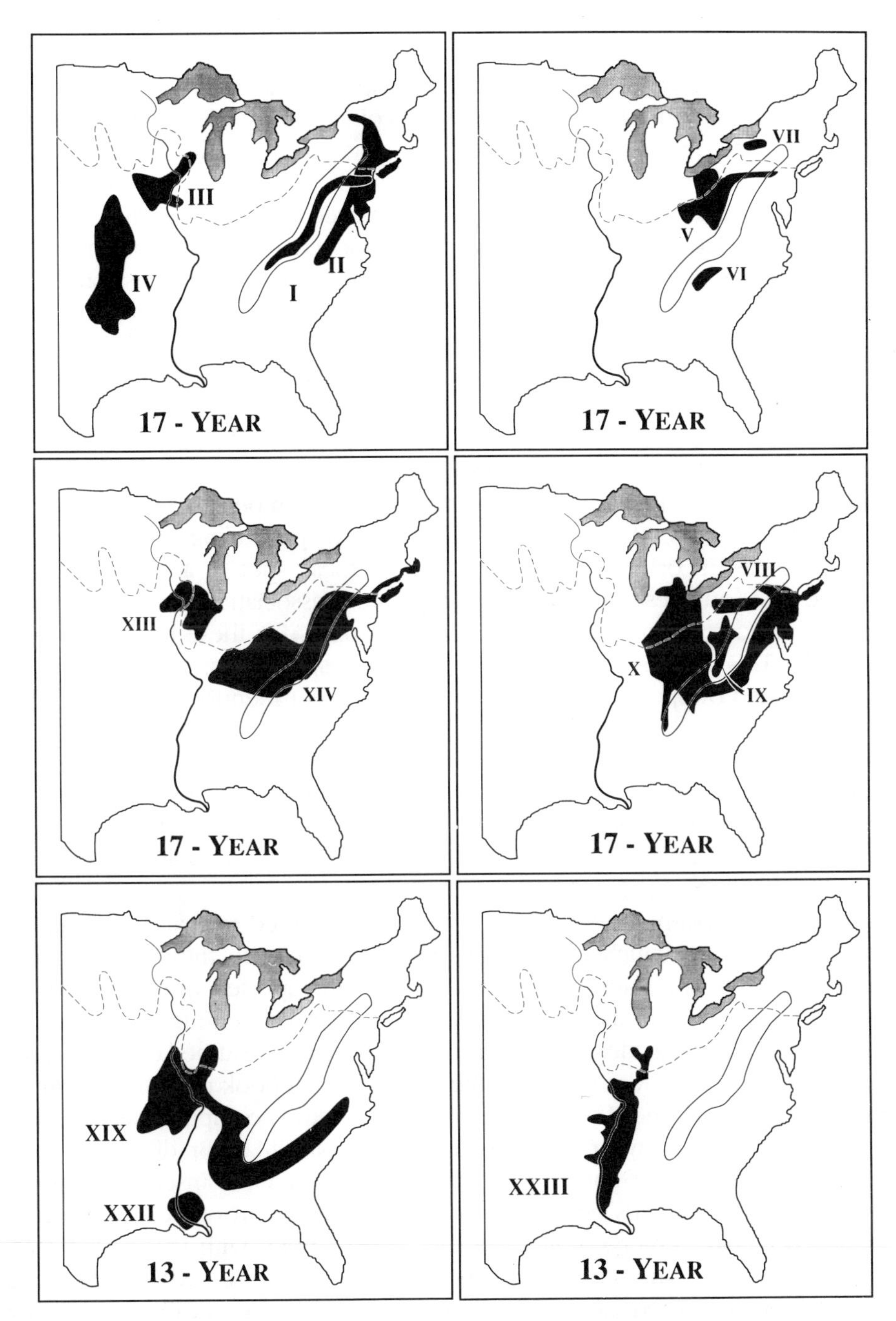

Figure 6 Generalized distributions of extant broods of periodical cicadas (*Magicicada*) in the eastern United States. All ranges include three species, except that of brood VII (*M. septendecim*, only). Upper four maps, 17-year cicadas (*M. septendecim, M. cassini, M. septendecula*); bottom two maps, 13-year cicadas (*M. tredecim, M. tredecassini, M. tredecula*). The Great Lakes, Mississippi River, Appalachian uplift area, and Wisconsinan (Pleistocene) glacial maximum boundary (dashed line) are also shown. (Modified from Alexander & Moore, 1962.)

have in fact moved further apart in nature with subsequent emergences, and that no sustained populations of intermediate emergence periods occur in regions of likely former overlap, all suggest that the natural experiments done on a grand scale whenever such 13- and 17-year broods have coincided as adults attest that viable adult offspring do not issue from such crossings. The handful of cross-matings reported within either 17-year or 13-year cicadas, in the hundreds of thousands of mating pairs observed, have always involved damaged individuals, have never been reported to have led to oviposition, and have not led to records of animals with intermediate songs or morphologies, suggesting insignificant introgression.

It might be expected that broods containing the species with the shorter life-cycle would out-reproduce 17-year cicadas wherever they overlapped, and spread more rapidly as forests expanded into new regions following Pleistocene glaciation, for in every 221-year period they would have four more generations. However, that has not been the clear case. One other expectation would be that if each of the 14 brood populations including three species remains isolated and persists long enough, sufficient differences will accumulate to make their members irreversibly separate from similar members of other broods, potentially producing 14 × 3 species of periodical cicadas. In order to understand where these species are today, and have been in the past (our best basis for predicting future associations), we need to show the locality and date records and the distribution maps for these broods by species (e.g. Fig. 7). This means using museum voucher specimens, tape-recordings, or records published by species and based upon competent identifications. Records and maps dealing with these six species as if they were one lead to a substantial loss of information and contribute to inadequate rigour in considering origins of broods. Comparisons of biochemical characteristics of less than all three species in each of the 14 extensive broods also are likely to contribute to incomplete or inadequate hypotheses (Simon, 1979, 1988; Simon & Lloyd, 1982; Cox & Carlton, 1988; Martin & Simon, 1988, 1990a, b; Cox, 1992).

There has always been concern about accepting presumed records, especially early records, for which no voucher specimens exist. Not only do these records usually lack precise locality and date, but also they are often reported from memory some time after the occurrence, usually by someone not familiar enough with cicadas to be certain even of generic identity. A prime example is the presumed earliest recorded emergence of 17-year cicadas (1633) in the first historical book published in America 36 years later (Morton, 1669:90–91). Morton (not Moreton) compiled his notes of the early 'New-Plimouth' colony largely from the journals of William Bradford and Thomas Winslow, noting that pages sometimes had fallen out, or were inserted in incorrect chronological order. Broods now known to have emerged in or near Massachusetts and expected to have emerged at about that period, none with vouchered records as far north-east as Plymouth (New-Plimouth), are XI (1631), XIV (1634), and II (1639). No other emergences are mentioned in Morton, specifically not for any of these years, nor for the next expected emergence in 1650 (or in 1651 – the date of the generally accepted first published emergence of 17-year cicadas (Anonymous, 1666; brood XIV) – if, as often assumed, the published date was incorrect by a year and the record represented brood XIV in 1634).

It is as likely that this 1663 record applies to *Okanagana rimosa* (Say), often confused for a 17-year cicada. It also is reddish and black, frequently occurs in very large populations at the same time of the year, sings loud songs in dense choruses

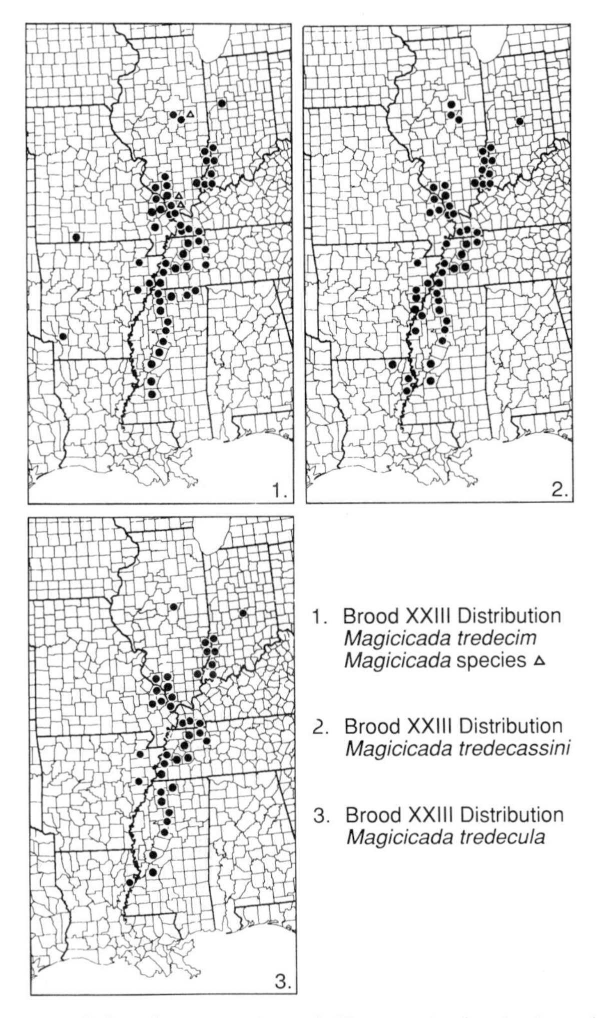

Figure 7 Distribution of the three species of 13-year cicadas in brood XXIII (*Magicicada*); consecutive emergence-years, 1976, 1989, 2002.

all day long, and has a fungus which rots off the tip of the abdomen, also characteristic of *Magicicada* species. *Okanagana rimosa* is known to occur from Long Island, New York, through Massachusetts, to the north-east, north, and westward through Canada and northern states to northern Iowa. We do not know a precise generation time for any *Okanagana* species; *O. rimosa* may average about 8–10 years in duration, and be variable.

Fidelity of *Magicicada* records, and their significance, has long been of concern (Marlatt, 1923; Alexander & Moore, 1962; Maier, 1985). Almost all records of adults emerging a few years before or a few years after an expected brood emergence-year (Fig. 7; cf. Alexander & Moore, 1962:50) are for either *M. septendecim* or *M. tredecim*, are within 4 years before or after a known major emergence and from known localities for existing broods, involve far fewer cicadas than a normal emergence, are eaten by birds or other predators within a week of emergence and before significant reproduction, and have rarely led to a corresponding emergence in more than one subsequent generation. These early or late adults are called stragglers. The suggestion that periodical cicadas could switch between 13- and 17-year cycles (Martin & Simon, 1988, 1990b), proposed for straggling *Decim* siblings, is not supported by robust evidence, and is countered by the vast preponderance of repeated emergences occurring as expected in all species throughout the extensive ranges of each brood. Suggestions that such episodes are likely events accounting for the origin of broods ignore the powerful countering fact that these broods contain three independent species, each responding in its own way to genetic or environmental factors, making it extremely unlikely that the same cluster of species would so extensively coincide in 14 brood populations. For similar reasons, it is not rigorous to discuss the origin of broods by dealing with them as if the broods consisted of a single species; and doing so can mislead others who are not specialists on cicadas and diminish the value of insights they might be able to provide (Cox & Carlton, 1988; Cox, 1992). It is also important to realize that periodical cicadas feed freely on roots and branches, and lay eggs successfully in twigs, of white pine (*Pinus strobus* L.) and of red cedar (*Juniperus virginiana* L.) in addition to a wide range of deciduous forest woody plants (Moore, unpublished).

The preoccupation with the origin of 13-year cicadas, or of broods of either 13- or 17-year cicadas, just from stragglers appearing 4 years early not only ignores the evidence that a single species is usually straggling, and that the population numbers are too small and survive for too short a time to produce another generation, but also the records for stragglers 3, 2 and 1 year(s) early, as well as those 1, 2, 3, and 4 years late. Data based on vouchers indicate that stragglers are nearly always *Decim* cognates for all extant broods, departing from only 1–7 emergence-years per brood (mean, 3), as follows: 17-year cicadas – two broods 6 years early, one 5 years early, five 4 years early, four 3 years early, five 2 years early, eight 1 year early, nine 1 year late, five 2 years late, four 3 years late, five 4 years late; 13-year cicadas – one brood 3 years early, two 2 years early, one 1 year early, one 1 year late, one 2 years late, one 3 years late, two 4 years late. The most common straggling period is 1 year early, or late, next most common is 4 years late, 2 and 3 years early or late are as common as 4 years early (Moore, unpublished).

Careful comparison of the 865 periodical cicada brood records for Indiana given by Kritsky (1987), none by species, shows that 85% of them are for 17-year cicada brood X; of these, 71 are straggler records, 63% 4 years late, 27% 4 years early, and the few others representing 2, 3 and 5 years late as well as 2 years early. He missed the emergence of brood XIX (Posey Co., Indiana, *M. tredecassini*, 1972). Contrary to Kritsky's emphasis on 4-year accelerations of emergence, it would seem that the more common straggling times should be at least as likely to initiate broods as any one of the rarer ones. Far rarer decelerations or accelerations of whole brood populations, or of all three species in major portions of broods, seem as likely as sources of broods or of variability within broods, as ascribing origin to one of the less common periodicities of principally single-species stragglers.

Lloyd, Kritsky & Simon (1983) assume hybridization between 17-year and 13-year cicadas created at least some broods of periodical cicadas, and suggest that both broods XIX and XXIII of 13-year cicadas uniquely occur sympatrically at Weldon Springs State Park, DeWitt County, Illinois, following a 1976 emergence at the park. This county is at the northern boundary of the range of brood XIX, the northernmost brood of all 13-year cicadas, and not far from forests known to be occupied by 17-

year cicadas. They made several predictions based on their analyses of presumed historical correlations. Important among these predictions were that (a) brood XXIII would emerge there again in 1989, and (b) would emerge particularly in the same woods where brood XIX would have emerged in 1985. No cicadas emerged there in 1985, nor in any year since then (including 1989), although in 1985 brood XIX emerged within a few miles, in 1972 in nearby Clinton, and in 1969 (stragglers of the 1972 emergence of brood XIX ?) in the village of De Witt (Moore, unpublished). The 1963 emergence Lloyd *et al.* cite from De Witt County, as well as the 1976 emergence 13 years later that prompted their paper, are at least as likely to be accounted for by stragglers of prior brood XIX populations, perhaps delayed in emergence by a series of unusually cold years such as those cited by Hughes (1979) for 1812–17, as they are to be accounted for by their more complicated scenario. Brood XIX would have been expected to emerge in 1816, the coldest of the summers listed by Hughes, just four emergence periods prior to the 1868 emergence cited by Lloyd *et al.* Emergence of brood XIX is known from the two adjacent counties analysed by Lloyd *et al.* for the years 1868, 1881, 1894, 1907, 1920, 1933, 1946, 1959, 1972 and 1985; while likely straggling 4-years-late populations of brood XIX are known from De Witt County only in 1885, 1898, 1911, 1963 and 1976, and from those same two adjacent counties in 1963 and 1976. By comparison, Alexander & Moore (1962) reported stragglers from four southern Ohio counties in 10 different years across a span of 18 consecutive years in a region in which just three well-studied broods of 17-year cicadas are known. The hybridization hypothesized to have occurred between cognate pairs of 13- and 17-year cicadas by Simon & Lloyd (1982) and Lloyd *et al.* (1983), also seems less likely in view of the interpretation of their own data by Martin & Simon (1990a), in which they suggest finding no evidence of gene flow between 13-year and 17-year cicadas.

Whatever the interactions of periodical cicadas at Weldon Springs, the populations occurring at 13-year intervals from 1885 to 1911, and in 1963 and 1976, have become extinct; and there is no compelling evidence of 17-year cicadas such as brood X (possibly present in 1868) or of other 13-year cicadas (such as brood XXIII), as presumed by Lloyd *et al.* (1983), ever having occurred there (Moore, unpublished). One must emphasize the unlikelihood of periodical cicada broods, including three independent species of the six periodical cicadas, changing simultaneously into another brood through climatic influence, or changing chronology simultaneously by all six species successfully hybridizing in large numbers. The suggestion of single-year weather fluctuations or hybridizations involving one cognate pair, followed by a single 4-year life-cycle change persisting through several slightly varying emergence years, and overlooking the fact that the changes must be simultaneous in three different species (e.g. Martin & Simon, 1990b), runs counter to the abundant evidence of multiple years of climatic variation as well as multiple year-change events in several broods, often in the same general locality. Hopefully, comparative studies involving all three species in each of the 14 large broods of periodical cicadas will eventually provide more robust and consistent evolutionary hypotheses for broods.

Nymphal growth rates have seldom been measured by species; we can only identify genera for nymphs. However, in the Ann Arbor, Michigan, area there is only one species of one brood: *M. septendecim*, brood X. In mid-June, 2 years after the 1970 emergence, all nymphs sampled from a mature second-growth forest were still in the first instar. By mid-summer, 4 years after emergence, all were in the second instar. In May, 8 years after emergence, 54% were in the third instar and 46% in the fourth, but by October of the same year only 21% remained in the third instar, while 79% were in the fourth. At 9 years, 9% were in fifth instar, and at 10 years from last emergence, 77% were in fourth and 23% were in fifth instar. By 13 years from last emergence, all were in fifth (final) instar, although none appeared mature enough to emerge. Other accounts of growth rates of nymphs of mixed, unknown species for both 17- and 13-year cicadas suggest, despite considerable variability in growth rates, that by the 8th or 9th year a significant percentage of nymphs have reached the fifth and final instar (Marlatt, 1923; White & Lloyd, 1975), which is not surprising, given the known pattern of straggling.

ORIGIN OF NORTH AMERICAN CICADAS

Quesada is a member of the subgroup of cicadas in which the tymbals are partially hidden from view by a dorsal abdominal flap, and in which the abdominal sternites are nearly transparent (see-through cicadas). *Tibicen* belongs to the subgroup of cicadas whose tymbals are completely concealed from view by a special dorsal abdominal flap (concealed-tymbal cicadas). *Magicicada* belongs to another subgroup in which the tymbals are completely exposed to view dorsally (exposed-tymbal cicadas). The exposed-tymbal condition is typical of the structure of most tymbals throughout the auchenorrhynchous Homoptera. One other subgroup, the wing-tapping cicadas, distinctive of western North America, has lost the tymbals while retaining an abdominal tracheal air chamber and ventral tympana typical of cicadas. Both males and females in this subgroup 'sing' by tapping their forewings to vibrate the wing surfaces.

All of these subgroups are known from the Tertiary of North America. *Lithocicada perita* Cockerell, a see-through cicada, is known from Miocene shales of Colorado; *Davispia bearcreekensis* Cooper, Palaeocene, Montana, and *Tibicen grandiosa* (Scudder), Miocene, Colorado, are both concealed-tymbal cicadas, similar to *Tibicen cultriformis* (Davis) or *Tibicen marginalis* (Walker); *Platypedia primigenia* Cockerell, Miocene, Colorado, is a wing-tapping cicada (Cooper, 1941). An undescribed rather complete fossil of *Okanagana* (Moore, unpublished), an exposed-tymbal cicada, is known from the Oligocene shales of the Ruby mountains of Montana (Becker, 1965). Thus, all of the major cicada subgroups now present in North America have been there for 25–60 million years. It seems likely that the six species of periodical cicadas (*Magicicada*) evolved and formed three-species broods sometime before the last Pleistocene glaciation. Seven 17-year broods, and two 13-year broods, have invaded, or re-invaded, these last glaciated areas (Fig. 6).

CONCLUSIONS

Cicadas are sometimes quite invariant morphologically and acoustically, despite great geographic ranges through many highly variable environments. Others vary morphologically, while showing almost no acoustic variation. The complex cluster of periodical cicadas (*Magicicada*) have evolved surprising developmental, biochemical and distributional variability, despite little morphological or acoustic variation among the cognate 13-year/17-year species pairs.

REFERENCES

ANONYMOUS, 1666. Some observations of swarms of strange insects and the mischiefs done by them. *Philosophical Transactions of London, 1(8):* 137.

ALEXANDER, R.D. & MOORE, T.E., 1962. The evolutionary relationships of 17-year and 13-year cicadas, and three new species. *Miscellaneous Publications of the University of Michigan Museum of Zoology, 121:* 1–59.

BECKER, H.F., 1965. Flowers, insects, and evolution. *Natural History, 74(2):* 38–45.

BRITTON, N.L., 1965. *Flora of Bermuda (illustrated).* New York: Hafner (reprinted from 1918).

COX, R.T., 1992. A comment on Pleistocene population bottlenecks in periodical cicadas (Homoptera: Cicadidae: *Magicicada* spp.). *Evolution, 46(3):* 845–846.

COX, R.T. & CARLTON, C.E., 1988. Paleoclimatic influences in the evolution of periodical cicadas (Insecta: Homoptera: Cicadidae: *Magicicada* spp.). *American Midland Naturalist, 120(1):* 183–193.

CLARIDGE, M.F., WILSON, M.R. & SINGHRAO, J.S., 1979. The songs and calling sites of two European cicadas. *Ecological Entomology, 4:* 225–229.

COOPER, K.W., 1941. *Davispia bearcreekensis* Cooper, a new cicada from the Paleocene, with a brief review of the fossil Cicadidae. *American Journal of Science, 239:* 286–304.

HAMILTON, W.D., 1971. Geometry for the selfish herd. *Journal of Theoretical Biology, 31:* 295–311.

HARMON, S.H., MITTERER, R.M., KRIAUSAKUL, N., LAND, L.S., SCHWARCZ, H.P., GARRETT, P., LARSON, G.J., VACHER, H.L. & ROWE, M., 1983. U-series and amino-acid racemization geochronology of Bermuda: implications for eustatic sea-level fluctuation over the past 250,000 years. *Palaeogeography, Palaeoclimatology, Palaeoecology, 44:* 41–70.

HUBER, F., KLEINDIENST, H.-U., SCHILDBERGER, K., WEBER, T. & MOORE, T.E., 1990. Acoustic communication in periodical cicadas: neuronal responses to songs of sympatric species. In F. G. Gribakin, K. Wiese & A. Popov (Eds), *Sensory Systems and Communication in Arthropods, Advances in Life Sciences:* 217–228. Basel: Birkhäuser.

HUGHES, P., 1979. 1816. The year without a summer. *Weatherwise 32(3):* 108–111.

KRITSKY, G., 1987. An historical analysis of periodical cicadas in Indiana (Homoptera:Cicadidae). *Proceedings of the Indiana Academy of Science, 97(1987):* 295–321.

LLOYD, M., KRITSKY, G. & SIMON, C., 1983. A simple Mendelian model for 13- and 17-year life cycles of periodical cicadas, with historical evidence of hybridization between them. *Evolution, 37(6):* 1162–1180.

MAIER, C.T., 1985. Brood VI of 17-year periodical cicadas, *Magicicada* spp. (Hemiptera: Cicadidae): new evidence from Connecticut, the hypothetical 4-year deceleration, and the status of the brood. *Journal of the New York Entomological Society, 93(2):* 1019–1026.

MARLATT, C.L., 1923. The periodical cicada. *U. S. Department of Agriculture, Bureau of Entomology Bulletin, 71:* 1–183.

MARTIN, A.P. & SIMON, C., 1988. Anomalous distribution of nuclear and mitochondrial DNA markers in periodical cicadas. *Nature (London), 336:* 237–239.

MARTIN, A. & SIMON, C., 1990a. Differing levels of among-population divergence in the mitochondrial DNA of periodical cicadas related to historical biogeography. *Evolution, 44(4):* 1066–1080.

MARTIN, A. & SIMON, C., 1990b. Temporal variation in insect life cycles. Lessons from periodical cicadas. *BioScience, 40(5):* 359–367.

MORTON, N., 1669. *New-Englands Memoriall: or, a Brief Relation of the Most Memorable and Remarkable Passages of the Providence of God, Manifested to the Planters of New-England in America; with Special Reference to the First Colony thereof, called New-Plimouth.* Cambridge, Massachusetts: Printed by S.G. and M.J. for John Usher of Boston.

SIMON, C., 1979. Evolution of periodical cicadas: phylogenetic inferences based on allozymic data. *Systematic Zoology, 28:* 22–39.

SIMON, C., 1988. Evolution of 13- and 17-year periodical cicadas (Homoptera: Cicadidae: *Magicicada*). *Bulletin of the Entomological Society of America, 34(4):* 163–176.

SIMON, C. & LLOYD, M., 1982. Disjunct synchronic populations of 17-year periodical cicadas: relicts or evidence of polyphyly? *Journal of the New York Entomological Society, 90(4):* 275–301.

VERRILL, A.E., 1903. The Bermuda Islands – their scenery, climate, productions, physiography, natural history and geology. *Transactions of the Connecticut Academy of Arts and Sciences, 11(2):* 413–957.

WHITE, J. & LLOYD, M., 1975. Growth rates of 17- and 13-year periodical cicadas. *American Midland Naturalist, 94:* 127–143.

WILLIAMS, K.S. & SMITH, K.G., 1991. Dynamics of periodical cicada chorus centers (Homoptera: Cicadidae: *Magicicada*). *Journal of Insect Behavior, 4(3):* 275–291.

CHAPTER

17

Speciation in insect herbivores – the role of acoustic signals in leafhoppers and planthoppers*

MICHAEL F. CLARIDGE

Introduction 286

Leafhoppers and planthoppers 287

***Oncopsis* leafhoppers** 287

***Nilaparvata* planthoppers** 288

Acoustic signals of rice-feeding *N. lugens* 290

Acoustic signals of *Leersia*-feeding *N. lugens* 293

Conclusions 294

Discussion 295

Acknowledgements 296

References 296

Keywords: Acoustic signals – biological species – host race – insect herbivore – *Nilaparvata lugens* – *Oncopsis* – reproductive isolation – speciation – specific mate recognition system.

Abstract

Leafhoppers and planthoppers are two species-rich groups of insect herbivores, related to the cicadas. Mate location and courtship in these insects are usually achieved by an exchange of acoustic signals between sexually receptive males and females when on the same or closely neighbouring plants. Few species have been studied in detail, but in those that have, patterns of amplitude modulation and rates of repetition of units of the calls have been shown to provide specificity.

Groups of closely related sympatric species usually differ most obviously in the patterns of their host-plant preferences. As in other groups of specialist feeders, this has led to suggestions that processes of sympatric speciation may have been the

*Expanded version of Presidential Address given to Linnean Society of London at Burlington House, 24 May 1991.

Evolutionary Patterns and Processes
ISBN 0-12-440895-8

dominant evolutionary mechanisms responsible for their diversification. However, where studied, such species of leafhoppers and planthoppers also differ clearly in patterns of acoustic signals. There is evidence of allopatric differentiation of such calls, suggesting the possibility of allopatric speciation. Detailed studies on *Oncopsis* leafhoppers and *Nilaparvata* planthoppers that show both host-plant and acoustic signal differentiation are discussed.

INTRODUCTION

Biological diversity on land is dominated by green plants and insects. More than 50% of all living species of macro-organisms are insects and at least another 20% are green plants (Southwood, 1978). Of the insects about half are estimated to be herbivores.

The enormous numbers of insect herbivores are mostly included in relatively few species-rich taxa. Often the most obvious features separating groups of allied species of these insects are specialized patterns of host-plant utilization. Indeed taxonomy is often very difficult and usually has to rely on small and not always very reliable morphological differences. Many biological species among such insects are now known that show little or no morphological differentiation – so-called sibling or cryptic species (Claridge, 1988).

It is hardly surprising that the evolutionary differentiation and speciation of the enormous numbers of extant insect herbivores should often have been thought to be primarily a product of host-plant shifts and adaptation to new hosts. Frequently large numbers of allied species of such insect taxa are found living together and clearly differentiated by patterns of host-plant specialization. Among modern workers, Guy Bush (*e.g.* Bush, 1975a,b; Bush & Diehl, 1982; Diehl & Bush, 1984, 1989; and Chapter 14) has been the most consistent advocate of speciation by a process of host shifts, host race formation and the emergence of new species, without any phase of isolation in space (allopatry). Arguments have raged over the years since the time of Charles Darwin and Alfred Russell Wallace as to the relative importance of allopatric and sympatric processes of speciation. In recent years some authors have suggested that speciation of parasites and all such specialist feeders is most usually by sympatric processes, while free-living organisms usually speciate by allopatric means (e.g. Bush, 1975a,b; Price, 1980; Strong, Lawton & Southwood, 1984). The evidence for such generalization is very sparse and unfortunately studies on insect herbivores and indeed other parasites have almost all concentrated on host differentiation and associated behaviour.

Adequate attention has rarely been given to mate recognition systems in speciation studies. In recent years Hugh Paterson has consistently advocated the central importance of distinct specific mate recognition systems (SMRS) in determining biological species boundaries (Paterson, 1985; Chapter 13). Thus, speciation is essentially the evolution of different SMRSs and these result in reproductive isolation between species. When such mate recognition systems are considered in the context of insect herbivore speciation, it is usually assumed that they are entirely associated with host-plant recognition.

In this chapter I shall consider two relatively well-worked examples from one group of insect herbivores, the leafhoppers and planthoppers, where studies on specific mate recognition signals have been made.

LEAFHOPPERS AND PLANTHOPPERS

Leafhoppers and planthoppers represent two species-rich and widely distributed families, the Cicadellidae and Delphacidae respectively, of the Homoptera Auchenorrhyncha. All of the Auchenorrhyncha, so far as known, are herbivores and feed by penetrating living host plants using specialized mouthparts and removing the fluid contents of plant cells. They may specialize, not only on particular host plants, but also on particular tissues of those plants. Thus, some leafhoppers and all known planthoppers feed on the nutrient-rich phloem sap, others feed exclusively on the extremely dilute xylem sap, and yet others on mesophyll and parenchyma cell contents.

Related to the leafhoppers and planthoppers are the cicadas (Cicadidae), well-known for their use of loud sound signals in communication and species recognition (Claridge, 1985a; Moore, Chapter 16). Until the middle of this century all other families of Auchenorrhyncha were thought not to use sounds in communication. Indeed early classifications divided these insects into the singing cicadas and the mute leafhoppers, planthoppers and other relatives (Claridge, 1985b). Ossiannilsson (1949) was the first to show that all families of Auchenorrhyncha possessed, at least in males, what appeared to be the analogue of the tymbal sound-producing mechanism, well-known in the Cicadidae. Also he showed, using quite primitive techniques, that many species do indeed produce very low intensity acoustic signals.

Following the development of modern electronic methods of recording, we now know that adult leafhoppers and planthoppers do produce low-amplitude vibratory signals. They are transmitted via the substrate on which the insects rest or walk, which is normally their host plant (reviews in Claridge, 1985a,b). In some examples closely related species have been shown to differ strikingly in patterns of amplitude modulation of the calls used in mate location and courtship. Thus these insects provide a group in which host-associated behaviour and specific mate recognition can both be investigated so that new sets of evidence may be considered when discussing the problems of speciation.

Here I shall consider two detailed examples: (1) leafhoppers of the genus *Oncopsis*, and (2) planthoppers of the genus *Nilaparvata*.

ONCOPSIS LEAFHOPPERS

Species of *Oncopsis* are all associated with broad-leaved trees in northern Holarctic forest and montane regions. In Britain, six species are generally recognized, but species taxonomy is very difficult and relies on small and often variable characters of the male genitalia (LeQuesne, 1965). The problem is further complicated because most species are polymorphic for adult and nymphal colour patterns (Claridge & Nixon, 1981). Host specificity is usually extreme, with each species restricted to one tree species or to a few closely related ones (Claridge & Reynolds, 1972; Claridge, Reynolds & Wilson, 1977). Life histories are completed in intimate association with the host tree. Eggs are laid precisely into particular plant tissues, characteristic for each species, where they overwinter (Claridge & Reynolds, 1972). Nymphs and adults feed from phloem tissue in the new growth during spring and early summer.

O. flavicollis (L.) is a common species associated with the two abundant native tree birches, *Betula pendula* Roth and *B. pubescens* Ehrh. in Britain. Claridge *et al.* (1977) showed that females prefer to oviposit in young shoots of the species of *Betula* from which they were collected. Also nymphs transferred from one species to the other tended to prefer their original host (Claridge & Reynolds, 1972). Populations of *O. flavicollis* from the two different birches also showed an apparent chromosome polymorphism (John & Claridge, 1974). In addition, the proportions of the different adult female colour morphs were significantly different from the two species of birches even in the same areas (Claridge & Nixon, 1981). Thus, the populations appeared to be good candidates for host races in the process of sympatric divergence (Diehl & Bush, 1984).

Acoustic signals are known to play an important part in courtship and mating in *Oncopsis* (Claridge & Howse, 1968). Closely related species differ strikingly in male calls. Claridge & Reynolds (1973) were able to confirm the biological species status of *O. subangulata* (Sahlberg), as separate from *O. flavicollis,* by analysis of male calls. Thus, Claridge & Nixon (1986) made a detailed study of populations of *O.flavicollis* from *B. pendula* and *B. pubescens* in southern Britain.

Multivariate analyses of morphometric data showed significant separations between populations from the two birch species in one area of South Wales, but no significant differences between populations from the same tree in different areas. In the males the most important characters responsible for the separation were measurements of internal apodemes associated with the sound-producing mechanism.

Acoustic studies (Claridge & Nixon, 1986) showed that males produced two distinct patterns of amplitude-modulated calls: (1) the calling song produced spontaneously and often in the absence of females, and (2) the courtship song only produced when either in close proximity or in physical contact with a female. Both courtship and calling songs were completely different in patterns of modulation in individuals from sympatric populations from the two tree species.

We thus concluded that the apparent host races of *O. flavicollis* in South Wales were in reality two very closely related sibling species. However, sampling further east in England showed an even more complicated situation with a further, third, sibling species associated with *B.pendula* and similarly identified by distinctive male calls and small differences in internal apodemes. In eastern areas, where all three species were found together, the specific *B. pendula*-associated form from western regions was also found on *B. pubescens* (Fig. 1)

It is clear now that where previously one biological species was suspected of showing incipient host race formation, in reality there are at least three different species which show geographical variation in host-plant exploitation. Thus, the assumption that sympatric host race formation will necessarily precede speciation cannot be made where a group of sibling species shows geographical variation in host-plant utilization.

NILAPARVATA PLANTHOPPERS

Nilaparvata is a small genus of largely tropical and subtropical planthoppers. *N. lugens* (Stål) is the well-known brown planthopper, a major pest of rice in Asia (Fig. 2) (Claridge & Morgan, 1987). It is widely distributed from Pakistan and India in the west to Fiji and the Solomon Islands in the east, and from Japan, Korea and northern

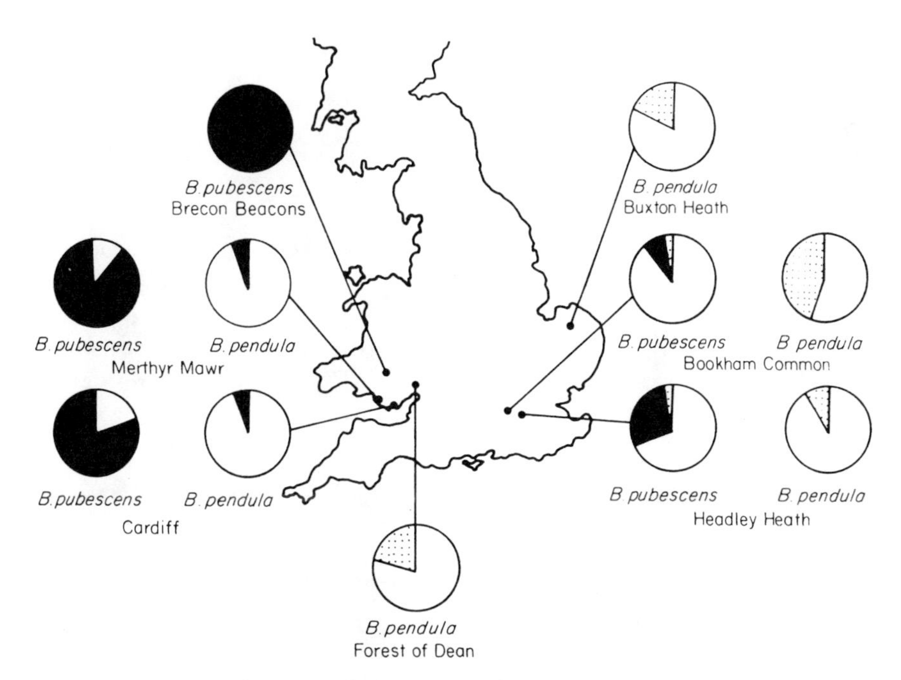

Figure 1 Sketch map to show sampling sites for the *Oncopsis flavicollis* complex from *Betula pendula* and *B. pubescens* in southern Britain. Pie diagrams show proportions in male samples of species 1 (solid), 2 (open) and 3 (stippled). (After Claridge & Nixon, 1986).

China in the north to Northern Australia in the south (Fig. 3) Across this large area of distribution the species shows little obvious variation in morphology. Its closest morphologically recognizable relative is *N. bakeri* (Muir), from which it differs in the structure of the male and female genitalia and with which it is sympatric over much of its range.

N. lugens is thought to feed only on rice, wild species of *Oryza* and cultivars of *O. sativa*. Nymphs and adults feed from the phloem and lose weight and die if transferred to other related grasses (Claridge, den Hollander & Morgan, 1985b). Some cultivars and wild species of rice show different levels of susceptibility to feeding by *N. lugens*. Also populations of the insect from different regions show different patterns of virulence to different varieties (see Claridge & den Hollander, 1982; Claridge, den Hollander & Furet, 1982; Claridge, 1990).

Over the past 10 years or so populations morphologically identical to *N. lugens* have been discovered widely in Asia and in Australia feeding on the semiaquatic grass, *Leersia hexandra* (Schwartz) (Claridge *et al.*, 1985b; Claridge, 1988, 1990). These insects do not survive if transferred from *L. hexandra* to rice. They frequently occur together in the field with the rice-feeding populations when their host plants grow in close proximity. Hybridization experiments between populations from rice and *Leersia* from the Philippines (Heinrichs & Medrano, 1984; Claridge *et al.*, 1985b, 1988) showed that F_1 and F_2 hybrids were easily obtained with little indication of hybrid inviability. It might be suggested, therefore, that the host-associated populations represent host races adapted to their specific food plants.

Hybridization experiments between geographically separated (allopatric) populations of rice-associated *N. lugens* showed varying degrees of success (Claridge, den

Figure 2 Macropterous adult male (right) and brachypterous female (left) of *N.lugens* on rice plants. Scale line approximately 1 mm.

Hollander & Morgan, 1985a). Replicate crosses were set up between individual virgin males and females from different geographically defined populations. Levels of success, measured by the presence of living sperm in the female spermathecae, were low in some crosses. For those crosses, the few pairs which mated successfully were allowed to lay eggs. These subsequently hatched and produced viable F_1 adults. So, despite definite barriers to hybridization, genetic differentiation between even the most distinct populations was not great. This suggested strongly that some premating barriers to interbreeding had evolved between at least some of the allopatric populations.

Acoustic signals of rice-feeding *N. lugens*

Males and females of *N. lugens* from rice are known to exchange acoustic signals during courtship and pair formation (Ichikawa & Ishii, 1974; Claridge *et al.*, 1985a). Unmated adult males and females when ready to mate may commence spontaneous calling. The female produces a regular drumming call while remaining motionless on the plant (Fig. 4A). Males produce a more complicated call, consisting of an irregular introductory phase, followed by a regular burst of rapidly reproduced pulses, and terminating in a series of more irregular pulses as a closing phase (Fig. 4B). Males call as they move on the plant and in response to a calling female.

Figure 3 Sketch map of Asia and part of Australia to show localities from which populations of the *N. lugens* complex have been sampled for study. The rice-associated species is indicated by solid circles and the *Leersia*-associated species by open circles. (After Claridge, 1990).

Likewise, females will call in response to a calling male within acoustic range. Thus courtship and mate location in *N. lugens* consist primarily of an exchange of acoustic signals (Fig. 4C). Pre-recorded calls may be played back into a plant to test the responses of individual insects to particular signals (Claridge *et al.*, 1985a).

Comparisons of male calls from allopatric populations of rice-feeding *N. lugens* showed differences in the pulse-repetition frequencies (PRF) in the main phase of the calls (Claridge *et al.*, 1985a). The most distinctively extreme populations in this respect were those from Australia and the Solomon Islands. In specific individual crosses between populations from those areas, very low rates of success in insemination were obtained (Claridge, den Hollander & Morgan, 1984). For those crosses where insemination occurred, the successful males were subsequently recorded. In all cases these individual males tended to be extreme variants for PRF from their own populations and nearest in PRF to that of the males of the female population with which they had mated. This strongly suggested that PRF of male calls is indeed an essential element of the specific mate recognition system of *N. lugens*.

Thus, within the geographical range of rice-feeding *N. lugens*, some allopatric

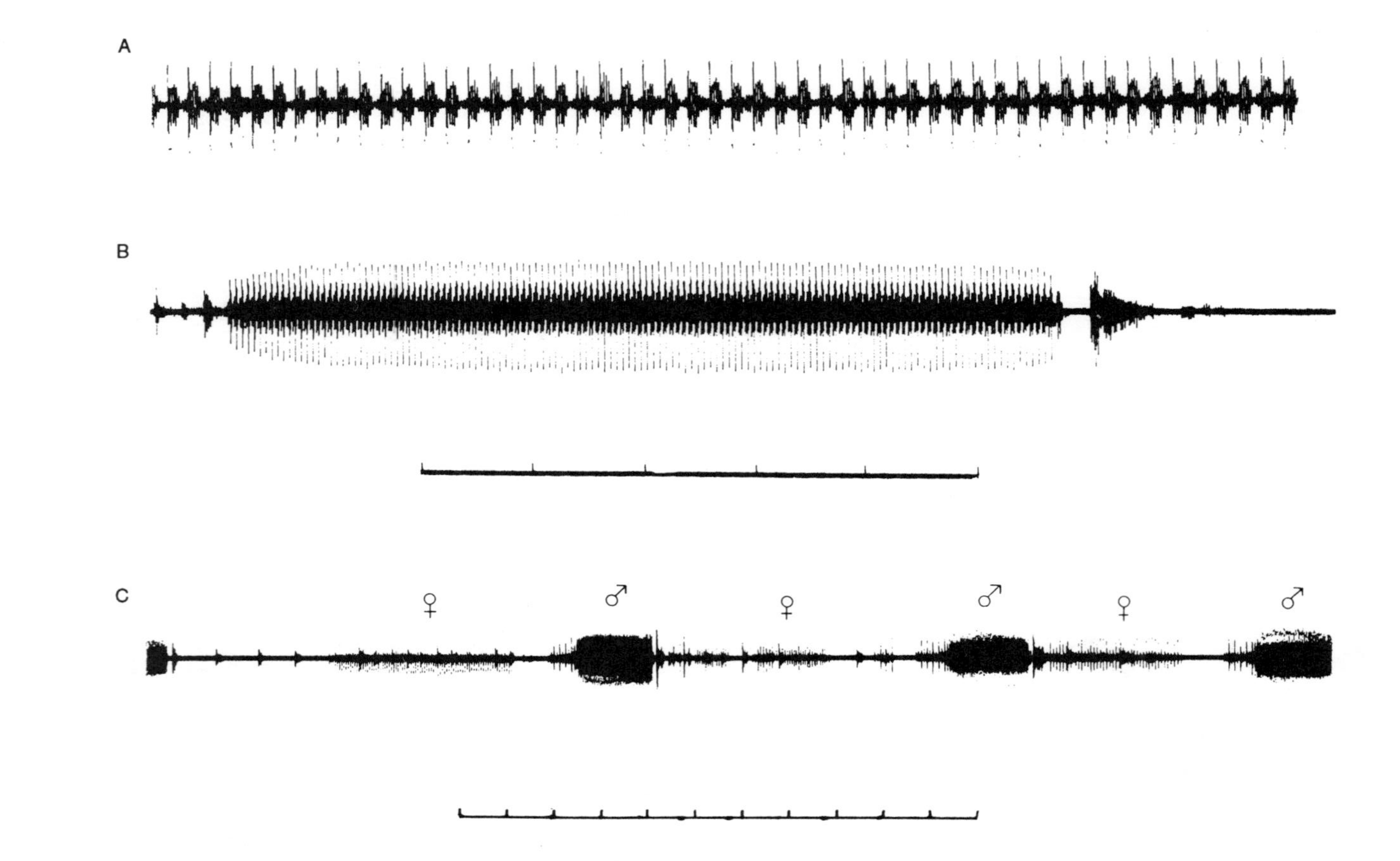

Figure 4 Oscillograms of parts of calls of adult *N. lugens* from rice. A: Female; B: male; C: male and female interaction. Time marks at 0.25 s intervals; A and B to same time scale. (After Claridge *et al.*, 1985a).

populations have diverged to such an extent in acoustic mate recognition systems that subspeciation or possibly speciation may already have occurred. Since the populations remain allopatric no absolute conclusions can be made.

Acoustic signals of *Leersia*-feeding *N. lugens*

Claridge *et al.*, (1985b, 1988) studied the acoustic signals of *Leersia*-associated populations of *N. lugens* from Asia and Australia and of hybrids with rice-associated populations. Males and females exchange qualitatively similar signals during courtship to insects from rice. However, both male and female calls of sympatric populations from the two hosts show clear and statistically significant differences in pulse-repetition frequencies (PRF) (Fig. 5). Laboratory-produced hybrids between the two population types produced calls characteristically intermediate in PRF (Fig. 5). Recordings from field samples taken from areas in Asia and Australia where the two host-associated populations occur in close proximity, showed no evidence of

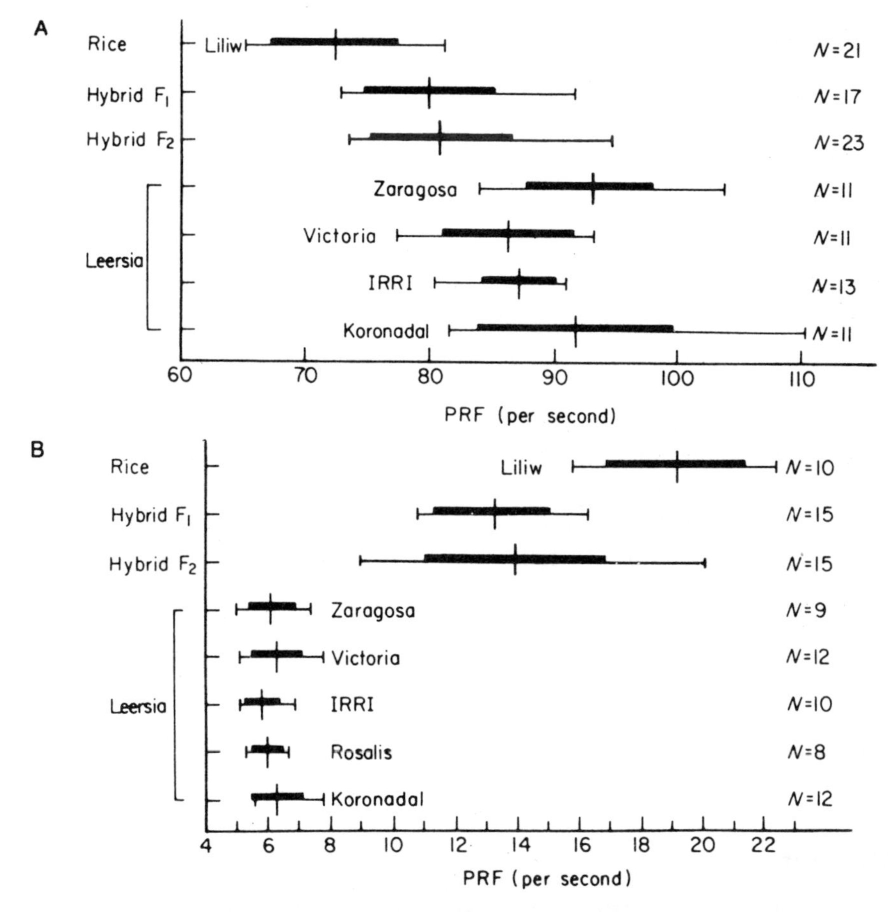

Figure 5 Pulse-repetition frequencies (PRFs) of male and female calls from sympatric rice-associated and *Leersia*-associated species of the *N. lugens* complex from the Philippines, and of F_1 and F_2 laboratory hybrids between them. A: Males; B: females. Vertical line represents mean for each population, thick bar 1 standard deviation on either side of the mean, thin bar total range. (After Claridge *et al.*, 1985b).

Table 1 Percentages of males and females of sympatric rice- and *Leersia*-associated species of the *N. lugens* complex responding to playback of pre-recorded calls of female and male calls respectively. Actual numbers of insects responding and total numbers given in parentheses.

	Percentage responding to	
	Rice male call	*Leersia* male call
Female response		
Rice females	93 (14/15)	13 (2/15)
Leersia females	27 (4/15)	87 (13/15)
	Percentage responding to	
	Rice female call	*Leersia* female call
Male response		
Rice males	77 (23/30)	30 (6/20)
Leersia males	17 (5/30)	90 (18/20)

After Claridge *et al.* (1985b).

intermediate calls and therefore of natural field hybridization. The two populations are reproductively isolated and retain both their host associations and their characteristic calls in nature. They are thus different biological species (Claridge *et al.*, 1985b, 1988).

Playback experiments were made in which both males and females from sympatric populations of each of the species were tested for their responses to prerecorded calls of the opposite sex of their own and the other species. These showed a very significant preference by both sexes for calls of their own population (Table 1). However, these results minimized the extent of the preferences. Insects responded much sooner and more vigorously and called for longer periods in response to calls from their own population than to those from the other. In addition, males began actively to search in response to calls from their own females, something they rarely did to calls from the other females.

Like populations of *N. lugens* from rice, those from *Leersia* also show geographical variation in PRF of both male and female calls (Claridge *et al.*, 1988). Similarly, the populations from Australia show the greatest differentiation from those from Asia. The patterns of geographical variation in both species are complex and not obviously correlated (Claridge, 1990).

Conclusions

Thus, as in the *Oncopsis* leafhoppers, what at first appeared to be sympatric host races of *N. lugens* associated with different host plants, are clearly different sibling species. They differ in acoustic signal patterns which have been demonstrated to function as specific mate recognition signals. Critical differences in PRF are sufficient to maintain reproductive isolation over large areas of Asia and Australia.

DISCUSSION

It is highly improbable that the enormous numbers of extant species of organisms with very diverse genetic systems should have evolved and diversified by the same processes. Many different mechanisms of speciation have been proposed and there is still certainly no unanimity of view (e.g. reviews in Otte & Endler, 1989). One of the problems is that "Although clearly a major focus of evolutionary biology, the study of speciation has never emerged as a coherent discipline" (Harrison, 1991). All authors are obviously influenced by their own particular interests and backgrounds and no general synthesis has been agreed.

The generalization already widely accepted and referred to above, that speciation of parasites and specialist feeders, such as insect herbivores, is likely to have occurred primarily by sympatric processes, suffers from just such problems. Most workers have been mainly interested in the insect/host-plant relationship and therefore regard it as inevitably central to the speciation process. Most such insects are relatively small (see Chapter 14) and few have been subjected to studies on courtship and mating behaviour. A superb example of the analysis of detailed interactions between a group of host-specialist insect herbivores in the context of speciation is provided by the work of Tom Wood and his colleagues (e.g. Wood, 1987, and Chapter 18) on the *Enchenopa binotata* treehopper complex in North America. This is widely regarded as one of the best-documented examples of sympatric speciation among insect herbivores (e.g. Strong *et al.*, 1984). However, no studies have been made on the specific mate recognition signals of these insects. Treehoppers are related to leafhoppers and planthoppers and they almost certainly use acoustic signals in similar ways to tree-associated leafhoppers, such as *Oncopsis*. An appropriate investigation might provide a whole new set of data to use in the discussion of *Enchenopa* speciation.

The importance of the study of mate recognition signals in species and speciation is obvious. Most insect herbivores probably communicate in mate location and recognition by the use of chemical signals. These are difficult to study in the field and few attempts have been made to do so in the context of speciation. The acoustic signal systems used predominantly by leafhoppers and planthoppers in species recognition have the advantage of being relatively simple to investigate. In the two examples discussed in detail above it is clear that acoustic signals do characterize different biological species even when genetic differentiation otherwise is slight. Even more important, these very signals are the ones that seem to be particularly prone to diverge in allopatry. Not only the two examples discussed here, but the few other relevant studies on these insects suggest similar allopatric divergence, whether or not host-plant divergence also occurs. These types of study suggest that the above generalization about parasite speciation is certainly premature. In these insects allopatric speciation by divergence of mate recognition signals seems as likely to account for the observed patterns of diversity as divergence by sympatric host race formation.

The above examples suggest that at least in leafhoppers and planthoppers premating isolation evolves more rapidly than postmating isolation. A wide-ranging review of speciation in *Drosophila* showed that this may be a more general phenomenon (Coyne & Orr, 1989). Various possible mechanisms that may drive the evolution of such premating isolation have been widely discussed. Perhaps the most generally favoured idea is the evolution of signals by divergent sexual selection in allopatric populations (e.g. West-Eberhard, 1983, 1984). This would at least be consistent with observed patterns of variation in leafhoppers and planthoppers. Whatever the processes, it is clear that no simple generalization will account for speciation in the most speciose groups of insect herbivores. Exciting problems still remain.

ACKNOWLEDGEMENTS

I am deeply indebted to my many colleagues and students for their assistance and interest over the years in species and speciation problems. Our work on leafhoppers and planthoppers has been generously supported by N.E.R.C., O.D.A. and E.C.

REFERENCES

BUSH, G.L., 1975a. Modes of animal speciation. *Annual Review of Ecology and Systematics, 6:* 339–364.

BUSH, G.L., 1975b. Sympatric speciation in phytophagous parasitic insects. In P.W. Price (Ed.), *Evolutionary Strategies of Parasitic Insects and Mites:* 187–206. New York: Plenum Press.

BUSH, G.L. & DIEHL, S.R., 1982. Host shifts, genetic models of sympatric speciation and the origin of parasitic insect species. In J.H. Visser & A.K. Minks (Eds), *Proceedings of the 5th International Symposium on Insect–Plant Relationships.* Wageningen.

CLARIDGE, M.F., 1985a. Acoustic signals in the Homoptera: behaviour taxonomy and evolution. *Annual Review of Entomology, 30:* 297–317.

CLARIDGE, M.F., 1985b. Acoustic behaviour of leafhoppers and planthoppers: species problems and speciation. In L.R. Nault & J.G. Rodriguez, (Eds), *The Leafhoppers and Planthoppers:* 103–125. New York: John Wiley.

CLARIDGE, M.F., 1988. Species concepts and speciation in parasites. In D.L. Hawksworth (Ed.), *Prospects in Systematics:* 92–111. Oxford: Oxford University Press.

CLARIDGE, M.F., 1990. Acoustic recognition signals: barriers to hybridization in Homoptera Auchenorrhyncha. *Canadian Journal of Zoology, 68:* 1741–1746.

CLARIDGE, M.F. & DEN HOLLANDER, J., 1982. Virulence to rice cultivars and selection for virulence in populations of the brown planthopper *Nilaparvata lugens. Entomologia experimentalis et applicata, 32:* 213–221.

CLARIDGE, M.F. & HOWSE, P.E., 1968. Songs of some British *Oncopsis* species (Hemiptera: Cicadellidae). *Proceedings of the Royal Entomological Society of London, A, 43:* 51–61.

CLARIDGE, M.F. & MORGAN, J.C., 1987. The Brown Planthopper *Nilaparvata lugens* (Stål), and some related species: a biotaxonomic approach. In M.R. Wilson & L.R. Nault (Eds), *Proceedings of the 2nd International Workshop on Leafhoppers and Planthoppers of Economic Importance:* 19–32. London: CABI Institute of Entomology.

CLARIDGE, M.F. & NIXON, G.A., 1981. *Oncopsis* leafhoppers on British trees: polymorphism in adult *O. flavicollis* (L). *Acta Entomologia Fennica, 38:* 15–19.

CLARIDGE, M.F. & NIXON, G.A., 1986. *Oncopsis flavicollis* (L): associated with tree birches (*Betula*): a complex of biological species or a host plant utilization polymorphism? *Biological Journal of the Linnean Society, 27:* 381–397.

CLARIDGE, M.F. & REYNOLDS, W.J., 1972. Hostplant specificity, oviposition behaviour and egg parasitism in some woodland leafhoppers of the genus *Oncopsis* (Hemiptera Homoptera: Cicadellidae). *Transactions of the Royal Entomological Society of London, 124:* 146–166.

CLARIDGE, M.F. & REYNOLDS, W.J., 1973. Male courtship songs and sibling species in the *Oncopsis flavicollis* species group (Hemiptera: Cicadellidae). *Journal of Entomology, B, 42:* 29–39.

CLARIDGE, M.F., REYNOLDS, W.J. & WILSON, M.R., 1977. Oviposition behaviour and food plant discrimination in leafhoppers of the genus *Oncopsis. Ecological Entomology, 2:* 19–25.

CLARIDGE, M.F., DEN HOLLANDER, J. & FURET, I., 1982. Adaptations of Brown Planthopper (*Nilaparvata lugens*) populations to rice in Sri Lanka. *Entomologia experimentalis et applicata, 32:* 222–226.

CLARIDGE, M.F., DEN HOLLANDER, J. & MORGAN, J.C., 1984. Specificity of acoustic signals and mate choice in the brown planthopper *Nilaparvata lugens. Entomologia experimentalis et applicata, 35:* 221–226.

CLARIDGE, M.F., DEN HOLLANDER, J. & MORGAN, J.C., 1985a. Variation in courtship signals and hybridization between geographically definable populations of the rice brown planthopper, *Nilaparvata lugens* (Stål). *Biological Journal of the Linnean Society, 24:* 35–49.

CLARIDGE, M.F., DEN HOLLANDER, J. & MORGAN, J.C., 1985b. The status of weed-associated populations of the brown planthopper, *Nilaparvata lugens* (Stål) – host race or biological species? *Zoological Journal of the Linnean Society, 84:* 77–90.

CLARIDGE, M.F., DEN HOLLANDER, J. & MORGAN, J.C., 1988. Variation in hostplant relations and courtship signals of weed-associated populations of the brown planthopper, *Nilaparvata lugens* (Stål), from Australia and Asia: a test of the recognition species concept. *Biological Journal of the Linnean Society, 35:* 79–93.

COYNE, J.A. & ORR, H.A., 1989. Patterns of speciation in *Drosophila. Evolution, 43:* 362–381.

DIEHL, S.R. & BUSH, G.L., 1984. An evolutionary and applied perspective of insect biotypes. *Annual Review of Entomology, 29:* 471–504.

DIEHL, S.R. & BUSH, G.L., 1989. The role of habitat preference in adaptation and speciation. In D. Otte & J.A. Endler (Eds), *Speciation and its Consequences:* 345–365. Sunderland, Massachusetts: Sinauer Associates.

HARRISON, R.G., 1991. Molecular changes at speciation. *Annual Review of Ecology and Systematics, 22:* 281–308.

HEINRICHS, E.A. & MEDRANO, F.G., 1984. *Leersia hexandra*, a weed host of the brown planthopper, *Nilaparvata lugens* (Stål). *Crop Protection, 3:*77–85.

ICHIKAWA, T. & ISHII, S., 1974. Mating signal of the brown planthopper *Nilaparvata lugens* Stål (Homoptera: Delphacidae): vibration of the substrate. *Applied Entomology and Zoology, 9:* 196–198.

JOHN, B. & CLARIDGE, M.F., 1974. Chromosome variation in British populations of *Oncopsis* (Homoptera: Cicadellidae). *Chromosoma (Berlin), 46:* 77–89.

LEQUESNE, W.J., 1965. Hemiptera Cicadomorpha (excluding Deltocephalinae and Typhlocybinae). *Handbooks for the Identification of British Insects: 2a.* London: Royal Entomological Society.

OSSIANNILSSON, F., 1949. Insect drummers. *Opuscula Entomologica Lund, Supplementum, 10:* 1–46.

OTTE, D. & ENDLER, J.A. (Eds). 1989. *Speciation and its Consequences.* Sunderland, Massachusetts: Sinaur Associates.

PATERSON, H.E.H., 1985. The recognition concept of species. In E.S. Vrba (Ed.), *Species and Speciation:* 21–29. Pretoria: Transvaal Museum.

PRICE, P.W., 1980. *Evolutionary Biology of Parasites.* New Jersey: Princeton University.

SOUTHWOOD, T.R.E., 1978. The components of diversity. In L.A. Mound & N. Waloff (Eds), *The Diversity of Insect Faunas:*19–40. London: Blackwell.

STRONG, D.R., LAWTON, J.H. & SOUTHWOOD, R., 1984. *Insects on Plants: Community Patterns and Mechanisms.* London: Blackwell.

WEST-EBERHARD, M.J., 1983. Sexual selection, social competition and speciation. *Quarterly Review of Biology, 58:* 155–183.

WEST-EBERHARD, M.J., 1984. Sexual selection, competitive communication and species-specific signals in insects. In T. Lewis (Ed.), *Insect Communication:* 283–324. London: Academic Press.

WOOD, T.K., 1987. Host plant shifts and speciation in the *Enchenopa binotata* Say complex. In M.R. Wilson & L.R. Nault (Eds), *Proceedings of the 2nd International Workshop on Leafhoppers and Planthoppers of Economic Importance:* 361–368. London: CABI Institute of Entomology.

CHAPTER

18

Speciation of the *Enchenopa binotata* complex (Insecta: Homoptera: Membracidae)

THOMAS K. WOOD

Introduction 300

The *Enchenopa binotata* species complex: biology and host-plant-related variation 302
- General life history 302
- Host-plant-associated variation 302
- Assortative mating and hybridization 303
- Host selection and the costs of inappropriate oviposition 304
- Allozyme variation 304
- Asynchronous life histories and mating 304
- Vagility 305
- Summary of evidence for reproductive isolation 306

Plant phenology hypothesis 306
- Evidence for plant mediation of life-history timing 307
- Experimental test of host-induced assortative mating 307

Phylogenetic relationships 307
- Geographic correlates of phylogeny 310
- Life-history timing, reproductive isolating mechanisms and fitness correlates 310

Experimental host-plant shifts 311
- Oviposition 'mistakes' 312
- Asynchronous egg hatch, adult maturation and mating 312

Summary 314

Evolutionary Patterns and Processes
ISBN 0-12-440895-8

Acknowledgements **315**

References **315**

Keywords: Speciation – sympatric – phenology – Insecta – Homoptera – Membracidae – *Enchenopa* – treehopper.

Abstract

The central question of this review is whether mechanisms which promote host-plant specialization initiate a process that ultimately results in genetic divergence of sympatric populations of phytophagous insects. The *Enchenopa binotata* species complex (Insecta: Homoptera: Membracidae) is used as a model to deduce, from extant plant–insect interactions, mechanisms which could have initiated sympatric divergence. Comparison of closely related species pairs, derived from an inferred phylogeny of extant *Enchenopa* species, suggests that asynchronous mating induced by differences in plant phenology may have initiated divergence. Although secondary contact cannot be excluded, geographic distributions of extant species are not allopatric, suggesting a possible sympatric origin of the species complex.

Experimental host-plant shifts demonstrate, in one generation, that phenologically novel host plants can result in asynchronous mating of *Enchenopa*. The existing data for the *Enchenopa* complex suggest that partial or complete mating asynchrony is achieved first through shifts to novel plants that disrupt extrinsically mating synchrony. Genetic divergence occurs secondarily through selection pressures imposed by novel host plants.

INTRODUCTION

The question of whether host-plant specialization of insects leads to speciation, in the context of sympatric models, has a long and controversial history in evolutionary biology (Mayr, 1947, 1963; Bush, 1975; Futuyma & Mayer, 1980; Wood, 1980; Jaenike, 1981; Templeton, 1981; Futuyma, 1986; Bush & Howard, 1986; Butlin, 1987; Barton, Jones & Mallet, 1988; Tauber & Tauber, 1989; Rice & Salt, 1990). Intrinsic (Bush, 1975) or gradual (Futuyma, 1986) sympatric models assume selection on insect genetic traits which differentially affect the fitness of individuals that colonize novel plant species (Rausher, 1984a,b). If a stable resource-associated genetic polymorphism is achieved, continued selection could lead to speciation (Maynard Smith, 1966). Bush (1975) and his colleagues (Feder, Chilcote & Bush, 1988; McPheron, Smith & Berlocher, 1988) have argued that intrinsic host recognition traits linked with traits influencing survival or the timing of the life history (Smith, 1988) are sufficient to promote divergence in the *Rhagoletis pomonella* complex (Feder *et al.*, 1988). However, experimental tests of some of these assumptions in other arthropods have generally not been supportive (Gould, 1979; Rausher, 1984a; Jaenike, 1989; Via, 1990; but see Diehl & Bush, 1989) but some models make sympatric speciation an attractive hypothesis while others do not (Maynard Smith, 1966; Pimm, 1979; Felsenstein, 1981; Rausher, 1984b; Diehl & Bush, 1989).

Most empirical research on insect sympatric speciation has concentrated on intrinsic genetic traits that influence diapause (Tauber & Tauber, 1977a,b; Tauber, Tauber & Masaki, 1986), ecological preference (Bush, 1975; Rice & Salt, 1988,

1990), or fitness (Rausher, 1984a) as mechanisms to disrupt a gene pool to allow genetic divergence. Invariably the insects chosen for study, such as the *Enchenopa* complex (Wood, 1980), are morphologically very similar and there is an a priori reason to suggest resource-based polymorphism. If the organisms in the complex are shown to be sibling or biological species, extant mechanisms that resolve the selection–recombination problem (Felsenstein, 1981) such as habitat preference (Diehl & Bush, 1989) or diapause (Tauber & Tauber, 1982), can be interpreted either as the cause (Tauber & Tauber, 1982), or the effect of speciation (Henry, 1982; Butlin, 1987). With the exception of the *Rhagoletis pomonella* complex, where host shifts are believed to have occurred during the last 200 years (Bush, Chapter 14) and reproductive isolation has not yet developed, there is little evidence that the postulated mechanisms effected divergence through sympatric host shifts.

Ideally, a case for sympatric speciation would involve mechanisms deduced from extant organisms, independent phylogenetic evidence, and an experimental test of the mechanism. Over the last 26 years, this has been the progression of my work on the eastern North American *Enchenopa binotata* (Fig. 1) species complex. In this review, I will describe evidence which suggests that divergence of this species complex could have occurred in sympatry.

Figure 1 An *Enchenopa* female on *Cercis canadensis* placing egg froth over an egg mass. Eggs are inserted in a mass under the bark of the branch. The horn-like structure is the pronotum.

THE *ENCHENOPA BINOTATA* SPECIES COMPLEX: BIOLOGY AND HOST-PLANT-RELATED VARIATION

General life history

All members of the *Enchenopa binotata* species complex are univoltine. Although there are host-plant related differences in timing, the general life history described here for *Enchenopa* on *Celastrus scandens* (bittersweet) is representative of the complex. In Delaware, egg hatch begins in late April or early May over a 7–10-day period. Essentially synchronous egg hatch results in a uniformly aged cohort which matures to the adult stage approximately 30 days later in early June. Adult maturation occurs within a 10-day period, where males eclose on the average 1–2 days before females. Mating begins 3–4 weeks later, around the first day of July, and is completed within 10–15 days. Males appear to mature sexually before females, since there is a 5–10-day span between the first attempted mating and the first copulation. Females mate once while males may mate several times. Under semi-field conditions most male mortality occurs before early August while females live well into November (Wood & Guttman, 1982).

Oviposition begins in the middle of July and is continuous through early November. Females deposit eggs in clusters or egg masses containing on average seven eggs (Wood, 1980). Eggs are inserted into woody stems where they contact the vascular tissues of the plant. Once an egg mass is deposited, the ovipositional wound is covered with a lipid-based secretion known as egg froth (Wood & Patton, 1971; Wood, 1982). During the prolonged oviposition, an average female may deposit 20 or more egg masses. When oviposition is completed in November, most of the egg masses are deposited in a relatively small proportion of the host's branches (Wood, 1980, 1982; Wood & Guttman, 1982).

The following points of the life history are essential to understanding the mechanisms which may have promoted divergence of the *Enchenopa binotata* species complex: (1) synchronous egg hatch, (2) uniform age structure, (3) temporal delay between adult maturation and mating, (4) limited period of mating, (5) females mate once, (6) males die substantially earlier than females, (7) placement of eggs in contact with plant vascular tissue, (8) prolonged oviposition, and (9) clustering of egg masses on branches.

Host-plant-associated variation

In eastern North America, *Enchenopa binotata* occurs on eight plant genera distributed among six different plant families (Table 1). With the exception of the *Enchenopa* on *Viburnum*, *Carya* and perhaps *Juglans*, the remaining *Enchenopa* are monophagous. Evidence that the six monophagous, two polyphagous and one possible oligophagous (on the two *Juglans* species) *Enchenopa* are biological species is presented later. Since the species in the *Enchenopa binotata* complex have not been formally named, they will be referred to by their associated host-plant genus or, in the case of the two on *Juglans*, by plant species.

The first suggestion that *Enchenopa binotata* was not a single polyphagous species was provided by host-associated variation in female pronotal colour, size and shape. This observation, combined with dramatic host-related differences in fifth instar nymphal colouration, aggregating habits, and feeding sites suggested considerable

Table 1 Host plants of the *Enchenopa binotata* species complex.

Genus	Species	Family
Ptelea	*trifoliata* (L.)	Rutaceae
Juglans	*nigra* (L.)	Juglandaceae
Juglans	*cinerea* (L.)	Juglandaceae
Carya	*illinoensis* (Wang) K. Koch *ovalis* (Wang) Sarg. *cordiformis* (Wang) K. Koch *laciniosa* (Michx.) Loud. *ovata* (Mill.) K. Koch	Juglandaceae
Celastrus	*scandens* (L.)	Celastraceae
Liriodendron	*tulipifera* (L.)	Magnoliaceae
Robinia	*pseudoacacia* (L.)	Leguminosae
Cercis	*canadensis* (L.)	Leguminosae
Viburnum	*cassinoides* L. *rufidulum* Raf. *lentago* L. *prunifolium* L.	Caprifoliaceae

intraspecific variation. Other characteristics such as diurnal differences in oviposition, seasonal timing of oviposition, oviposition sites, egg froth composition, and egg mass size which varied with host species also suggested distinct species or races in the *E. binotata* complex (Wood, 1980).

Assortative mating and hybridization

To determine whether *E. binotata* is a complex of host races or cryptic biological species, males and females from six host species were placed in a common field cage where they were free to select mates and host plants. Although some females dispersed to inappropriate hosts and formed mixed precopulatory pairs, mating in general occurred on the female's natal host and only 6 of 103 matings were between individuals from different hosts. Although other pre-mating barriers could not be excluded, the observed assortative mating appeared to be the result of temporal mating asynchrony related to host-plant origin (Wood, 1980). These results supported the model of Maynard Smith (1966), in which a stable resource-associated polymorphism could develop through assortative mating, effected by host fidelity during mating.

Three attempts to hybridize six host-associated *Enchenopa* have been made by transferring males from one host and confining them with females on their host. Males from four hosts, when confined to *Ptelea* and *Juglans*, died before mating. In other transfers to inappropriate hosts, male mortality was high but mating did occur. The viability of eggs from these matings could not be established as they were accidentally destroyed the following year (Wood, unpublished).

Male and female genitalia of all *Enchenopa* in the *binotata* complex, regardless of host-plant origin, are remarkably similar and are of no taxonomic value in species recognition (Pratt & Wood, unpublished) and preliminary karyotype analysis failed to detect any differences in chromosome number (Greene & Wood, unpublished).

The lack of genitalic differences and similar chromosome numbers suggest no mechanical impediments to prohibit dispersing males from successful mating with females on other host species. However, high male mortality on inappropriate hosts and tight host-associated assortative mating suggests that little or no gene flow occurs between *Enchenopa* utilizing different hosts through male dispersal.

Host selection and the costs of inappropriate oviposition

When females from a host species were given free access to oviposit on seven plant species in a cage, they made few or no mistakes in selecting the appropriate host species. None of six host-associated *Enchenopa* female cohorts were ambivalent in host selection for oviposition even though individuals were observed on inappropriate hosts prior to oviposition (Wood, 1980).

Although female host selection is tight, another experiment was done to test whether gene flow among *Enchenopa* on different hosts could occur through the dispersal of mated females. When females associated with each of six plant genera were forced to oviposit on inappropriate *Enchenopa* hosts, females either: (1) died rather than oviposit, (2) deposited few egg masses, (3) produced few hatching eggs, or (4) produced no adult offspring (Wood & Guttman, 1983).

These results suggest that extensive gene flow does not occur among *Enchenopa* on different host-plant species through the movement of mated females and pointed to considerable genetic divergence in host preference and the ability to survive on inappropriate host species. The coupling of host selection fidelity during oviposition, fitness costs of inappropriate oviposition, host-associated mating, and assortative mating suggest that resource-associated divergent selection could operate as described by Maynard-Smith (1966).

Allozyme variation

Initial allozyme analysis of seven plant-associated *Enchenopa* demonstrated fixed or nearly fixed differences between four, while frequency differences separated the remaining three (Guttman, Wood & Karlin, 1981; Wood & Guttman, 1985). Although sample sizes and number of polymorphic loci were low, *Enchenopa* populations were from habitats where introgression was possible. Since the respective gene pools were distinguishable, *Enchenopa* on each host were considered to be either a race or biological species. Fixed allelic differences disappeared when six *Enchenopa* were collected over broader geographic ranges, but differences in allelic frequencies were maintained within and between geographic areas (Guttman & Weigt, 1989; Pratt, Wood & Datz, unpublished). Evidence that *Enchenopa* on *Carya* and *Juglans cinerea* are biological species is presented in Pratt *et al.* (unpublished).

Asynchronous life histories and mating

Assortative mating of this univoltine complex of species was hypothesized to be the result of asynchronous mating associated with different host species (Wood, 1980). In common garden experiments, seven of the host-associated *Enchenopa* species confined to separate cages differed in the timing of egg hatch, seasonal time of mating, diurnal mating patterns, and periods of flight activity. For *Enchenopa* on an array of host species, asynchrony of egg hatch interacting with apparent host effects on development time resulted in asynchronous restricted seasonal mating where females mate once. The longevity of males was substantially shorter than that of females, with most mortality occurring before and during the limited interval of female sexual receptivity. Because of host-associated differences in life-history timing, male mortality was also asynchronous (Wood, 1980, 1988; Wood & Guttman, 1982, 1985). The relationship between the timing of egg hatch and flowering phenology of the host plants suggests that the consequence of utilizing phenologically different host species was asynchronous mating (Wood, 1980; Wood & Guttman, 1982, 1985). Thus, host-associated differences in life-history timing and limited mating periods appear to augment genetic differences in host preference and survival to isolate extant *Enchenopa* species.

Vagility

For *Enchenopa* there is only a relatively short period in the life history where dispersal is possible. Eggs are in plant tissue for as long as 10 months and nymphs cannot, under normal circumstances, leave the host. Nymphs in the *Enchenopa binotata* complex interact facultatively with ants which appears to favour low female vagility and persistent yearly colonization of trees. Attracting and maintaining ant attendance is a function of nymphal aggregation size on branches. Large nymphal aggregations are attended earlier by more ants than smaller aggregations, and as a result have higher survival. Nymphal survival on individual trees is related to the number of nymphs on the tree and whether nymphs are ant attended (Wood, 1982). Persistent colonization of individual trees appears to be related to the yearly predictability of ant mutualists. Since a female's fitness is directly related to offspring being tended by ant mutualists, this suggests female dispersal should only occur at high densities (Wood & Guttman, 1981) and that genetic traits associated with philopatry should be favoured.

Nymphal aggregations are established the previous year by ovipositing females and adult aggregations are loosely maintained until males become sexually active. Flight occurs before and during the pre-mating period but decreases as mating continues (Wood & Guttman, 1981). Once oviposition begins, the ovipositional attractant(s) in egg froth placed over egg masses (Fig. 1) attracts females to a limited number of branches within the host and there is a rapid decline in male and female flight. At the peak of oviposition, on some host species, aggregations of females are common and persistent for weeks unless disturbed. At the completion of oviposition on some host species, several hundred egg masses can be deposited in a branch 1–2 feet long (Wood, 1980, 1982).

When males and females were marked, dispersal between trees of the same species during the pre-mating stage was low and dispersal to different *Enchenopa* host species was not detected at all. Although marked insects declined on the original host during the pre-mating period (Wood & Guttman, 1981), the decline corresponded to high mortality in caged control populations (Wood & Guttman, 1982). At the end of mating, few marked males were recaptured on their original host which also coincided with high mortality of caged populations. Females were observed on their original host up to 2 months after being marked. Recaptures adjusted for non-predator related mortality indicated that 35–41% of the females remained on their original host for their entire lifespan (Wood & Guttman, 1981). The few recaptured dispersed marked males were recovered on their appropriate hosts within 18 m of the release host (Wood & Guttman, 1981) while in another study all but two marked females were recaptured on adjacent touching trees of the same species (Guttman, Wilson & Weigt, 1989).

Low *Enchenopa* dispersal between trees and apparent philopatric tendencies suggest that gene flow within a microhabitat is limited. If true, then a reasonable prediction would be the existence of statistical differences in allozyme frequencies of *Enchenopa* among trees of the same species (Guttman *et al.*, 1981). Guttman *et al.* (1989) examined three *Enchenopa* species and found evidence of among-tree differences in allozyme frequency for *Enchenopa* on *Cercis*, *Juglans* and *Ptelea*. In summary, an important consequence of a female's gain in fitness from ant mutualists are genetically based ovipositional behaviours that increase the probability that offspring will be ant attended. These same behaviours reduce dispersal by mated females (Wood & Guttman, 1982).

Summary of evidence for reproductive isolation

At present I consider the seven *Enchenopa* associated with seven plant genera and the two on different *Juglans* species as biological species. Nymphal morphology provides evidence that each of the nine *Enchenopa* are distinct species. Differences in colouration, setal morphology, shape of the pronotal horn, and the length and shape of the dorsal abdominal scoli of fifth instars can be used to distinguish the nine species (Pratt & Wood, in press). Adult female pronotal shape and size can also be used to distinguish the nine species but the differences are not discrete (Wood, Datz & Pratt, unpublished).

Asynchronous and host-associated assortative mating (Wood, 1980), host selection for oviposition (Wood, 1980), and the inability of females to produce viable offspring on inappropriate host species (Wood & Guttman, 1983) supports the species status of seven *Enchenopa*. Consistent allozyme frequency differences within and among geographic regions (and habitats) where introgression was possible indicate all nine host-associated *Enchenopa* are isolated divergent gene pools (Guttman *et al.*, 1981; Wood & Guttman, 1985; Guttman & Weigt, 1989; Pratt *et al.*, unpublished). The allozyme and morphological variation among species in the *E. binotata* complex are in the range typically used to distinguish geographic races of other insects. This suggests either a recent historical origin or that these characters have not been subject to divergent selection, or a combination of both.

PLANT PHENOLOGY HYPOTHESIS

Recognition that *Enchenopa binotata* is a complex of biological species presents the dilemma of whether the asynchrony of life histories which facilitates reproductive isolation is the cause or the effect of speciation. Extant host–insect interactions suggest that shifts to phenologically novel hosts disrupted mating synchrony which permitted genetic divergence. If life-history timing is mediated extrinsically by the host plant rather than an intrinsic insect genetic characteristic, then recombination would have no effect on mating asynchrony. Rapid selection for host-plant adapted genotypes is far more plausible if the nullifying effects of recombination do not affect the mechanism that maintains mating asynchrony. In this hypothesis, the phenology of the host plant operates in a manner analogous to an allopatric barrier.

Evidence for plant mediation of life-history timing

In extant *Enchenopa* there is a relationship between the timing of egg hatch and the flowering phenology of the host plant (Wood, 1980; Wood & Guttman, 1982, 1985). The first direct evidence that the timing of egg hatch is mediated by the host plant rather than by an intrinsic genetic trait in the insect was the observation that eggs deposited on an inappropriate host hatched at a different time to those on appropriate host species (Wood & Guttman, 1983). Eggs are inserted under the bark in contact with the vascular tissue of the stem (Wood & Patton, 1971; Wood, 1980) and they require liquid water from the plant to begin development (Wood, Olmstead & Guttman, 1990). Asynchronous egg hatch can be induced experimentally by manipulating the time at which water is added to excised branches with eggs. Since six of the host species differ in branch water content during the late winter and early spring, temporal plant differences in the ascent of sap were hypothesized to be the factor initiating asynchronous egg hatch. Thus, a critical step in initiating asynchronous *Enchenopa* life histories appeared to be under extrinsic mediation (Wood *et al.*, 1990).

For host-induced differences in life-history timing to be the mechanism initiating divergence, there must be year to year consistency in patterns of egg hatch. Over 3 years of study, eggs on three hosts always hatched earlier than those on three others. Within each of these two groups, differences in development time to mating and diurnal mating patterns result in asynchronous mating (Wood & Guttman, 1982).

Experimental test of host-induced assortative mating

Although extant *Enchenopa* species appear to mate assortatively through asynchronous mating (Wood, 1980), this is not a direct test of whether host-induced asynchronous mating could initiate divergence. Assortative mating could be the result of other pre-mating barriers. However, experimental manipulation of one species (*Enchenopa* on *Celastrus*) demonstrated that plant-induced differences in life-history timing can result in asynchronous mating through uniform female maturation and asynchronous male mortality. Thus, assortative mating as a consequence of host-plant effects on the timing of egg hatch and development appear to be a mechanism (Wood & Keese, 1990) to minimize the potential effect of recombination.

PHYLOGENETIC RELATIONSHIPS

Consistent yearly patterns of mating asynchrony interacting with host effects on fitness (Wood, 1980; Wood & Guttman, 1982, 1983) appear to be the requisites for divergence. If this hypothesis is correct then phylogenetic patterns should reflect these mechanisms. To attempt a historical reconstruction of divergence, phylogenies were inferred using three independent data sets and techniques. For each of the analyses, *Campylenchia latipes*, the only North American species closely related to the *Enchenopa binotata* complex was used as an outgroup.

Sixteen polymorphic allozyme loci from 3725 *Enchenopa* and 77 *C. latipes* produced essentially the same tree regardless of whether the data were treated as distances (Biosys-1, Swofford & Selander, 1989) or characters (PAUP, Swofford, 1985). The distance Wagner tree (Fig. 2A) places *Enchenopa* from *Robinia, Liriodendron* and *Carya* at the base of the tree. *Enchenopa* on *J. nigra* and *J. cinerea* are more derived but placed together. Of these two, the *Enchenopa* on *J. cinerea* is the most derived. The remaining *Enchenopa* form a separate group, with those on *Cercis* and *Viburnum* being more plesiomorphic to those on *Ptelea* and *Celastrus*. Of this group, the *Enchenopa* on *Ptelea* is the most derived. The lack of fixed differences as well as

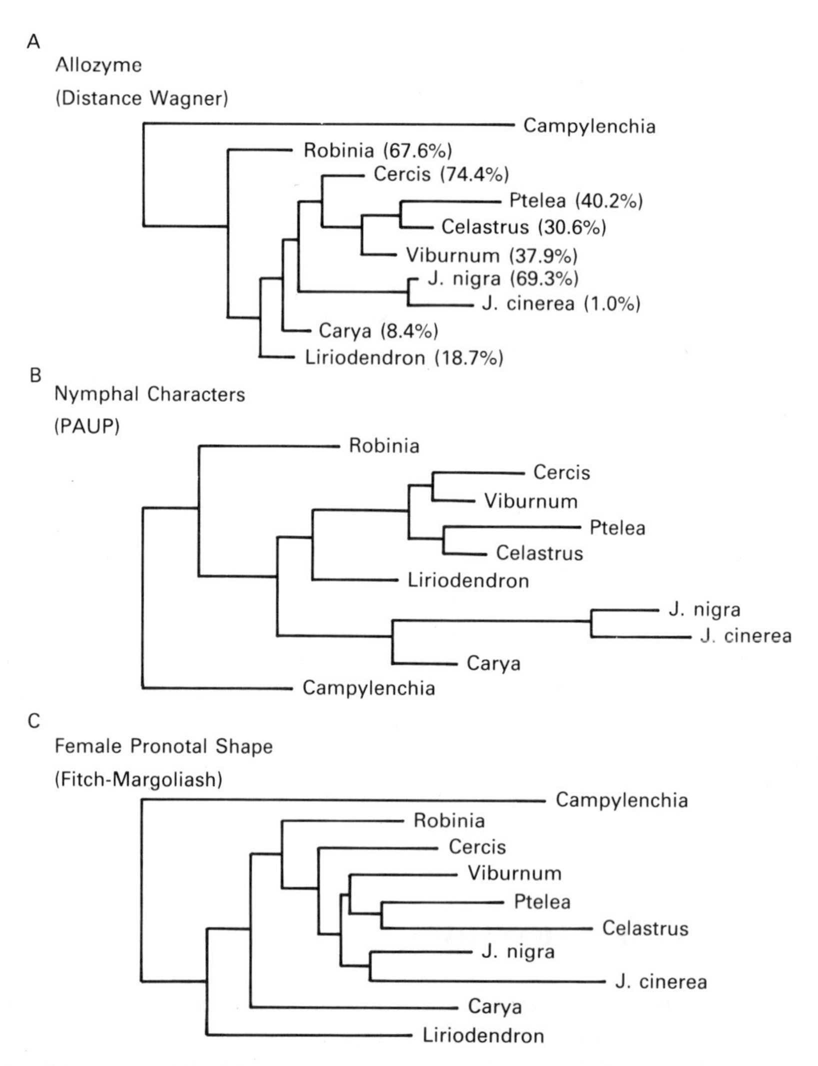

Figure 2 Dendrograms suggesting relationships among nine host-plant associated *Enchenopa* species, using *Campylenchia latipes* as an outgroup. In part A the percentages indicate the proportion of the host plant's range occupied by *Enchenopa* on that host.

small genetic distances among the *Enchenopa* species indicates that divergence has been rather recent (Pratt *et al.*, unpublished).

A more traditional phylogenetic analysis (PAUP, Swofford, 1985) using 33 nymphal characters results in a tree (Fig. 2B) very similar to the allozyme derived distance Wagner tree (Fig. 2A). The few differences that do exist are compatible with the genetic distance data and the topology of the distance Wagner tree (Pratt & Wood, in press).

Pronotal colour, shape and size differs among the *Enchenopa* associated with different host plants. Pronotal shape (Fig. 1) can be measured and analysed using canonical discriminate analysis (Wood & Pesek, 1992). The pronota of more than 2400 females were traced with a digitizer. Digitized data ($X - Y$ coordinates) were used to derive 300 variables that reflect pronotal shape of each individual. Canonical discriminate analysis was used to remove correlations among these variables and to generate a Mahanolobis distance matrix for the *Enchenopa* species and *Campylenchia*. The dendrogram (Fig. 2C) from a Fitch–Margoliash analysis (Felsenstein, 1986; Phylip, ver. 2.7) using the female distance matrix, is in general agreement with phylogenies inferred from allozyme and nymphal characters.

Although there are differences in topology, all three trees (Fig. 2A–C) are in general agreement as to which *Enchenopa* are plesiomorphic and which are apomorphic. For the following discussion, the allozyme distance Wagner tree and genetic distances were used to define 12 closely related pairs of *Enchenopa* species (Table 2). Some pairs are not, in a strict cladistic sense (Wiley, 1981), sister species but will be referred to as such in the remainder of this paper.

If speciation of the *Enchenopa* complex was a result of sympatric shifts to phenologically novel plants that imposed differential selection pressures, then closely related *Enchenopa* should be geographically sympatric, differ in critical life-history

Table 2 'Sister' pairs of *Enchenopa* species associated with nine species of host plants. 'Sister' species were defined by genetic distance and topology of the Distance Wagner tree (Fig. 2A).

Enchenopa 'sister' host-associated species	% Geographic sympatry	Egg hatch	Development time to mating	Asynchronous mating	Diurnal mating preference
Robinia/Liriodendron	38.6	+	+	+	?
Liriodendron/Carya	92.4	+	+?	+	?
Carya/J. nigra	29.6	+	?	?	?
J. nigra/J. cinerea	1.5	0	?	?	?
J. nigra/Cercis	88.5	+	+	+	+
Cercis/Viburnum	75.6	+	0	0	+
Viburnum/Celastrus	72.9	+	+	+	+
Celastrus/Ptelea	54.8	+	+	+	0
Viburnum/Ptelea	63.7	+	+	0	+
Robinia/Cercis	81.6	+	+	+	+
Cercis/Carya	17.7	+	?	?	?
Viburnum/Liriodendron	39.1	+	−?	+	?

Data from Pratt *et al.*, unpublished; Wood & Guttman, 1982, 1985 were used to determine percentage geographic sympatry and indicate statistically significant differences in the timing of egg hatch, development time to mating, asynchronous mating and diurnal mating preference.
+ = significant difference; 0 = no difference; +? = inferred difference; and ? = no available data.

characteristics and in their ability to use the host plant of a 'sister' species. A consistent pattern would indicate that these factors contribute to present-day ecological isolation and could be those that initiated divergence and speciation.

Geographic correlates of phylogeny

Since phytophagous insects are dependent on their host plant, one potential indirect measure of the historical length of the association could be the amount of the host geographic area that is occupied by the insect. A high degree of distributional similarity between both could indicate concomitant historical dispersal, especially into glaciated regions of eastern North America. With two under-sampled *Enchenopa* exceptions (on *Carya* and *Liriodendron*), there appears to be a correspondence between host area occupied and phylogenetic position (Fig. 2A). Generally, more plesiomorphic *Enchenopa* species occupy a larger proportion of their host plant's range than do apomorphic species, suggesting either a long host association or greater dispersal ability. Although this relationship must be considered tentative, pending more sampling, the extensive distribution of both host and its associated *Enchenopa* in glaciated regions indicates colonization during the last 20 000 years. The lack of fixed allelic differences among the *Enchenopa* complex and relatively low host geographic areas occupied by apomorphic species implies that speciation and dispersal could have been recent (Pratt *et al.*, unpublished).

Of the 12 potential 'sister' species pairs listed in Table 2, none presently show classic allopatric geographic distribution. Although secondary contact cannot be excluded, seven of the pairs share 54.8 – 92.4% of the geographic range and must be considered sympatric. Three pairs share 29.6 – 39.1% of their geographic range, however these 'sister' pairs involve *Enchenopa* on either *Carya* or *Liriodendron* where adequate distributional data are not available. At this point I would consider these pairs sympatric. Only two pairs suggest parapatric distributions in that they share less than 17.7% of the geographic area at the extreme of the 'sister' species distribution.

Life-history timing, reproductive isolating mechanisms and fitness correlates

The plant phenology hypothesis suggests that divergence was initiated by host characteristics which disrupted mating synchrony. Phylogenetic support for the hypothesis would be life-history differences between sympatric 'sister' *Enchenopa* species. Of the 12 pairs where adequate experimental data are available, there is a remarkable congruence with this prediction (Table 2). Ten of the 12 pairs differ in the timing of egg hatch and 8 of 9 differ in the length of development to mating. Seven of 9 are asynchronous in mating and the two pairs that are not, differ in the time of day they mate (morning v. evening).

If there was selection for host-adapted genotypes and speciation was recent, then latent genetic variation in the ability of extant *Enchenopa* to utilize different host species could reflect the sequence of host shifts suggested by their phylogeny. When first instar nymphs from each of six *Enchenopa* species were transferred to non-natal host plants (Wood & Datz, unpublished), *Ptelea* and *Juglans* were completely unacceptable hosts (100% mortality) for all *Enchenopa* nymphs except those native to them. Although mortality varied, all *Enchenopa* nymphs survived on the remaining four host species. *Robinia* is the most acceptable host plant for all *Enchenopa* nymphs regardless of their host origin, since they have the same or higher survival compared to that on their natal host. The remaining host species (*Celastrus, Viburnum* and *Cercis*) had more variable effects on nymphal survival of *Enchenopa* from different host-plant species. In general, nymphal survival was equal to or lower than that on the natal host species (Wood & Datz, unpublished).

Survival to reproduction and the ability to oviposit is the most critical step in promoting successful host-plant shifts. Of the 30 non-natal transfer combinations, only 8 had females that survived to oviposit (Table 3). Five of these combinations were *Enchenopa* transferred to *Robinia*. In all but one of the 8 combinations, the number of egg masses deposited per female was substantially lower than comparable females raised on their natal host. The lower number of egg masses represents an interaction between lower survival during oviposition and a reduction in fecundity. In two combinations, where sufficient egg masses were deposited to permit comparison, the number of eggs in each mass was substantially lower than those from females raised on their natal host (Wood & Datz, unpublished).

Of the six *Enchenopa* studied, non-natal host-plant effects on survival, reproduction and fecundity are reflected by their phylogenetic position. The host of the most plesiomorphic *Enchenopa* is *Robinia* where all *Enchenopa* survived to reproduce, although fecundity was lower than on their respective natal hosts. At the opposite extreme, the most apomorphic *Enchenopa* are on *Ptelea* and *J. nigra*. These hosts resulted in 100% nymphal mortality for all non-natal *Enchenopa*. *Enchenopa* from *Celastrus* lived to reproduce on *Viburnum* and *Cercis*, while those on *Ptelea* lived to reproduce only on *Cercis*. In both of these cases, reduced but successful reproduction occurred on hosts with more plesiomorphic *Enchenopa* (Wood & Datz, unpublished). The pattern described here is consistent with the Futuyma & McCafferty (1990) hypothesis which suggests host shifts should occur from chemically or nutritionally less difficult hosts to those that are more difficult.

EXPERIMENTAL HOST-PLANT SHIFTS

Survival of extant *Enchenopa* species to reproduction on non-natal hosts, and the isolating mechanisms discussed above are compatible with the hypothesized phylogeny.

Table 3 The number of egg masses per female for females that survived from the first instar to reproduction on their natal and non-natal hosts. Zeros indicate either no survival to reproduction or failure to deposit egg masses.

Original host	Transferred to: *Robinia*	*Cercis*	*Juglans*	*Viburnum*	*Ptelea*	*Celastrus*
Robinia	4.9	0	0	0	0	0
Cercis	0.8	5.9	0	0	0	0
J. nigra	2.2	0	4.3	0	0	0
Viburnum	0.4	0	0	4.5	0	0
Ptelea	1.8	0.5	0	0	24.0	0
Celastrus	0.9	8.5	0	1.3	0	5.2

Overall, the pattern suggests that speciation in this complex of treehoppers was initiated through host shifts to phenologically different host plants that altered life-history timing and imposed selection favouring extant genotypes capable of surviving and reproducing on novel hosts. Although secondary contact cannot be excluded, the geographic distributions of 'sister' taxa provide no evidence to suggest classic allopatry, supporting the hypothesis that speciation could have occurred in sympatry or parapatry.

The plant phenology hypothesis is parsimonious, in that oviposition mistakes on phenologically novel host plants should sufficiently disrupt life-history timing to allow for assortative mating. Host effects on development time, survival and fecundity should favour host adaptive traits and host preference. Direct evidence for the hypothesis would be to induce experimentally sympatric shifts to novel host plants that, over time, resulted in life-history asynchrony and detectable genetic divergence in fitness traits. In the following two sections I provide the first experimental evidence for this hypothesis.

Oviposition 'mistakes'

A host shift can only be made through the movement of mated females that make ovipositional mistakes. Such 'mistakes' could be the result of genetic variation in host preference (Bush, 1975), overcrowding or death of a host plant. Sympatric host shift models (Bush, 1975) predict that sufficient genetic variation exists within a population to permit successful colonization of new host species. A corollary is that some genetic traits permit successful colonization on some hosts but not on others. In the initial stage of a host shift, the models predict differential mortality on females during oviposition and their offspring. To test this hypothesis, females collected from a single *Viburnum lentago* population were forced to oviposit on an array of eight novel native and introduced *Viburnum* species and hybrids (Table 4). With the exception of one *Viburnum* species normally utilized in the eastern United States, female mortality was higher during a 7-day period and fecundity lower on novel *Viburnum* than on *V. lentago* (Wood, Pratt & Greene, unpublished).

Asynchronous egg hatch, adult maturation and mating

To test the hypothesis that the phenology of the above eight novel hosts disrupts life-history synchrony of offspring, the timing of egg hatch, adult maturation and mating were determined on these plants grown on the same greenhouse bench. If egg hatch is correlated with plant flowering phenology (as on extant *Enchenopa* host plants), manipulated shifts to *Viburnum* that flower earlier should result in significantly earlier egg hatch. As predicted, the temporal sequence of egg hatch (Table 4) was earlier on the novel plant species than on the parent host (*V. lentago*) which is the last to flower in the spring (Wood *et al.*, unpublished). The chronology of adult maturation did not reflect the temporal pattern of egg hatch (Table 4). Adults on novel hosts matured earlier, at the same time, and later than on *V. lentago*. Thus, the interaction between host effects on the timing of egg hatch and nymphal development time produced novel *Enchenopa* populations that were not synchronized with those on the parent host.

To determine host effects on the chronology of mating, males and females were

Table 4 Chronological patterns of egg hatch, adult maturation and mating of *Enchenopa* on nine species or hybrids of *Viburnum*.

Host	Chronological pattern of Egg hatch ± S.E.	Adult maturation ± S.E.	Mating ± S.E.
V. rhytidophyllum	7.2 ± 0.2 (21)		
V. opulus	7.2 ± 0.1 (280)	57.0 ± 0.5 (51)	107.0 ± 0.0 (1)
V. dentatum	7.3 ± 0.9 (58)	51.4 ± 0.4[a] (66)	105.0 ± 5.9[a] (10)
V. utile	7.5 ± 0.1 (199)	48.9 ± 0.2 (252)	91.4 ± 1.2 (52)
V. burkwoodii	7.5 ± 0.1 (260)	52.1 ± 0.3 (154)	95.3 ± 1.5 (44)
V. lantana	7.5 ± 0.1 (693)	54.4 ± 0.1 (891)	109.7 ± 0.8 (245)
V. rhytidophylloides	8.0 ± 0.1 (260)	53.7 ± 0.5 (30)	99.7 ± 2.4[a] (12)
V. prunifolium	9.2 ± 0.0 (1179)	53.7 ± 0.1 (1835)	104.9 ± 0.6[a] (512)
V. lentago	9.9 ± 0.1 (425)	51.2 ± 0.1[a] (488)	104.1 ± 1.2[a] (100)

A Kolmogorov–Smirnov two-sample test and a '*t*'-test were used to determine if the frequency distributions and the means on novel hosts differed from those on *V. lentago*. Both tests provided identical results and the same superscripts indicate no difference ($P<0.05$) from that on *V. lentago*. Numbers in parentheses indicate number of individuals. Low numbers of matings on *V. opulus* precluded comparison.

confined to cages on their new host. The chronology of mating generally reflected that of adult maturation. The mean day of mating (Table 4) on two novel hosts was earlier than on *V. lentago*. The 12.5- and 8.8-day differences between the mean day of mating on these hosts and *V. lentago* demonstrates that plant phenologies promoting early egg hatch combined with minimal host effects on development time do result in early asynchronous mating. The mean day of mating on three novel hosts did not differ from that on *V. lentago*. Although *Enchenopa* on these hosts hatched earlier, host-plant effects on development time brought them into mating synchrony. On one novel host the mean day of mating was later than *V. lentago*. The 5.6-day difference between the means was the result of increased nymphal maturation time. The 18.3-day difference between the mean day of mating on two novel hosts represented almost complete allochrony in mating. The magnitude of this difference attests to the power of interacting effects from plant phenology and apparent nutritional quality, causing asynchrony in the life histories of these sap-feeding insects in only one generation (Wood *et al.*, unpublished).

When *Enchenopa* on all novel hosts were compared to those on *V. lentago* to determine if the proportion of observed matings differed, all differed. Although the occurrence of unobserved matings cannot be excluded, it is possible in the early stage of a host shift that a significant proportion of females go unmated. Unmated females may be those that were not capable of obtaining sufficient nutrition to mature sexually. If confirmed, unmated females, in addition to mortality, represent a powerful selective pressure during the first generation of a host shift (Wood *et al.*, unpublished).

As expected, some novel hosts imposed high mortality on nymphs, with extinction occurring on one novel host 15 days after egg hatch. During the mating period, most novel hosts had no effect on mortality compared to that on *V. lentago*. On one novel host, male mortality was lower while on two it was higher. Female mortality on two hosts was higher than on *V. lentago*. Shortly after mating was completed an additional extinction occurred. The lifetime survival distributions of adult males differed from that on *V. lentago* on all but two novel hosts. On all but one novel host female lifetime survival distributions differed from *V. lentago* (Wood *et al.*, unpublished).

The number of eggs in each egg mass was greater on three novel hosts than on *V. lentago*. Only *Enchenopa* on one novel host did not differ from *V. lentago*. Although individual clutch size appears to be the same or higher on novel hosts, overall lifetime fecundity of these populations was lower than on *V. lentago*. On two novel hosts the number of egg masses deposited was extremely low (there were not enough for destructive sampling), suggesting only a small proportion of females on these hosts were capable of egg production (Wood *et al.*, unpublished).

This experiment demonstrates that the essential predictions of the intrinsic and plant phenological hypotheses of sympatric host shifts were met by the *Enchenopa* system. Forced oviposition 'mistakes' were costly in terms of initial female mortality and fecundity. The life-history synchrony of offspring was disrupted through the interaction of host effects on the timing of egg hatch and development time, resulting in asynchronous mating on some novel hosts relative to the parent population, and there were also fitness effects (Wood *et al.*, unpublished). Thus the original parent population of *Enchenopa* on *V. lentago* contained sufficient genetic variation to permit successful colonization of novel host species.

Although mating synchrony was disrupted, the question of how effective asynchronous mating is as a pre-mating barrier to dispersing males and females remains to be determined for these *Viburnum Enchenopa*. Other studies (Wood & Keese, 1990) suggest that, in the absence of spatial and genetic considerations, assortative mating should occur between some combinations of these experimentally created *Viburnum* asynchronous host 'races'. The ultimate rate of sympatric divergence of these 'races' should be a function of the degree of mating asynchrony, the intensity of selection, genetically based behavioural tendencies toward philopatry (Wood & Guttman, 1981; Guttman *et al.*, 1989; Wood, 1982) and microhabitat influences on gene flow between *Enchenopa* on parent and novel hosts.

SUMMARY

Although far more and better evidence for sympatric speciation has been produced since the early critiques of Mayr (1947, 1963), an unequivocal case for sympatric speciation among extant species is difficult to make. In reality, as with most speciation models, nobody will know whether the extant *Enchenopa binotata* complex actually speciated in sympatry. The best that can be done is to demonstrate a plausible mechanism by which populations can differentiate in sympatry (Futuyma & Mayer,

1980; Butlin, 1987). In my estimation, the present data for the *Enchenopa* complex are more consistent with a sympatric model than any variant of allopatric models. My interpretation of the existing *Enchenopa* data is that, if sufficient genetic variation exists within a population to permit successful colonization of novel hosts, partial or complete mating asynchrony can be induced extrinsically through shifts to phenologically novel plants. Life-history asynchrony then permits genetic divergence to occur through selection pressures imposed by novel host-plant species. If novel hosts impose high fitness costs, as with extant *Enchenopa*, divergence will be rapid and selection for traits associated with host preference intense. Ultimately, only long-term experiments using the *Enchenopa* on novel *Viburnum* species can test stringently whether or not sympatric host shifts can lead to sufficient genetic differentiation for biological species formation. Only this kind of documentation will resolve the debate for the *Enchenopa* complex.

ACKNOWLEDGEMENTS

I appreciate the very helpful critical manuscript reviews by Drs G. Bush, L. Hurd, G. Pratt and D. Tallamy. Appreciation is extended to the National Science Foundation for supporting the initial *Enchenopa* research and the Competitive U.S.D.A. Grants Program for supporting other aspects. This is paper no. 1440 of the Delaware Agricultural Experiment Station and contribution no. 639 of the Department of Entomology and Applied Ecology, and no. 166 of the Life Science Ecology program at University of Delaware, Newark.

REFERENCES

BARTON, N.H., JONES, J.S. & MALLET, J., 1988. No barriers to speciation. *Nature, 336:* 13–14.

BUSH, G.L., 1975. Modes of animal speciation. *Annual Review of Ecology and Systematics, 6:* 339–364.

BUSH, G.L. & HOWARD, D.L., 1986. Allopatric and non-allopatric speciation: Assumptions and evidence. In S. Karlin & E. Nevo (Eds). *Evolutionary Processes and Theory:* 411–438. New York: Academic Press.

BUTLIN, R., 1987. A new approach to sympatric speciation. *Trends in Ecology and Evolution, 26:* 310–311.

DIEHL, S.R. & BUSH, G.L., 1989. The role of habitat preference in adaptation and speciation. In D. Otte & J. Endler (Eds), *Speciation and its Consequences.* Sunderland, Massachusetts: Sinauer Associates.

FEDER, J., CHILCOTE, C.A. & BUSH, G.L., 1988. Genetic differentiation between sympatric host races of the apple maggot fly *Rhagoletis pomonella. Nature, 336:* 61–64.

FELSENSTEIN, J., 1981. Skepticism towards Santa Rosalia, or why are there so few kinds of animals? *Evolution, 35:* 124–138.

FELSENSTEIN, J., 1986. PHYLIP-Phylogeny Inference Package. Version 2.7. Seattle, Washington: University of Washington.

FUTUYMA, D.J., 1986. *Evolutionary Biology*. Sunderland, Massachusetts: Sinauer Associates.

FUTUYMA, D.J. & McCAFFERTY, S.S., 1990. Phylogeny and the evolution of host plant associations in the leaf beetle genus *Ophraella* (Coleoptera, Chrysomelidae). *Evolution, 44:* 1885–1913.

FUTUYMA, D.J. & MAYER, G.C., 1980. Non-allopatric speciation in animals. *Systematic Zoology, 29:* 254–271.

GOULD, F., 1979. Rapid host range evolution in a population of the phytophagous mite *Tetranychus urticae* Koch. *Evolution, 33:* 791–802.

GUTTMAN, S.I. & WEIGT, L.A., 1989. Macrogeographic genetic variation in the *Enchenopa binotata* complex (Homoptera: Membracidae). *Annals of the Entomological Society of America, 82:* 156–165.

GUTTMAN, S.I., WOOD, T.K. & KARLIN, A.A., 1981. Genetic differentiation along host plant lines in the sympatric *Enchenopa binotata* Say complex (Homoptera: Membracidae). *Evolution, 35:* 205–217.

GUTTMAN, S.I., WILSON, T. & WEIGT, L.A., 1989. Microgeographic genetic variation in the *Enchenopa binotata* complex (Homoptera: Membracidae). *Annals of the Entomological Society of America, 82:* 225–231.

HENRY, C.S., 1982. Reply to Tauber and Tauber's 'sympatric speciation in *Chrysopa*: further discussion'. *Annals of the Entomological Society of America, 75:* 3–4.

JAENIKE, J., 1981. Criteria for ascertaining the existence of host races. *American Naturalist, 117:* 830–834.

JAENIKE, J., 1989. Genetic population structure of *Drosophila tripunctata*: Patterns of variation and covariation of traits affecting resource use. *Evolution, 43:* 1467–1482.

McPHERON, B.A., SMITH, D.S. & BERLOCHER, S.H., 1988. Genetic differences between host races of *Rhagoletis pomonella*. *Nature, 3363:* 64–66.

MAYNARD SMITH, J., 1966. Sympatric speciation. *American Naturalist, 100:* 637–650.

MAYR, E., 1947. Ecological factors in speciation. *Evolution, 1:* 263–288.

MAYR, E., 1963. *Animal Species and Evolution*. Cambridge, Massachusetts: Belknap Press of Harvard University Press.

PIMM, S.L., 1979. Sympatric speciation: a simulation model. *Biological Journal of the Linnean Society, 11:* 131–139.

PRATT, G. & WOOD, T.K., in press. A phylogenetic analysis of the *Enchenopa binotata* complex (Membracidae: Homoptera) using nymphal characters. *Systematic Entomology*.

RAUSHER, M.D., 1984a. Tradeoffs in performance on different hosts: Evidence from within- and between-site variation in the beetle *Deloyala guttata*. *Evolution, 38:* 582–595.

RAUSHER, M.D., 1984b. The evolution of habitat preference in subdivided populations. *Evolution, 38:* 596–608.

RICE, W.R. & SALT, G.W., 1988. Speciation via disruptive selection on habitat preference: experimental evidence. *American Naturalist, 131:* 911–917.

RICE, W.R. & SALT, G.W., 1990. The evolution of reproductive isolation as a correlated character under sympatric conditions: experimental evidence. *Evolution, 44:* 1140–1152.

SMITH, D.S., 1988. Heritable divergence of *Rhagoletis pomonella* host races by seasonal asynchrony. *Nature, 336:* 66–67.

SWOFFORD, D.L., 1985. PAUP-Phylogenetic analysis using Parsimony. Version 2.4. Champaign, Illinois: Illinois Natural History Survey.

SWOFFORD, D.L. & SELANDER, R.B., 1989. Biosys-1. Release 1.7. Champaign, Illinois: Illinois Natural History Survey.

TAUBER, C.A. & TAUBER, M.J., 1977a. Sympatric speciation based on allelic changes at three loci: Evidence from natural populations in two habitats. *Science, 197:* 1298–1299.

TAUBER, C.A. & TAUBER, M.J., 1977b. A genetic model for sympatric speciation through habitat diversification and seasonal isolation. *Nature, 268:* 702–705.

TAUBER, C.A. & TAUBER, M.J. 1982. Sympatric speciation in *Chrysopa*: Further discussion. *Annals of the Entomological Society of America, 75:* 1–2.

TAUBER, C.A. & TAUBER, M.J., 1989. Sympatric speciation in insects perception and perspective. In D. Otte and J. Endler (Eds), *Speciation and its Consequences*. Sunderland, Massachusetts. Sinauer Associates.

TAUBER, M.J., TAUBER, C.A. & MASAKI, S., 1986. *Seasonal Adaptations of Insects:* 411. New York: Oxford University Press.

TEMPLETON, A.R., 1981. Mechanisms of speciation – A population genetic approach. *Annual Review of Ecology and Systematics, 12:* 23–48.

VIA, S., 1990. Ecological genetics and host adaptation in herbivorous insects: The experimental study of evolution in natural and agricultural systems. *Annual Review of Entomology, 35:* 421–446.

WILEY, E.O., 1981. *Phylogenetics: The Theory and Practice of Phylogenetic Systematics.* New York: John Wiley & Sons.

WOOD, T.K., 1980. Intraspecific divergence in *Enchenopa binotata* Say (Homoptera: Membracidae) effected by host plant adaptation. *Evolution, 34:* 147–160.

WOOD, T.K., 1982. Ant attended nymphal aggregations in the *Enchenopa binotata* complex (Homoptera: Membracidae). *Annals of the Entomological Society of America, 75:* 649–653.

WOOD, T.K., 1988. Consequences of membracid life histories mediated by plant phenology. In C. Vidano & A. Arzone (Eds), *Proceedings of the 6th Auchenorrhyncha Meeting, Turin, Italy, 1987.* Silvestrelli & Cappelletto di S. Rosa - Clot e C, Torino.

WOOD, T.K. & KEESE, M., 1990. Host plant induced assortative mating in *Enchenopa* treehoppers. *Evolution, 44:* 619–628.

WOOD, T.K. & GUTTMAN, S.I., 1981. The role of host plants in the speciation of treehoppers: An example from the *Enchenopa binotata* complex. In R.F. Denno & H. Dingle (Eds), *Life History Patterns: Habitat and Geographic Variation:* 39–54. New York, Heidelberg, Berlin: Springer-Verlag.

WOOD, T.K. & GUTTMAN, S.I., 1982. Ecological and behavioural basis for reproductive isolation in the sympatric *Enchenopa binotata* complex (Homoptera: Membracidae). *Evolution, 36:* 233–242.

WOOD, T.K. & GUTTMAN, S.I., 1983. The *Enchenopa binotata* complex: Sympatric speciation? *Science, 220:* 310–312.

WOOD, T.K. & GUTTMAN, S.I., 1985. A new member of the *Enchenopa binotata* Say complex on tulip tree (*Liriodendron tulipifera*). *Proceedings of the Entomological Society of Washington, 87:* 171–175.

WOOD, T.K. & PATTON, R., 1971. Egg froth distribution and deposition by *Enchenopa binotata* (Homoptera: Membracidae). *Annals of the Entomological Society of America, 65:* 1190–1191.

WOOD, T.K. & PESEK, J., 1992. Pronotal shape: A source of confusion or panacea in systematic studies of treehoppers (Homoptera: Membracidae). In J.T. Sorensen & R. Foottit (Eds), *Ordination in the Study of Morphology, Evolution and Systematics of Insects.* Amsterdam: Elsevier.

WOOD, T.K., OLMSTEAD, K.L. & GUTTMAN, S.I., 1990. Insect phenology mediated by host-plant relations. *Evolution, 44:* 629–636.

Index

Acoustic signals
  in cicadas 269–84
    and morphological variability 270–82
    similarity of 271
  in leafhoppers 287–8
  in planthoppers 287, 290–4
    Leersia associated 293–4
Adaptation
  and evolutionary change 141
  of invading species 110–11
  and natural selection 210–12
  and preadaptive features 47
  in punctuational evolution 14–15
Adaptive divergence 46–7, 49, 55
Adaptive radiation 45–58
  and comparative cladistics 51–3
  diagnostic tests for 51–4
  meaning and usage 46–8
  and microevolution 54–5
  and non-adaptive radiation 49–51
  and stratigraphical range analysis 53–4
  testing for 48–9
Alleles
  and genetics of advance 113–14
  postglacial gene flow 101
Allopatry
  and frogs: mate recognition in 252
  speciation, and GISH 200
Allopolyploids, origins of, and GISH 197–9
Allozyme variation in treehoppers 304
Alps 112
Amazonia: frogs in 253–4
Anagenesis
  in natural selection 5, 7–8
  and teeth, of mammoth, evolution of 80
Animal species
  diversity of 210
  and sexual selection 208–9
  small 229–49
Antlers, of moose, evolution of 74–6
Ants, and vagility of treehoppers 305–6
Arthropods, in Cambrian explosion 34–44
Artificial selection 211
Asia: planthoppers in 288–91
Assemblage, in Holocene studies 126, 129
  proposal for study 131–3
Associative mating 239
  in frogs 252
Assortative mating
  host-induced 307
  and hybridization 303–4
Asymmetry, fluctuating
  and evolutionary change 148
  and stress 142–4
Australia: hybrid zones in 100–1

Balkans 112
Baupläne
  of Cambrian radiation 34
  and cladistic analysis 35–41
  and disparity analysis 41–3
  in punctuational evolution 5–6
  appearance 236
  recognizing 34–41
Beetles 27
  in postglacial advance 108–9
  rate of spread of 109
Behaviour and body size in animals 232
Benthic foraminifera 65–7
Biases
  and evolutionary trends 88–9
  in microevolutionary patterns 20–1
Binomial taxonomy and microevolutionary patterns 20–1
Biological species concept 242–3
Bivalves 27
Body size in animals 231–2
  and behaviour 232
  and developmental complexity 233–6
  and ecological isolation 237–42
  and mate recognition 232
  and species problem 243
Bone diameters in mammals 84–5
Bryozoans 27

Builth trilobites 21–2, 25–6

Calls, and mate recognition in frogs 254–6
  interspecific preferences 256–8, 259–63
  intraspecific preferences 258–9, 264–5
Cambrian explosion 33–44
  cladistic analysis of 35–41
Carbon isotopes
  in radiation of planktonic foraminifera 60–1, 64–8
  and variations in earth's orbit 98–9
Cenozoic era 27
Central America: frogs in 253–4
Cereals, and crop domestication 180
Chelicerata, in Cambrian explosion 34–5
Chromatin 190
Chromosomes
  banding 164
    and chromosomal diversity 193–5
  differences in hybrid zones 100
  diversity
    and chromosome banding 193–5
    and genomes 190–6
    and karyotypes 190–2
    meiosis, analysis of 192–3
    and molecular cytogenetics 195–6
  and GISH 199
  and molecular cytogenetics 195–6
  pairing
    and fertility in × *Festulpia* 163–4
    and GISH 202
    in hybrids 159–60
  recombination of, and meiosis 192–3
Chronospecies
  of mammoth 78–9
  of moose 74
Cicadas
  acoustic signals in 269–84
    similarity of 271
  defences of 275–6
  evolutionary relationships 273–5
  morphological variability of 270–82
  phylogenetic relationships 275
  records of emergence of 278–81
Cladistic analysis
  and Cambrian radiation 35–41
  comparative, and adaptive radiation 51–3
Cladogenesis
  and adaptive radiation 48, 50
  antlers of moose, evolution of 76–7
Colonization
  and adaptive radiation 47
  genetics of advance 113–16
  by invading species 104, 110
  and postglacial advance 108–9
Competition
  and extreme environments 140
  and natural selection 211–12
  and sexual selection 215–16
Compound radiations 51
Convergence
  and correlation 4–6
  and evolutionary change 141
  in punctuational evolution: De Vries 12–13
Correlation
  and convergence 4–6
  life history timing, reproductive isolation and fitness 310–11
Courtship 217
Cretaceous/Teritiary (K/T) boundary 60, 63–8
Crop domestication, rapid evolution in 179–81
Crustacea: in Cambrian explosion 34–5
Cues
  for adaptive radiation 46–7
  for mate recognition in small animals 232

Data amalgamation and microevolutionary patterns 24–5
Depth habitats of planktonic foraminifera 60–3
Desiccation, and stress 150–1
Developmental complexity in animals 233–6
  hierarchical organization 233–5
  modification of 235–6
Differential reproduction 211–12
  and sexual selection 216
Differential survival 211
  and sexual selection 216
Discontinuities, and evolutionary change 147–50
Disparity
  and Cambrian radiation 34
  cladistic analysis of 35–41
  quantifying 41–3
Dispersal
  and genetics of advance 113–14
  and microevolution 55
  pioneer, in hybrid zones 104
Divergence
  in adaptive radiation 47
  and fertilizer in plant populations 174–6
  of invading species 110–11
  in plant populations and heavy metals 173–5
  in plants 173–6
Diversity of animal species 210
  and body size 231–2
  and insect–plant interactions 240
DNA
  ancient 132–3
  and chromosomal diversity 190–2
  and GISH 196, 198–9
  hybridization of, and molecular cytogenetics 196
  satellite, and chromosome banding 195
Dollo's Law 21–2

Domestication of crops 179–81
Dwarfing of mammals on islands 84–5

Earth's orbit: variations in 98–9
Ecological isolation 237–42
  and non-allopatric race formation 239–40
  role of 240–1
Ecophenotypic variations 25
Eemian interglacial 82, 84–6
Egg hatch, asynchronous, in treehoppers 312–14
Elk *see* moose
Endler's diffusion rule 101, 104
Energy requirements, and stress 144–6
Environment
  and body size in animals 231–2
  and evolutionary divergence 173–6
  extreme, and evolution 140
  interpretation, in Holocene studies 126
  and morphological variability 141
  and speciation and extinction 150–1
  and stasis 26–8
  and stress 139–56
  see also palaeoenvironment
Eocene era: planktonic foraminifera in 63, 66–8
Epigenetic development in animals 235
Euploid gametes, in polyploid introgression 166–7
Eurasia
  mammals in 72–3
  mammoth in 78–81
  moose in 74–8
  red deer in 82–7
  water voles in 81–2
Europe
  advance in Southern edge 112–13
  hybrid zones in 107–8
  postglacial advance 108–9
Evolution
  and Cambrian explosion 33–44
  karyotypes in 189–206
  polyploid, rare events in 157–69
  of polyploid complexes 164–6
  potential for in plant populations 172–3
  rapid
    in crop domestication 179–81
    in plant populations 171–88
    in time 178–9
  rates of change 87–8
  and stress 139–56
  trends in 88–9
  see also gradualistic evolution; punctuational evolution
Evolutionary change
  selection and discontinuities 147–50
  and stress 147–53

Extinction
  mass, and adaptive radiation 47
  and stress 150–3

Female choice 220
  in fruit flies 221–2
  in sticklebacks 219–20
  in widowbirds 223
Fertility
  and chromosome pairing in × *Festulpia* 163–4
  in hybrids 159
Fertilization, and genetical species 212–14
Fertilizer and divergence in plant populations 174–6
Fitness
  and fluctuating asymmetry 142–4
  life history timing and reproductive isolation correlates 310–11
  and stress, and reductionism 146–7
Flora, hybridization in 159
Founder effects
  and ecological isolation 241
  in natural selection in plants 172
Frequencies of cicadas' songs 273–4
Freshwater ostracods 125–37
  proposals for study 131–3
Frogs
  mate recognition in 252–3
  species groups of study 253–4
Fruit flies
  sexual selection in 220–2
  stress and evolution in 142–4, 145–52
Functional approach to evolutionary morpology 6
  and planktonic foraminifera 62

Gene flow
  and ecological isolation 241
  and extreme environments 140–1
  in plant populations and heavy metals 174
  postglacial 114–15
  postglacial, in hybrid zones 101–3
  and selection in plant populations 174
Gene pools
  and genetical species 212
  and natural selection in plants 173
Genes
  and chromosome banding 193–5
  and evolutionary change 148–50
  mapping of, and molecular cytogenetics 195–6
  in punctuational evolution 11–12
  and reproductive isolation in plants 177–8
Genetic isolation 54–5
Genetic uniformitarianism 158

Genetic variation
  and natural selection in plants 172–3
  within populations 181–3
Genetical species 212–14
  and recognition concept 213–14
Genomes
  and chromosomal diversity 190–6
  and GISH 198–9
  and hybrid zones 100–1
  and karyotype studies 189–206
  and meiosis, analysis of 192–3
  range shifts, repeated 117–19
  reassortment 116–17
Genomic *in situ* hybridization (GISH) 196–202
  allopolyploids, origins of 197–9
  and homology 200–2
  and identical karyotypes 199–200
Geographic correlates of phylogeny in treehoppers 310
Geographic range, and songs of cicadas 270–1, 271–82
Germ line determination in animals 234
Germ line sequestration in animals 235
Globigerine planktonic foraminifera
  depth habitats of 60–3
Gradualistic evolution
  in antlers of moose 76–7
  and microevolutionary patterns 19–20, 22
  and rates of change 87
  in teeth of mammoth 78–81
Grasshoppers: in hybrid zones 101–5, 112

Habitats
  and moose antlers, evolution of 78
  and non-allopatric race formation 239–41
  of planktonic foraminifera 60–3
Heavy metals
  and evolutionary divergence in plants 173–6
  plant tolerance and genetic variation 172–3
  and rapid evolution 178–9
  and reproductive isolation in plants 176–8
Herbicide resistance 178
Heterochromatin 192
Holocene era
  microevolutionary studies in 128–30
  paleoecology of, molluscs and ostracods in 125–37
  planktonic foraminifera in 61
  population data in 129
  suitable groups for study 129–30
Homology
  and convergence 4–6
  and GISH 200–2
Host race formation
  in leafhoppers 288
  and speciation in insects 241
Host selection
  and inappropriate oviposition 304
  plant phenology hypothesis 306–7
Hybrid zones 99–101
  adaptation and divergence 110–11
  contact and mixing in 103–5
  genome reassortment 116–17
  movement and position of 106–8
  in postglacial advance 108–9
  postglacial reinforcement 105–6
  rate of spread 109–10
  secondary contact in 103–5
  width of 101–3
Hybridization
  and assortative mating 303–4
  failure, and GISH 200–2
  in *Festuca* and *Vulpia* 161–2
  in language of evolution 218
  in planthoppers 289–90, 293–4
  in plants 158–60
  see also genomic *in situ* hybridization (GISH)

Ice ages
  and earth's orbit: variations in 98–9
  and hybrid zones 99–101
  postglacial reinforcement 105–6
Idea-systems, convergence and correlation in 4–6
Insects
  body size, life history traits, and sympatry 231–2
  host-plant, experimental shifts 311–14
  host-plant associated variation 302–3
  host-plant interactions 295
    plant phenology hypothesis 306–7
  plant interactions with 240
  speciation in 285–97
    treehoppers 299–317
  see also leafhopper; planthopper; treehopper
Interglacial periods 84–6
Introgression, polyploid 166–7
Invasion
  adaption and divergence in 110–11
  and adaptive radiation 47
Islands
  dwarfing of mammals on 84–5
  and evolutionary change 149
  isolation on 82–5
Isolation
  genetic 54–5
  on islands 82–5
  see also ecological isolation; reproductive isolation
Isotopes, in planktonic foraminifera studies 59–70

Jersey: deer on 82–7

K/T boundary, and planktonic foraminifera evolution 63–8
Karyotypes
  bimodal 190–2
  and chromosomal diversity 190–2
  and genome organization 189–206
  and GISH 197
  and homology and GISH 201–2
  identical, and GISH 199–200
Key evolutionary innovation (KEI) 51–3

Lacustrine molluscs 125–37
  proposals for study 131–3
Lacustrine ostracods 125–37
  proposals for study 131–3
Leafhoppers 287–8
  acoustic signals in 288
Life history
  body size, and sympatry 231–2
  of treehoppers 302
    asynchronous, and mating 304–5
    plant mediation in 307
    timing correlates 310–11
    vagility 305–6
Limulids 27
Lingulid brachiopods 27

Mammals, of Quaternary era 72–3
  mammoths 78–81
  moose 74–8
  rates of change in 87–8
  red deer 82–7
  water voles 81–2
Mammoth, evolution of 78–81
Mate choice, and non-allopatric race formation 239
Mate recognition
  and body size of animals 232
  calls, in frogs 254–6
  and ecological isolation 237
  in frogs 251–67
  see also specific mate recognition systems
Maternal predestination 235
Mating, of treehoppers 304–5
  asynchronous 307–8
  and asynchronous egg hatch 312–14
Maturation, of treehoppers 312–14
Mediterranean Sea 112
Meiosis
  and chromosomal diversity 192–3
  and polyploid evolution 163
Metabolic rate
  fitness and reductionism 146–7
  and speciation and extinction 150
  and stress 144–6
Microevolution
  and adaptive radiation 54–5
  Holocene era, studies in 128–30
  patterns of 19–31
    biases in 20–1
    and different environments 26–9
    and random walks 24–5
    reversals in 21–6
  population data in 129
Milankovitch theory 98
Molars *see* teeth
Molecular cytogenetics 195–6
Molluscs
  conventional use in Holocene palaeoecology 126–8
  Holocene, evolution of 125–37
  infraspecific change in 130–1
  as suitable groups for study 129–30
Moose, evolution of 74–8
  antlers 74–6
Morphogenesis in fruit flies 236
Morphological change and reversals in microevolutionary patterns 26
Morphological variability
  and cicada's songs 270–82
  in leafhoppers 288
  in planthoppers 289
  quantifying disparity 41–3
  and stress 141–4
Mutation
  and stress 142–3
  theory in punctuational evolution 9, 12–13
Mutations
  and developmental complexity in animals 233
  and preformistic development in animals 235

Natural selection
  and adaptation 210–12
  and adaptive radiation 46
  convergence and correlation in 4–6
  De Vries on 10–12
  and genetic variation in plants 172–3
  hierarchical theory of 7–8
  microevolutionary patterns in trilobites 22–4
  punctuational paradigm for 3–18
  and speciation 210
Neocene era: planktonic foraminifera in 61
Niche separation 54–5
Non-adaptive radiation 49–51
Non-allopatric race formation 239–40
Non-allopatric speciation 242–5

North America
  cicadas in 270
    distribution of 277
    origins of 282
  hybrid zones in 107–8
  postglacial advance 108–9
Nucleating points in idea-systems 5–6
Nucleolus organizing regions (NORs) 200
Nymphs
  and vagility of treehoppers 305–6

Ocean Drilling Program 59–60
Oceans, and planktonic foraminifera 59–70
Ordovician trilobites
  descriptive bias 20–1
  different environments 26–9
  reversals in 21–6
Ostracods 27
  conventional use in Holocene palaeoecology 126–8
  Holocene, evolution of 125–37
  infraspecific change in 130–1
  as suitable groups for study 129–30
Oviposition in treehoppers
  and inappropriate selection of host 304
  mistakes 312
Oxygen isotopes
  in radiation of planktonic foraminifera 60–1, 63–4, 68
  and variations in earth's orbit 98–9

Palaeocene: radiation of planktonic foraminifera 59–70
Palaeoenvironment, and teeth of mammoth, evolution of 80–1
Pangenes 11
Parapatric speciation 105–6
Park Grass experiment 174–6
Phalanxes in hybrid zones 103–4
Phenology
  of plants, and treehoppers 306–7
    correlates of 310–11
Phenotypes, variability of and stress 146–7
Photic zones, planktonic foraminifera in 64
Phylogenetic relationships
  cicadas 275
  insects–plant interactions 240
  treehoppers 307–11
Phylogeny
  and chromosomal diversity 190–6
  geographic correlates in treehoppers 310
Pioneers
  dispersal, in hybrid zones 104
  and genetics of advance 113–14
Planktonic foraminifera 28
  depth habitats of 60–3
  evolution after K/T boundary 63–8
  keeled forms of 61–2
  Palaeocene radiation of 59–70
  shell shapes of 62–3
  and variations in earth's orbit 98–9
Planthoppers 287–94
  acoustic signals in 287, 290–4
  Leersia associated 293–4
Plants
  and body size of animals 231
  and heavy metal tolerance 172–3
  host
    associated variation 302–3
    experimental shifts 311–14
    insect interactions 295
    plant phenology hypothesis 306–7
    and speciation 300–1
  insect interactions with 240
  rapid evolution in 171–88
  variability within 181–3
Pleistocene era 27
  ice ages in 98–9
  mammals in 74, 78, 81–2, 84
Pliocene era: mammals in 74, 78
Pollen 108–9
  in Holocene studies 127, 130
Polyploidy
  complexes, evolution of 164–6
  in *Festuca* and *Vulpia* 161–2
  introgression 166–7
  in plants 160–1
  see also allopolyploids
Population data in Holocene evolutionary studies 129
Post-mating isolation 54
Postglacial advance 108–9
  rate of spread 109–10
Preadaption in organisms 47
Preformistic development in animals 234–5
Principal component analysis of disparity 41–3
Pronota, shape in treehoppers 308–9
Pulse-repetition frequencies, in planthoppers 291, 293–4
Punctuated equilibrium 6–8
Punctuational evolution
  in antlers of moose 76–7
  basic empirics of 7
  biases in 88
  De Vries on 9–15
  in dwarfing of deer 86
  logical implications of 7
  and microevolutionary patterns 19–31
  paradigm for 3–18
  and punctuated equilibrium 6–8
  and rates of change 87
Pyrenees: hybrid zones in 101–5, 112

Quaternary era 72–3
  Holocene evolutionary studies 128–9

Race-breeding in punctuational evolution 13
Race formation, non-allopatric 239
Radiation
  in Cambrian era 33–44
  of planktonic foraminifera in Palaeocene 59–70
  see also adaptive radiation; non-adaptive radiation
Random walks
  and adaptive radiation 48
  in evolutionary patterns 22–4
Range shifts, and speciation 118–19
Recognition concept of animal species 210, 212–14, 217
  and sexual selection 224–5
Recombination
  and stress 142
  in treehoppers 306–7
Reductionism, stress and fitness 146–7
Refugia, and hybrid zones 112–13, 116–18
Reinforcement, postglacial 105–6
Reproductive isolation
  and meiosis, analysis of 192–3
  in planthoppers 293–4
  in plants 176–8
  in treehoppers 306
    concept of, in animal species 210
    correlates of 310–11
    post-mating 54
Resources, and body size in animals 231–2
Reversals in microevolutionary patterns 21–6
  consequences of 24–6
  and morphological change 26
  sampling and data amalgamation 24–5
  in teeth of water voles 82–3
  and timespans 25
Rudist bivalves 53

Sallian glacial 82, 85–6
Saltation in punctuational evolution 9, 12
Sampling and microevolutionary patterns 24–5
Secondary somatic differentiation in animals 235
Selection
  and crop domestication 179–80
  and evolutionary change 147–50
  and gene flow in plant populations 174
  host, and inappropriate oviposition 304
  and reproductive isolation in plants 176–8
  see also artificial selection; natural selection; sexual selection
Sexual selection
  in animal species 209–8
  case histories 218–3
  and evolution 214–18
  in fruit flies 220–2
  in language of evolution 216–18
  and natural selection 210–12
  and recognition concept 224–5
  in sticklebacks 218–20
  in widowbirds 222–3
Shatsky Rise (North Pacific) 64–5, 68
Songs *see* acoustic; calls
South America: cicadas in 270
Speciation
  and adaptive radiation 50–1
  allopatric, and GISH 200
  and biological species concept 242–3
  and developmental complexity in animals 233
  and ecological isolation 237–42
  and genomes, influenced by 189–206
  and host-plant specialization 300–1
  in insect herbivores 285–97
  of mammoth 78–80
  of moose 74–6
  non-allopatric 242–5
  and non-allopatric race formation 239–40
  parapatric, and reinforcement 105–6
  by polyploidy 210
  and range shifts 118–19
  and sexual selection 224–5
  and stress 150–3
  sympatric, among insects 295
  of treehoppers 299–317
Species
  advance
    genetics of 113–16
  problem of 242–5
  recognition 218
  selection
    and adaptive radiation 50
    and different environments 28
    in punctuational evolution 13–14
Specific mate recognition systems (SMRSs) 50–1, 54, 295
  in fruit flies 221–2
  and genetical species 213–14
  in language of evolution 216–18
  in planthoppers 291–3
  and range shifts 118
  in sticklebacks 219–20
Stasis
  and different environments 26–8
  and microevolutionary patterns 20, 24
  in punctuational evolution 9–10
  and rates of change in evolution 87
Sterility
  in hybrids 159
  in × *Festulpia* 163
Sticklebacks: sexual selection in 218–20

Stratigraphical range analysis in adaptive radiation 53–4
Stress
  and energy requirements 144–6
  and evolution 139–56
  and evolutionary change 147–53
  fitness, and reductionism 146–7
  and morphological variability 141–4
  speciation and extinction 150–3
Structural approach to evolutionary morpology 6
Sulphur dioxide 178–9
Sustained metabolic rates (SusMR) 144
Suture zones 107
Sympatric speciation 238
  and host-plant specialization 300–1
Sympatry
  and body size in animals 231–2
  in planthoppers 289

Teeth
  of mammoth, evolution of 78–81
  of water voles, evolution of 81–2, 83
Tension zones 106
Teretiusculus Shales 20
Terrestrial molluscs 125–37
  proposals for study 131–3
Tertiary era: planktonic foraminifera in 60, 63–8
Thermoclines, and planktonic foraminifera 61
Timescale
  and crop domestication 180–1
  and evolutionary change 141
  in Holocene evolutionary studies 128
  and rapid evolution 178–9
Timespans and reversals in microevolutionary patterns 25
Treehoppers
  host-plant, experimental shifts 311–14
  host-plant-related variation 302–6
  phylogenetic relationships 307–11
  plant phenology hypothesis 306–7
  speciation of 299–317
Trilobites
  in Cambrian explosion 34–5
  computer simulation of 22–3
  and different environments 26–9
  microevolutionary patterns in 19–31
Tundra: postglacial advance of 108–9, 112

Uniramia: in Cambrian explosion 34–5

Vagility, of treehoppers 305–6
Variety-testing in punctuational evolution 13
Vicariance radiation 49, 55

Wagner parsimony 35
Wales: leafhoppers in 288
Walvis Ridge (South Atlantic) 64, 66–7
Wheat, rapid evolution in 180–1
Widowbirds: sexual selection in 222–3

Younger Dryas period 105, 108–10

Reports of Linnean Symposia

Speciation in Tropical Environments (Lowe-McConnell)	*Biological Journal of the Linnean Society* Vol. *1* 1969 pp. 1–246
New Research in Plant Anatomy (Robson, Cutler & Gregory)	Supplement to the *Botanical Journal of the Linnean Society* Vol. *63* 1970
Early Mammals (Kermack & Kermack)	Supplement 1 to the *Zoological Journal of the Linnean Society* Vol. *50* 1971
The Biology and Chemistry of the Umbelliferae (Heywood)	Supplement 1 to the *Botanical Journal of the Linnean Society* Vol. *64* 1971
Behavioural Aspects of Parasite Transmission (Canning & Wright)	Supplement 1 to the *Zoological Journal of the Linnean Society* Vol. *51* 1972
The Phylogeny and Classification of the Ferns (Jermy, Crabbe & Thomas)	Supplement 1 to the *Botanical Journal of the Linnean Society* Vol. *67* 1973
Interrelationships of Fishes (Greenwood, Miles & Patterson)	Supplement 1 to the *Zoological Journal of the Linnean Society* Vol. *53* 1973
The Biology of the Male Gamete (Duckett & Racey)	Supplement 1 to the *Biological Journal of the Linnean Society* Vol. *7* 1975

Continued as the Linnean Society Symposium Series

No. 1 **The Evolutionary Significance of the Exine** (Ferguson & Muller) (1976)

No. 2 **Tropical Trees. Variation, Breeding and Conservation** (Burley & Styles) (1976)

No. 3 **Morphology and Biology of Reptiles** (Bellairs & Cox) (1976)

No. 4 **Problems in Vertebrate Evolution** (Andrews, Miles & Walker) (1977)

No. 5 **Ecological Effects of Pesticides** (Perring & Mellanby) (1977)

No. 6 **The Pollination of Flowers by Insects** (Richards) (1978) (*Botanical Society of the British Isles Conference Report, No. 16)*

No. 7 **The Biology and Taxonomy of the Solanaceae** (Hawkes, Lester & Skelding) (1979)

No. 8 **Petaloid Monocotyledons** (Brickell, Cutler & Gregory) (1980)

No. 9 **The Skin of Vertebrates** (Spearman & Riley) (1980)

No. 10 **The Plant Cuticle** (Cutler, Alvin & Price) (1982)

No. 11 **Ecology and Genetics of Host-Parasite Interactions** (Rollinson & Anderson) (1985)

No. 12 **Pollen and Spores** (Blackmore & Ferguson) (1986)

No. 13 **Desertified Grasslands Their Biology and Management** (Chapman) (1992)